MODERN PETROLEUM

A Basic Primer of the Industry

MODERN PETROLEUM

A Basic Primer of the Industry

Third Edition

By Bill D. Berger and Kenneth E. Anderson, Ph.D.

PennWell Books

PennWell Publishing Company

Tulsa, Oklahoma

Copyright © 1992 by
PennWell Publishing Company
1421 South Sheridan/P.O. Box 1260
Tulsa, Oklahoma 74101

Berger, Bill D.
    Modern petroleum: a basic primer of the industry / by Bill D. Berger
and Kenneth E. Anderson. — 3rd ed.
        p. cm.
    Includes bibliographical references and index.
    ISBN 0-87814-386-6
        1. Petroleum engineering.   I. Title.
    TN870.B457   1992   92-26928
        665.5—dc20   CIP

Printed in the United States of America

2 3 4 5 96 95 94

I claim that I did invent the driving pipe and drive it in and without that they could not bore on the bottom lands when the earth is full of water. And I claim to have bored the first well that was ever bored for petroleum in America . . . If I had not done it, it would not have been done to this day.

(Statement attributed to Col. Edwin L. Drake late in his life.)

Contents

List of Tables

List of Figures

Acknowledgments

A book seldom results from the labors of the authors alone. A project as ambitious as *Modern Petroleum* could only come into being through the efforts of the many—family, friends, colleagues, and total strangers—who contributed so much to this book. To them we humbly say "Thank You."

Jim Potts and Gail Hernandez, Yale Gas Company. Gary Gibson, F/64 Studio. Peter R. Rano, Arabian CBI Ltd. and G. Graham Harper and Ann Weidart, of the Public Relations Department of CBI Industries, Inc., Don Krone, Krown Systems Inc., David L. Baldwin, Baldwin and Associates. William Rhey, U.S. Environmental Protection Agency. Al Kite, Eastman Christensen. Rick Bloxom, Applied Automation Hartmann & Braun. Tina T. Underwood, Intergraph. Linda Scott, Imperial Oil Limited. Lynn Railsback, Phillips Petroleum. Robert A. Hefner III, The GHK Company. Professor Thomas Gold FRS, Cornell University. Carl Balcer, Mertz Inc. Bob Buchanan and Anne McNamara, The Petroleum Resources Communication Foundation. M. Christine Wignall, The Canadian Petroleum Association, and John L. Kennedy, *Oil & Gas Journal.*

Will L. McNair, Texas Engineering Extension Service, The Texas A & M University System. The Hydril Company. Ed Moerbe, Dresser Security. David Gilmore, The Foxboro Company. The Balcones Research Center, Division of Continuing Education—PETEX, The University of Texas at Austin. Jean H. Smith, Schlumberger Well Service, Nigel D. Muir, British Petroleum, Cecil Hunt, Oil Field Division of Lufkin Industries, Inc., and the Halliburton Services Company.

A special thanks to American Petroleum Institute, 1220 L Street, Northwest, Washington, D.C. 20005, for the use of their specification tables throughout the book.

To all the others who were so generous in their assistance but whose names we may have inadvertently omitted, we offer our humble apologies.

Closer to home, we owe thanks to Delinda McDonald who did much of our word processing. We are also indebted to our mentor at PennWell Books, Don Karecki, whose good humor, enthusiasm for our work, and friendship we have enjoyed for more than a decade.

Last, but by no means least, to Twylla and Kathy who have endured yet another edition of *Modern Petroleum*, thank you.

Yale, Oklahoma 1992   Bill D. Berger
Oklahoma City, Oklahoma 1992   Ken Anderson

Introduction

In writing this third edition of *Modern Petroleum*, the authors faced two challenges. The first was to keep this truly a petroleum book for the non-technical person. While the first edition back in 1978 was adopted as a basic petroleum text at more than 23 universities, tens of thousands of copies were also purchased by people who were not engineers or geophysicists or chemists or geologists; but salesmen, secretaries, bankers, lawyers, investors, roughnecks, accountants, government officials, land and mineral owners, truck drivers, housewives and many other people who had a desire to learn what the petroleum industry was all about. To make our book as interesting and understandable as possible, we have italicized those words and terms which many of our readers may be encountering for the first time. A large number of these terms appear in the glossary, which we heartily encourage everyone to use.

We do, of course, hope that our technical friends and colleagues will buy and read the new *Modern Petroleum* and use it in in their work and in their classrooms. However, since we did write it for a wide audience, we deliberately omitted as much as possible any need for the reader to be skilled in science and mathematics.

While we did "begin at the beginning" and present the story of petroleum in a logical sequence of events, we also employed a modular approach in that each chapter stands alone. Thus the reader who has only an interest in say, marketing, or in land operations, can read and understand those chapters without first having to read the chapters on history or drilling.

The second challenge that we faced was that in the 13 years since we wrote the first edition, there have been more technological advances in the petroleum industry than there were in the previous 100 years. We have included as many of these as possible, yet they continue on a daily basis. There came a point where we simply had to stop cataloging and begin writing. And, the reader must also accept that this is a *primer*—a brief overview of as many aspects of the total petroleum industry as is possible in one volume.

Therefore, we would like the reader to consider the following points as he or she reads the story of the petroleum industry.

The petroleum story is really one of a never-ending cycle of change.

Oil is really too valuable to burn.

Today, man is both master and slave of petroleum.

Oil has played a major, if little noticed role in the sweeping geopolitical changes of the past 10 years.

Over the past decade, computers have had a tremendous impact on every phase of the petroleum industry. What will the next decade bring?

Carbon is lacking on the other planets. Can it be the source of life on Earth?

1

How It All Began

Oil.

Petroleum.

Black gold.

Formed under the surface of the earth millions of years ago; man has long been aware of the existence of oil, yet, it has only been in the last hundred years or so that he has fully realized its value and usefulness. In hardly more than a century, our modern society has become totally dependent upon the men and women and dreams and realities of the petroleum industry. Today we depend upon petroleum products not only for transportation, heating, and to generate electricity, but also for fertilizers and fabrics, plastics and pantyhose, munitions and medicines, paints and pesticides, and thousands of other items we take for granted every day. Other fuels, such as coal and uranium, would not be available without the diesel fuel and petroleum-based explosives required to mine, transport, and process them, or the billions of cubic feet of natural gas necessary to manufacture the Portland cement needed to build the generating stations they fuel. In short, as the past decade has painfully taught us, the world's economy is based not on gold or political philosophies, but rather on the price of a barrel of crude.

Man first became acquainted with petroleum through natural "seeps," or spots on the earth's surface where shallow deposits of crude oil—and often natural gas—oozed upwards into pits, creeks, and marshes, or along beaches and bays. Petroleum, as the "rock oil" was called by the ancients, is referred to not only in the Bible, but in man's earliest recorded history. Oil seeps and bitumen pits furnished the pitch and asphalt used in the mortar that built the Tower of Babel, the walls of Ninevah, and Solomon's temple. It was from "slime pits" near the Dead Sea, that the Egyptians obtained the bitumen they used to embalm their dead. In 3000 B.C. the Sumerians were using oil-burning lamps; a practice adopted in Greece around 250 B.C. In the East, sometime after 100 B.C., the Chinese were digging wells hundreds of feet deep to obtain salt water from which they extracted the sodium chloride to season and preserve their food. Occasionally these wells also produced natural gas, which the Chinese learned to use not only for heating and cooking, but also to pipe through bamboo tubes and burn to heat the well water to hasten its evaporation. By 615 A.D., the Japanese were digging wells nearly 1,000 ft deep in attempts to obtain "burning water," a practice which was also occurring in Burma and India at about the same time. And, along the banks of the Caspian Sea, men were digging shallow wells by hand to obtain oil to light their lamps. As man found more and more of this strange substance, he slowly began to learn that it could be used in many ways. In 671 A.D., Kallinikos of Byzantium invented a primitive missile which he called Greek Fire, which carried an incendiary payload composed of petroleum, sulfur, resin, and rock salt, which the Greeks used against their enemies, the Arabs, during the siege of Constantinople. The Greeks also used petroleum in war at sea, pouring it on the water and igniting it, a fearsome weapon against wooden ships. Even earlier, the legionnaires of ancient Rome won a battle by setting oil-soaked pigs aflame and driving them into the ranks of the approaching enemy.

Gradually, the use of petroleum began to spread across parts of Asia and Europe. But, in order for any new industry to come about, three factors must be present: a widely demonstrated need for the product provided, a reliable and economical source of supply, and an established market price that would guarantee a return on investment sufficient to overcome the risks inherent in any new venture. Thus, the time was not yet right for the birth of the worldwide petroleum industry, and it would not be until the Europeans ventured west that the sleeping giant would begin to awaken.

The New World

When the first explorers ventured into the New World, they found petroleum marshes as well as huge asphalt pools in Trinidad, Venezuela, and later, in California. As more and more adventurers made the long voyage across the Atlantic, they learned to come to these places, beach their sea-weary ships and caulk their leaks with boiled-down petroleum residues. Not only did this seal the hulls, it also protected them against teredos, small, worm-like marine mollusks that inhabit tropical waters and bore into wooden ships, destroying them. In the years to come, this practice of ship coating would grow into an industry of its own and create one of the first economic demands for petroleum products.

Modern Petroleum

Latin America

The Indians in what is now Mexico, called this tarry substance *Chapapote*, while in Venezuela and other parts of South America it was called *Mene*. The pre-Colombian peoples used chapapote or mene—which had worked its way to the earth's surface—as medicine in the form of liniments and ointments, to coat footwear, boats, and roofing, as glue, for illumination, and as incense. In 1579, Commander Melcor de Alfaro Santa Cruz had written home to Spain describing the many uses of chapapote in Mexico, which included toothpaste and chewing gum. However, Venezuela may have become the world's first petroleum exporting country when, in 1539, several barrels of mene were shipped to Spain, apparently in response to an urgent request for a "miracle drug" to cure the painful gout afflicting Emperor Charles V; rumors having reached the royal ears of the great medicinal powers attributed by the Indians to a mysterious substance found around Lake Maricaibo. Thus the origins of today's petroleum industry are closely entwined with the history of Latin America, and many Latin nations have contributed to the industry that has changed the world. (Figure 1–1).

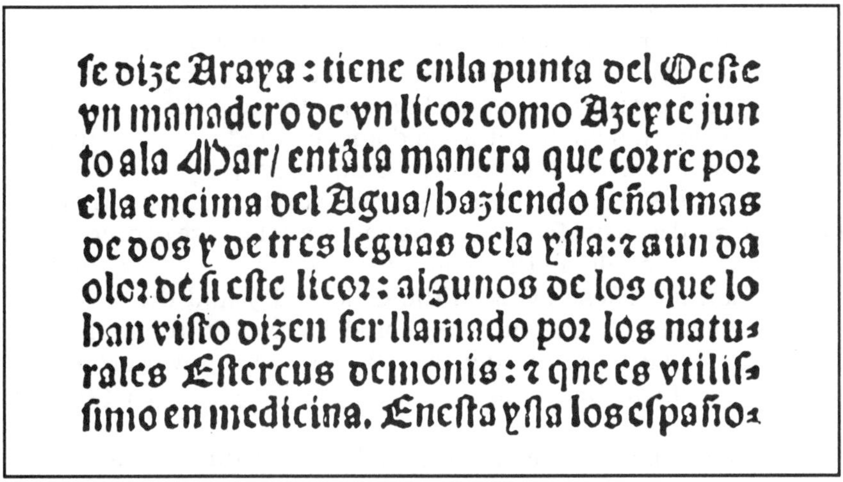

fe dize Araya : tiene enla punta del Oeſte
yn manadero de yn licoʒ como Azeʈte jun
to ala Albar/ entâta manera que coʒrc poʒ
ella encima del Agua/baʒiendo ſeñal mas
de dos y de tres leguas dela yſla: z aun da
oloʒ de ſi eſte licoʒ: algunos de los que lo
ban viſto dizen ſer llamado poʒ los natu-
rales Eſtercus demonis: z qne es vtiliſ-
ſimo en medicina. Eneſta yſla los eſpañoᴣ

Figure 1–1 One of the first known references to petroleum in Venezuela was made by Gonzalo Fernandez de Oviedo in his *General and Natural History of the Indies*, written in 1535.

North America

Later, as Europeans penetrated the northern portion of the hemisphere, they found that the Indians from California to what is now Pennsylvania and Canada were also familiar with this strange black substance that came from within the earth, and had developed many practical uses for it. In southern California, they applied asphalt to baskets or cloth to make them waterproof. These baskets were used to carry water since they were not only lighter than pottery, but unbreakable as well. Asphalt was also used to caulk their boats and waterproof their roofs. In 1788 Peter Pond, exploring western Canada in what is now Alberta, reported that the Cree Indians were using tar from the Athabasca River to caulk their canoes and also as medicine.

Today, one of the most widely used skin ointments in the world is known in English as "Indian Petrolatum," or, "petroleum jelly," a scientific name derived from the Greek which obscures its American Indian invention. In making this nearly colorless gelatinous material of olefin hydrocarbons and methane, the Indians found one of the first practical uses of petroleum. They applied it to human and animal skins to protect wounds, stimulate healing, and to keep the skin moist. They also used it to lubricate the moving parts of tools. Today, petroleum jelly has found its way to the most remote parts of the earth, including the Sahara where nomadic tribesmen smear it on their skin to protect themselves from the relentless sun, dry wind, and pounding sand.

The Indians in Pennsylvania used a number of open pits as oil wells, a fact noted by the white settlers of the Quaker State who, in the nineteenth century, not only launched the American petroleum industry from this spot, but in so doing, created the impetus that would move the world into the petroleum society.

As the settlers began to move westward into the interior of this vast land, the need was not for oil, but, as had been the case in other parts of the world, for salt, so they could preserve their food for the long winter months. Those living in the Ohio Valley and the westward slope of the Alleghenies found they could drill wells through the rock down into subsurface brine deposits. When the brine was brought to the surface, it was boiled down and the resulting salt used to cure their meat.

But some of these crude, hand-drilled wells produced oil as well as salt. Since the oil contaminated the salt, it was considered a nuisance by the colonists who either drained it into nearby creeks or into sump pits where it was burned. However, plantation owner George Washington considered his Virginia estate's "bitumen spring" to be of value. Harking back to the days of the Indian healers, a few entrepreneurs bottled this residue and sold it as a widely acclaimed "cure-all" through itinerant wagon shows and local druggists.

The Search for Light

Samuel M. Kier, a Pittsburgh druggist who owned some brine wells near Tarentum, Pennsylvania, was one of these. In addition to salt, Kier's wells also produced much petroleum. From this "useless" by-product, Kier bottled his oil, labeled it "Pennsylvania Rock Oil," and advertised it far and wide. But, it failed to find a ready market. Undaunted, Kier then devised a crude still to convert his petroleum into a lamp oil. His "rock oil" burned, but it had a bad odor, and produced a heavy, black smoke. Although Kier was not successful, suddenly the first birth pains of the oil industry were felt because man began to realize there was a widespread need for petroleum.

By mid-nineteenth century, the idea of popular education had spread widely; more and more people could read and write. And, there was more to read since this was also a great period of growth for the newly popular newspapers and magazines which were now beginning to be published for the masses instead of for an elite few. But for the most part, the world was still largely an agrarian society. On farms, and in the cities which were becoming industrialized, people worked a minimum of twelve hours a day, or from dawn to dark. This made reading a leisure time activity, which had to be done at night, after the day's labors were completed. Therefore, in order to read, people needed light. Light was also needed by the new factories of the Industrial Age if they were to be able to operate during the short and dark winter days and survive economically. Many used tallow candles which were both expensive and ineffective, or, sperm oil from whales, which though relatively clean burning, was becoming more and more expensive because even then the great herds of whales had been hunted to the point of extinction.

In both America and Europe a search was on for a source of light that could be manufactured to sell at a reasonable price, burn cleanly, and provide effective illumination.

In a few large cities, plants had been built to process artificial gas from coal, or to furnish natural gas from wells located in the towns themselves. Gas lights illuminated the streets at night, and indoors "jets"—thousands of times brighter than candles—not only brightened homes, but also made the forerunners of the modern factory possible for the first time. Gaslight quickly became enormously popular, but the piping was crude and expensive to install, which meant that gas was available only in metropolitan areas. So the search, including attempts to produce a light source from petroleum, continued.

In 1849, James Young, a Scotsman, obtained a patent for processing cannel coal (or more likely, cannel shale which is prevalent in

Scotland), which he distilled into what he called coal oil. It became popular almost immediately, and the world's first refining industry sprang up in Britain as Young issued licenses on his patent for the production of coal oil in both Great Britain and the United States.

Canada

A brilliant and innovative Canadian was responsible for the next step in the birthing. Abraham Gesner, a medical doctor who was also an amateur geologist and self-taught chemist from Nova Scotia, began experimenting in the 1840s with producing lamp oils from both asphalt and coal. Using a process similar to Young's, he became the first in North America to distill a hydrocarbon into lamp fuel. In 1853, he moved his operation to New York and began producing coal oil. The following year he introduced a new product called "Kerosene," from the Greek words for oil and wax. Even though coal oil and the early kerosene were smelly, smoked badly, and cost dearly, they immediately became far more popular than whale oil, which had risen in price to $2.50 per gallon, which placed it far beyond the reach of most consumers. Gesner began to issue licenses under his American patent, and by 1860, there were more than 70 plants in the United States producing an estimated 23,000 gals of coal oil per year from natural asphalts, soft coal, and shale.

All of this would last only a few short years however, for it was to be kerosine refined from petroleum, not coal oil, that would light the world. Up until the mid-1850s all crude oil had either been skimmed from seeps or produced as an unwanted by-product of brine wells, and was considered by many to be good only as a medicine show novelty. But two brothers in a tiny village in Canada West, as Ontario was called in 1854, would change all that and in so doing, change the world. Henry and Charles Nelson Tripp of Enniskillen Township on the north shore of Lake Ontario, a metropolis inhabited by 37 settlers, 34 cows, and 16 hogs, founded the world's first incorporated oil company, International Mining and Manufacturing. Their interests were not in lamp oil, but in possible industrial uses for the black, tar-like substance from gum beds located on their property. In 1850 a chemist with the Canadian Geological Survey had pointed out that the gum had possible commercial use as paving material, a sealant for ship bottoms, and as raw material for the manufacture of illuminating gas. By 1854 they were digging the bitumen by hand, and boiling it down in open cast iron kettles. Their major thrust was to sell their product for paving and for marine use as ship coating. Even though the Hamilton Gas Company reported that gas made from the bitumen produced far more illumination than the gas produced from coal, the Tripps ignored this market and by 1856 their company failed and was forced to sell its gum beds. It was to the buyer of their 600 ac of potential oil land that fame as the founder of the North American oil industry would come.

James Miller Williams was a carriage maker from Hamilton, Ontario. In 1857 he set up a simple refinery on Black Creek and set out to drill for oil, correctly assuming that if he penetrated below the surface of the asphalt beds he would find more oil. His first well was abandoned after the pipe broke off at 27 ft, but in 1858 he successfully produced oil from a depth of 49 ft. Whereas the Tripps had mined the bitumen from its surface beds, Williams began pumping crude oil from a reservoir beneath them, which he then refined into lamp oil. In 1860 he reincorporated as the Canadian Oil Company. This was North America's first integrated oil company, since its activities included exploration, production, refining, and marketing. That same year, nearly 100 wells were drilled in the Oil Springs area with the successful ones producing 12 to 23 B/D (barrels per day). Then, on January 16, 1862, the greatest gusher the world had ever seen blew in at Oil Springs. Spouting crude more than 20 ft into the air, it flowed at an astonishing 2,000 B/D. Refineries quickly sprang up across Ontario, but almost as quickly, the momentum of the growing industry flowed south across the border (Figure 1–2).

Figure 1–2 James Miller Williams (courtesy Imperial Oil Archives).

The United States

In 1857, Col. A. C. Ferris, a major supplier of lamp oil, saw samples of Samuel Kier's rock oil in use. He purchased some and ran it through his processing plant. It produced such a superior lighting oil that he sent agents far and wide buying crude for $20/bbl, a price the world would not see again for more than a century. Some of this oil he purchased from Williams in Ontario, although most of his supply came haphazardly from brine wells, oil springs, or oil seeps. Suddenly, the pieces of the puzzle began to come together; there was a growing need for the product, a firm market price had been established, so all that remained was the development of a reliable supply of raw material in quantity, in this case, crude oil.

The Birth of the Industry

James Miller Williams may well have been the father of the modern oil industry, but history, fickle as always, has chosen to award that honor to an American, Col. Edwin Drake. A former railway conductor, the "colonel" was an affectation adopted to impress the townspeople of Oil Creek, Pennsylvania, where Drake, as representative of the Pennsylvania Rock Oil Company, spent three unsuccessful years trying to skim oil in marketable quantities from the same springs the Indians had used years before. When the company failed, Drake organized the Seneca Oil Company to try again. This time he looked at the brine wells that had been drilled at nearby Tarantum and made the momentous decision to try to obtain oil in quantity by drilling for it—a decision that brought him ridicule and derision from the local population.

He assembled a steam-powered cable-tool drilling rig much the same as the ones used by the salt water drillers and began operations. That he was successful was due in no small part to the skill and dedication of his driller, William A. "Uncle Billy" Smith, a blacksmith and experienced brine well driller. After penetrating 30 ft of rock, Drake struck oil at a total depth of 69 1/2 ft. The well was not a gusher—the oil had to be pumped to the surface, but it was the first "oil well" in the United States and the lamp oil producers quickly flocked to the site to buy Drake's oil for $20/bbl. That day, August 27, 1859, is noted as the birthday of the oil industry, for Drake had proven that it was possible to obtain oil in quantity by drilling for it through rock. At last the combination had come together; the need for oil, an established market and price, and a method of obtaining oil in quantity (Figure 1–3).

Figure 1-3 The search for light helped give birth to the industry. Col. Edwin Drake (in stovepipe hat) speaks to an associate in front of his famous well (courtesy API).

Almost overnight Titusville and the Oil Creek area became the world's first oil boom town as buyers, would-be producers, and lessors swarmed to the area, giving western Pennsylvania the title "Cradle of the Oil Industry." Although many men in many countries around the world were seeking oil, Drake just happened to be the first to find it in quantity by drilling through rock, and he was far more fortunate in his selection of a site than anyone at the time realized.

By pure chance the Drake well was located in an area where the pay sands were close to the earth's surface and could easily be reached with the primitive equipment at hand. In fact, the Oil Creek field was the shallowest and most productive ever discovered. This permitted large-scale production in a short time. The area was also accessible to the available transportation facilities. In addition to the railroad, Oil City, south of Titusville, was located on the banks of the Allegheny River which joined both the Monongahela and the Ohio at Pittsburgh. Thus, needed supplies could be brought in and the oil shipped out to the markets of the rapidly industrializing north. As a final blessing, the oil was of high gravity and sulfur-free. In fact, for decades to come Pennsylvania-grade crude was the world's standard against which all other oils were measured. It could easily be refined into a high-quality kerosine with the unsophisticated refining processes that were then in use. Also, since it had a paraffin base, it could easily be made into lubricants with the available technology. If Drake's oil had been heavy, sour crude with a high sulfur content, the pioneer refiners would not have known what to do with it and the birth of the industry would have been delayed.

New Uses for Petroleum

Dozens of rigs soon covered the Pennsylvania landscape. Oil was produced so rapidly and in such great quantities that the existing storage facilities soon proved woefully inadequate and thousands of barrels spilled out on the ground and into rivers and creeks. The price rapidly began to drop, but despite the initial overproduction and a lowered price structure, the infant industry continued to grow and prosper as more and more uses were discovered for petroleum. At first, the only needs were for lamp oil and lubricants; anything else that came out of the barrel including naphtha, benzene, and gasoline, was considered dangerous waste which was either burned or dumped into the nearest creek.

The new industry was still only a two-year old infant when the horror of the Civil War burst across America, unleashing a new technological age. No longer were ships powered by sail alone—steam engines propelled supplies across the Atlantic, and the first iron warships and steamboats up and down the mighty rivers. Their engines, as well as the machinery in the factories that turned out cannons, repeating firearms, telegraph wire, barbed wire, and tents and clothing all needed lubrication, and lamp oil was needed to keep them working through the night, and to illuminate the field hospitals and headquarters tents.

The Move Westward

By the time the last Confederate general, Stand Watie, a Cherokee Indian, surrendered in Texas, half the nation lay in ruins. All of the railroads in the south, together with most of the port and manufacturing facilities, had been destroyed. A massive job of rebuilding was needed, one that called for more and more petroleum products. No sooner than this was underway, than men began to turn their ambitions westward. The War had given them the tools: barbed wire, steamboats, railroads, and repeating firearms, with which they could finally conquer the American West. The need for oil had turned into a never-ending demand.

Not only was there a great demand for oil, there were great profits to be made in the ever-growing industry, a fact which did not escape the notice of John D. Rockefeller, who, in 1870, founded the Standard Oil Company. There were also great profits to be made by those who found new sources of oil to meet these increasing needs. This was still an era when little was known about reservoir mechanics, and there were fears that the Pennsylvania fields were becoming totally depleted. So as Americans moved westward, so did the search for oil. New fields were discovered in the Ohio Valley and in Illinois and Kentucky, although much of this proved to be heavier and more sour than the Pennsylvania grade crude. The search continued however, even into Wyoming, although these early wildcatters could not know that an American and two Germans,Thomas A. Edison, Gottlieb Daimler, and Carl Benz were on the verge of perfecting inventions that would change the oil industry—and the world— forever.

Winners and Losers

In early part of the nineteenth century the Cherokee Indians in the southeastern United States became known as one of the Five "Civilized" Tribes because they had adapted so well to the white man's society. Many owned large estates and lived in fine homes whose furnishings often included extensive libraries. They published their own newspaper in the Cherokee language, and they sent their children to universities in the north and in Europe to be educated. Suddenly however, disaster struck. Gold was discovered on the Cherokee lands in northern Georgia. Overnight, greedy gold seekers evicted them from their homes and confiscated their lands. Those who escaped the volunteer "militia" were driven westward at bayonet point, and hundreds died along the shameful Trail of Tears. Eventually, the survivors were resettled in the vast Cherokee Nation in Indian Territory. Very few, if any at all, ever got to enjoy any of the mineral wealth of their ancestral home.

Removal, an integral part of the "taming of the west," was the government's policy of moving as many Indians as possible onto reservations, many of which were located in Indian Territory. A few years after the Cherokees, it came the Osages' turn to go. They pointed out to the government that there was no land left that was not already assigned to a tribe. The terse response was, "Find some land, buy it, and go—or else." What they found was 1.8 million ac of hilly, rocky scrub, little suited for farming or much else, at the northern end of the Cherokee Nation. The Cherokees were more than happy to unload it on the hapless Osages for 75¢/ac. Happy that is, until the Nellie Johnstone #1 was drilled in 1897.

The first producing well in the Territory, it signaled the opening of the vast Oklahoma oil fields and for a time made the Osages the richest per capita nation in the world. But it was during a few hours on the afternoon of January 10, 1901, on a marshy bit of waste land near Beaumont, Texas, that the world was changed forever.

Spindletop

Spindletop—to oilmen the name still evokes feelings akin to Columbus' first sighting of the New World or Neil Armstrong first setting foot on the moon. For it was on this worthless bit of Texas swamp that the first American gusher blew in with a mighty roar that was heard around the world. For if Drake's well signaled the birth of the oil industry, Spindletop gave birth to Age of Liquid Fuel, and in so doing changed the lives of everyone on Earth. Now there was no looking back.

Spindletop—although that is not the name they would have chosen—was the culmination of the dreams of two highly unlike, but equally colorful oilmen: Captain Anthony Lucas, a transplanted Slav mining engineer, and Patillo "Bud" Higgins, a one-armed Texas Hellraiser turned Sunday School teacher.

As a youth, Higgins had lost an arm as the result of a prank. One evening he and a group of companions had tossed a hornet's nest into a tent where a Baptist revival meeting was taking place. An overly zealous deputy sheriff pursuing the group fired a shot that stuck young Higgins in the arm. The wound became infected and his arm had to be amputated. Higgins was a carousing brawler however, and to him the loss of a limb seemed to mean little. He headed for the timber country where he proceeded to show the other loggers he could outfight, outdrink, and outwork anyone with two arms and fists. His reputation grew until the townspeople dreaded to see him come in out of the woods on Saturday nights. But on one such evening fate stepped in again. He was passing another Baptist tent meeting much like the first one, when he paused for a moment, stayed to listen, and as a result "got religion."

One of the pillars of the church, George W. Carroll, was deeply impressed by Higgins' conversion and offered him a job in his land and real estate business. Higgins accepted, and quickly became a successful and respected member of the community. One day, after a heavy rainstorm, Higgins happened to notice an unusual deposit of red clay that contrasted sharply with the black soil of the area. He had a sample analyzed and found it was the type of clay from which bricks could be made.

Sensing a ready market for bricks, Higgins had a group of investors join him in building a kiln. The business showed a profit, but was inefficient, so Higgins went north to study established kiln operations. In Pennsylvania and Ohio he learned the kilns were fired by natural gas and by oil, more stable and efficient fuels than wood or coal. In seeking more information, he learned that many of the geological characteristics that oilmen looked for in their search for petroleum were present in a small mound that rose from the marshes south of Beaumont. When he returned to Texas, he studied the area carefully and became convinced that the mound was sitting above oil—much oil.

He approached George Carroll with his theory. Carroll agreed to support him, and in 1892 they formed the Gladys City Oil, Gas, and Manufacturing Company. Together with his dream of oil, Higgins had another dream—that of a whole new planned city. One with oil wells of course, but so laid out that there would be orderly growth and beauty as well as industry. For five years, using crude water well equipment, the company sought oil with no success. Finally, Higgins sold his interest to Carroll, and spent three years trying to form another group to drill on his land.

Deeply in debt, and with his remaining friends thinking him more than a little strange, Higgins, as a last resort, advertised for help in a trade journal. The man that answered the ad was as colorful as Higgins himself, but between them they would usher in the Age of Liquid Fuel. (Figure 1–4).

Figure 1–4 Patillo Higgins (courtesy *Amoco Torch*).

Anthony Luchich was born on the island of Hvar, in the Adriatic Sea. After graduating from the Graz School of Mining Engineering and being commissioned in the Austrian Navy, he came to the United States for a visit. Liking what he saw, he applied for citizenship, changed his name to Lucas, married an American girl and settled in Louisiana where he worked as a mining engineer (Figure 1–5).

Figure 1–5 Captain Anthony F. Lucas (courtesy *Amoco Torch*)

After gaining experience in working with subsurface structures, Lucas formed a theory that gas, oil, and sulfur would accumulate under salt domes. Since he had visited Beaumont, he was familiar with the mound or dome, that Higgins talked about in his ad. Sensing an opportunity to test his beliefs, he decided to join Higgins, and the two were soon at work on yet another attempt to find oil at Gladys City, as Higgins called his dream town.

However, this time there was a difference. Lucas decided to use steam-powered rotary drilling rigs. Rotary rigs were certainly not new, the first having been patented in 1833, and by 1901 more than 100 wells in Texas alone had been drilled with them. But most of these were crude devices powered by nothing more than a mule walking in a circle, so rotaries had not gained wide acceptance in the industry. Lucas was to put together all the best elements of rotary drilling—ones that in one form or another are still used today—the boiler, the engine, drawworks, rotary pumps, swivel, drillpipe, crown block, traveling block, and bits. And, as a final touch, he used mud

Figure 1–6 Spindletop—America's first "gusher" (courtesy *Amoco Torch*)

to lubricate his drillstring. Mud, he probably knew, had been tested in Louisiana, but he had none. So, the legend goes, he created his own by flooding a field and driving a herd of cattle back and forth through it.

On January 10, 1901, drilling had reached 1,020 ft. Lucas had gone into town for supplies leaving the three-man drilling crew—Peck Byrd and Al and Curt Hamil in charge. Later, no one remembered who heard the noise first; it began with a low, menacing rumble that shook the earth. Then, as the crew ran for their lives, a geyser of mud shot forth carrying 700 ft of heavy drilling pipe up through the earth and into the sky, taking most of the drilling rig with it. The noise ceased as suddenly as it began; the only sound was that of the twisted pieces of steel crashing back into the mud surrounding the site. The men timidly approached what was left of the rig, now standing in a sloppy pool, and with uncommon good sense, shut down the boiler fires. No sooner had they done so, than the rumblings began again; once more they ran as fast as they could as more mud came boiling out of the hole only to be blown aside by a column of gas, which was immediately followed by the loudest roar yet: a solid spout of heavy green crude oil that reached 200 ft into the air (Figure 1–6).

The three drillers gaped in amazement at a sight never before seen—the world's first true gusher. For nine days the well continued to spew until more than 800,000 bbl of oil formed a lake that spread nearly three-quarters of a mile from the well. Finally it was capped by a collection of valves and fittings—the forerunner of today's Christmas tree—that Al Hamil devised and gingerly lowered into place over the monster. News of the discovery spread almost as quickly as the free-flowing oil, and within days Higgins's dreams for Gladys City disappeared forever under a maze of wooden derricks and the frame and canvas shanties of the wheeler-dealers, con artists, and pimps and prostitutes that flocked to Spindletop in hopes of finding an easy fortune (Figure 1–7).

Drake's well gave birth to the industry because the time and place were right. Now the time was right for Higgins and Lucas to bring it to maturity. Edison's electric light had been successful, and the demand for kerosine was already on the wane. Daimler and Benz had brought forth the world's first successful automobiles, a fact not ignored by Henry Ford in this country, and, these new devices were powered by gasoline, the oil industry's largest waste by-product. In addition, there is the sheer number of firsts that Spindletop repre-

Figure 1-7 Instead of homes, "Gladys City" sprouted a jungle of derricks. This is Boiler Avenue at Spindletop (courtesy API).

sented: the first true gusher, the proving of Lucas's theory about oil and gas forming under salt domes, the first large-scale success of rotary drilling rigs, and the first successful use of drilling mud. The world would never be the same.

Chapter References

Bill D. Berger, *Facts About Oil,* (Stillwater: Oklahoma State University, 1975), p. 2.

Bill D. Berger and Kenneth E. Anderson, Gustavo Pena, and the editors of *Petróleo Internacional* editors, *Petróleo Moderno Introducción básica a la Industria Petrolera,* (Tulsa: PennWell Books, 1980), pp. 1–6.

A Brief History of the Petroleum Industry in Canada (Calgary: The Canadian Petroleum Association, November, 1980), p. 1.

John Feehery, "Spindletop...Birthplace of A New Era," *Amoco Torch,* Vol. 4, No. 6, November/December, 1976. pp. 8–9.

Jack McIver Weatherford, *Indian Givers: How the Indians of North America Transformed the World,* (New York: Crown Publishers, 1988), pp. 48–49.

Anne McNamara, *Our Petroleum Challenge: The New Era* (Calgary: Petroleum Resources Communication Foundation, 1986), p. 13.

2

The How and Where of Petroleum

The Beginnings

To understand the origins of oil and gas, where they are most often found, and how we find them, one must first have some knowledge of the Earth itself.

Our Earth, one of the terrestrial, or inner planets of the solar system, is thought to be about 4.5 billion years old. The inner planets, so-called because they are the ones nearer to the Sun, also include Venus, Mercury, and Mars. They differ from the outer planets, Jupiter, Saturn, Uranus, Neptune, and Pluto, in that they are believed to have been composed mostly from rock-forming substances with little gases or liquids. Today, there is little specific knowledge of the Earth's early history, but it is believed that these constituents, which contained solid or nonvolatile molecules, adhered, and then combined, condensed, and compressed to form the sphere upon which we live.

Geology

Today, most of our knowledge of the Earth is based upon its past 450,000 years, or, about 10% of its total history. Geology is the science of the Earth, its composition, structure, and history, and is partly descriptive and partly historical. As our knowledge has increased, and our scientific tools have grown more sophisticated, other fields of learning such as geophysics, astrophysics, and seismology have all contributed to what we know about the Earth.

We do know that the Earth is not a static body; it has a record of past and present change. Through time, all its features have been altered by the interplay of both external and internal forces.

The Interior of the Earth

Our planet is approximately 24,900 mi (40,079 km) in circumference around the equator. The surface area is comprised of approximately 70.8% water and 29.2% land. Its approximate volume is 260 billion mi^3 ($1.08 \times 10^{12}km^3$). The volume of its water is 330 million mi^3 ($1.370 \times 10^6 km^3$).

Most of what we know about the interior of the earth has come to us from secondary evidence gathered from studying rocks and minerals, seismic data, heat flow emanating from the Earth, study of the Earth's gravity and magnetic fields, and comparisons with other bodies in our solar system. From all this, it has become apparent that there is a heavy inner core surrounded by an outer core, a lower mantle, an upper mantle, and finally, a relatively thin crust (Figure 2–1).

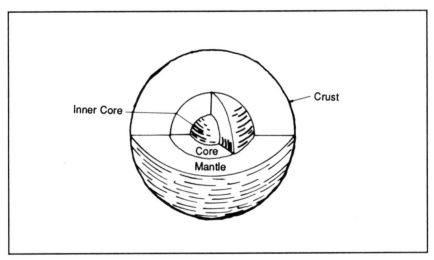

Figure 2–1 The Earth's crust, mantle, core, and inner core.

The crust, or skin, has a volume of 2 billion mi^3 ($6,210 \times 10^6 km^3$), and is 10 to 30 mi in thickness. This accounts for about 1% of the total Earth by volume and about .4% by weight. The volume of the mantle is 216 billion mi^3 ($898,000 \times 10^6 km^3$), or, about 84% of the total volume, and 67% of the total weight. It is some 1,800 mi thick. The volume of the core is 42 billion mi^3 ($175,500 \times 10^6 km^3$) and this equals 15% of the total volume and 32% of the total weight of the Earth. The core, composed of the heaviest material, is 4,400 mi in diameter.

So far, man has only been able to drill a few miles into the crust. With our current technology, it is not likely that a hole could be drilled all the way to the core. However, it is believed possible, as tools are developed to withstand higher and higher temperatures and pressures, to drill through the crust into the upper reaches of the mantle.

Minerals

In our study of petroleum our basic interest is in the fundamental building blocks of the Earth's crust, which are minerals. Minerals are chemical elements and chemical compounds that have been formed by inorganic processes. There are more than 2,000 known minerals on Earth. All of their properties are determined by the composition and internal atomic arrangement of their elements. Every mineral has a definite weight per cubic inch, and has a crystalline structure which does distinguish it from other minerals.

Of all the known minerals, only a few are rock-forming and more than 90% of these are *silicates*. These silicates are compounds containing silicon and oxygen and one or more metals. The Earth's crust is divided into two layers. The upper one, called the *sial*, is composed mainly of minerals containing silicon and aluminum, and the lower one, called the *sima*, contains minerals composed mostly of silicon and magnesium. Minerals then, are the units from which the rocks of the Earth's crust are built. These rocks are divided into three general groups: *igneous, metamorphic,* and *sedimentary.*

Igneous Rocks

Igneous rocks were once a hot, molten, liquid-like mass known as magma that cooled into firm, hard rock. These fire-formed rocks are the ancestors of all other rocks. The chief characteristics of igneous rocks are: color, specific gravity, texture, and mineral composition.

Metamorphic Rocks

Metamorphic rocks occur when sedimentary rock is exposed to great heat and pressure. This may take place in areas where the crust of the earth warps downward due to large-scale shifting deep below the surface. Metamorphic rocks are those formed from original rocks by Earth pressure, heat, and chemically active fluids beneath the Earth's surface.

Sedimentary Rocks

Sedimentary rocks are made up of particles derived from the breakdown of preexisting rocks. The most characteristic feature of sedimentary rocks is the layering of the deposits which make them up. These deposits are particles that were transported to new locations by ice, wind, or water where they settled and became hardened. Later, we shall see why in recent years, sedimentary rocks have been of particular importance to petroleum geologists.

As the Earth began to assume its present shape, its crust buckled and shrank to form a rough surface of igneous rock, which developed from the molten form called magma. Rain fell from the moisture condensing in the new atmosphere, and as it struck the rough surface it collected and flowed downhill into the low places to form the primeval rivers and oceans. As the water ran downhill toward the depressions, it carried along small particles of rock, a process we call erosion. Another process, sedimentation, occurred when this combination of water and rock particles reached a quiet body of water, and the particles, or sediment, settled to the bottom. While erosion is often thought of as being the result of the washing action of flowing water, it can also be caused by wind, waves, and freezing, and well as by moving ice (glaciers). The first particles to erode were, of course, from igneous rock, since that is all that existed in the beginning. As time went on, sediments were deposited on top of sediments. The earlier ones were compressed by the weight of the succeeding layers into sedimentary rocks. These rocks can again erode and produce sediments and the cycle of erosion and sedimentation occurs over and over again.

As layer after layer of sediments were buried, they were compressed by the weight of the layers above them. Pressure and heat, as well as chemical, bacterial, and radioactive action changed them into sedimentary rock.

Sedimentary rocks may occur as loose mud and sand or be hard and compacted, depending upon how much pressure has been exerted and how old the layers are. The rocks are made up of clastic material, chemical precipitates, and organic debris. The clastic fragments are made up of broken and worn particles of other minerals, rocks, and shells that have been moved into place by erosion. Chemical precipitates were formed in place by the action of dissolved salts or the evaporation of entrapped seawater. Organic debris is composed of shells and plant and animal remains that accumulated in one place, such as a coral reef or peat bog.

The Rock Cycle

If sedimentary rocks are exposed to great heat, they may be transformed into metamorphic rock, which can melt and become magma. This molten rock may be forced to the surface where, as it cools, it again becomes igneous rock which of course, is subject to erosion. Thus a definite relationship exists among igneous, sedimentary, and metamorphic rocks because any one of these rocks may be changed into some other form with time and changing conditions (Figure 2–2).

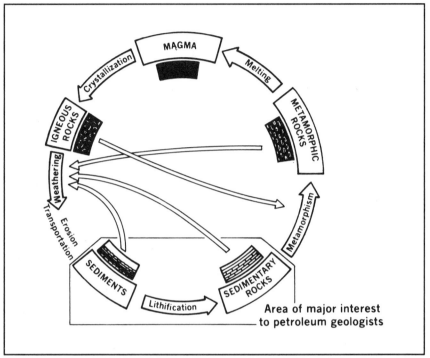

Figure 2-2 The "Rock Cycle." The emphasized portion is the area of major interest to petroleum geologists.

Inside the Mantle

Many geologists view the mantle as the source of both the crust and the core of the earth: its lighter parts moved upwards to form the crust, and the heavier parts gravitated toward the center to form the core. This belief gave rise to speculation that in its history, the mantle could have been subject to heat-driven convection currents that could have resulted in the continents being formed over the descending currents, and the deep ocean basins formed over rising currents. This in turn, lead to conjecture that convection currents might have even lead to the splitting apart of land masses and ocean bottoms and their movement to new locations.

Plate Tectonics

As was mentioned earlier, our Earth, or rather its crust upon which we live, is not a static platform. In reality:

> ...the outer shell is fragmented into large and small plates, all moving in relation to all the others with steady velocities that reach 5 in. (13 cm) per year—pulling apart here, slipping past one another there, sliding beneath another somewhere else, and in other places colliding slowly to build some of our most dramatic mountain ranges.[1]

Plate tectonics ("plate" is the basic unit of the system, and "tectonics" comes from a Greek word, "tekton," which means builder) is the study of the processes and products of motion which takes place within the Earth. The concept and recognition of such phenomena developed slowly, but has come to the fore over the past 30 years as the science of paleomagnetism has expanded and added evidence to support the idea of continental movement (Figure 2–3).

Knowledge that continents can, and have moved, explains why today glacial deposits are found in the tropics, fossilized tropical remains in arctic regions, and seabed debris atop mountain peaks.

The Earth's crust is broken into seven very large plates consisting of both continental and oceanic portions, as well as a dozen smaller plates. Each is about 50 mi (80 km) thick and has a shallow part that deforms either by elastic bending, or by breaking, and a deeper,

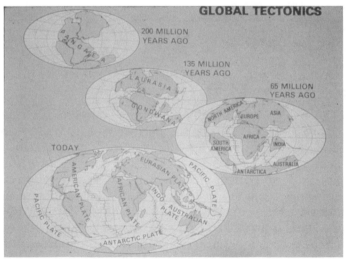

Figure 2–3 Plate tectonics (courtesy API).

Modern Petroleum

yielding part. The plates lie over a viscous layer upon which they slide. Their internal portion is rigid with most action taking place around the edges.

When plates come together, the oceanic one tends to tip down and slide beneath (subduct) the continental plate. At sea or on land, plates may also slide along strike-slip faults, such as the famous San Andreas fault, where the Pacific and North American plates are sliding past each other at about 1.5 in. (3.8cm) per year.

It is such forces that occur beneath the surface of the Earth which have brought about the conditions necessary for the entrapment of petroleum in reservoirs from whence it may be recovered.

The Origins of Oil and Gas

Many theories have been advanced as to the origin of petroleum and natural gas. Yet is has not been possible to determine the exact place or materials from which any particular reservoir originated. The two most prevalent theories today are the *organic*, or *bionic*, and the *inorganic*, or *abionic*. It is important that both of these be mentioned and briefly described before we proceed further with our discussion about the origins of petroleum.

The Organic Theory of the Origin of Oil and Gas

The organic theory of the origin of petroleum began to come to the forefront around the turn of the century as the oil and gas industry grew in size and economic importance, and as geologists were called upon to accurately locate new and larger deposits. It became the most widely held theory, and is still so today, although recently it has been facing renewed challenges as new tools and techniques for the study of the Earth's history are developed.

Simply stated, the organic theory is that the carbon and hydrogen necessary for the formation of gas and oil came from the early forms of life on Earth, or, in short, from biological origins. The remains of these plants and animals were caught up in the process of erosion and sedimentation and carried down the rivers to the seas which then covered large portions of the Earth's surface. They and their accompanying mud and silt were spread along the shoreline where they were covered and compressed by the weight of many more succeeding layers building up on top of them. In time, these layers became sedimentary rock. Today, these sedimentary rocks, sandstone, shale, and dolomite are often where deposits of petroleum are found (Figures 2–4 through 2–8).

Figure 2-4 Organic theory of origin of gas and oil. Some scientists believe petroleum formation began millions of years ago, when tiny marine creatures abounded in the seas (courtesy API).

Figure 2-5 The marine plants and animals, held in clay, sand, and silt, were changed to gas and oil, probably by gradual decay, heat, pressure, and possibly bacterial and radioactive actions (courtesy API).

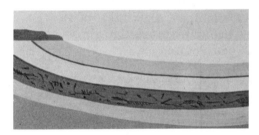

Figure 2-6 As millions of years passed, pressure compressed the deeply buried layers of clay, silt, and sand into layers of rock (courtesy API).

Figure 2-7 Earthquakes and other earth forces buckled the rock layers (courtesy API).

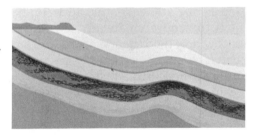

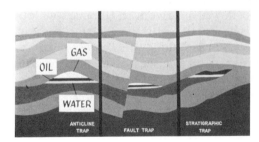

Figure 2-8 The petroleum migrated upward through porous rock until it became trapped under nonporous rock (courtesy API).

The Inorganic Theory

The inorganic theory that hydrocarbons come from deep in the Earth, from materials that were incorporated as the Earth was formed, was devised early in the 1800s, when scientists came to believe that petroleum was a residue left from the formation of the Solar System or, as a result of chemical actions deep within the Earth. Dimitri Mendeleev, the distinguished Russian chemist who discovered the periodic nature of the elements, believed that petroleum emanated from deep within the Earth, from inorganic origins rather than from organic sediments. His view was that occurrences of petroleum seemed to be controlled more by the large scale features of the crust, like the mountain ranges and the great valleys, than by the details of sedimentary deposits laid down over the years. Since then, other scientists have adopted the same view. Today, these proponents of the inorganic theory claim that data acquired by modern technology has strengthened their claims while at the same time weakening the case for a purely biological origin of hydrocarbons.

In defense of their views, those who argue in favor of an abionic origin, observe that:

(1) Petroleum and methane (natural gas) are frequently found in geographic patterns of long lines or arcs which are related more to deep-seated, large-scale structural features of the crust, rather than to smaller-scale sedimentary deposits.

(2) Hydrocarbon-rich areas tend to be rich at many levels and extend down to the crystalline basement that underlies the sediment.

(3) Some petroleum from deeper and hotter levels almost completely lacks biological evidence.

(4) Methane is found in many areas where a biogenic origin is improbable.

(5) Hydrocarbon deposits of a large area often show common chemical features independent of the varied composition or geological ages of the formations in which they are found.

(6) The regional association of hydrocarbons with the inert gas, helium, and a higher level of natural helium seepage in petroleum-bearing regions has no explanation in the theories of biological origin.

They also maintain that there are far greater amounts of hydrocarbons to be found deep within the Earth than have previously been believed to exist, and that organic debris cannot possibly account for all the Earth's petroleum.

The Validity of the Theories

It is not the intent of the authors to attempt to provide all of the evidence available supporting either side of the question, but rather to make the new student of petroleum aware that two differing theories, which are not always mutually exclusive, do exist. Many distinguished scientists and researchers on both sides have published their views and findings on this subject, and readers who are interested in exploring the topic further should seek out their books and journal articles.

Occurrence and Entrapment

If we are to accept the organic theory, which is prevalent today, that once-living organisms are the basis for oil and gas that resulted from the remains of these organisms being subjected to pressure, temperature, and chemical and bacterial actions as they were buried under the silt of the ancient seas, then we must also accept that the search for oil is confined to areas containing thick beds of sedimentary rocks, since it was in such rocks that the source materials were buried.

A common form of rock in which oil or gas may be found is *sandstone*, which is composed of grains of sand mixed with particles of clay and shale. Petroleum also occurs in porous *limestones* and *dolomite*. The oil migrates from its place of formation through *pores* (tiny spaces) which occur between the particles in the sandstone, or between the pores and cracks which appear in dolomite and limestone. These openings form the reservoirs in which oil and gas accumulate.

Entrapment

Oil and gas are usually not found where they were formed. Source rocks, in which the original organic material was trapped, are fine-grained and relatively impervious. They rarely hold oil and gas in anything but small quantities. Instead, the oil and gas moves from the source rock upwards toward the surface. Some escapes through faults to the surface where the gas disperses into the atmosphere, the lighter oil eventually evaporates, and a tar-like deposit of bitumen is left. Usually however, much of the oil and gas does not reach the surface. It migrates upward until its progress is blocked by an impermeable barrier or cap rock where it accumulates in place to form a reservoir. The barrier and the resulting reservoir are called a *trap* (Figure 2–9).

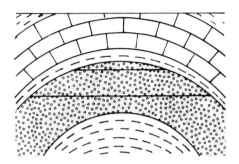

Figure 2–9 Anticlinal trap. The gas, oil, and water are prevented from further migration by an impermeable layer of rock (courtesy Anne McNamara, Petroleum Resources Communication Foundation).

This upward movement is also accompanied by a separation of the oil, gas, and water. The oil and gas rise as they displace the sea water that originally filled the pore spaces of the sedimentary rock. When they reach the impenetrable barrier, the materials separate. If you were to place equal parts of salt water, natural gas, and oil into a sealed glass container, you would note that they would separate themselves into three more or less distinct layers—the gas at the top, the oil in the middle, and the water at the bottom. This same separation occurs in the formation—gas is found in the highest part, then oil, and finally, water at the bottom.

Not all of the salt water is displaced from the pore spaces, however. Often they contain from 10 to more than 50% salt water in the gas and oil accumulation. This remaining water, called *connate water*, fills the smaller pores and also coats the surfaces of the larger openings. The principal search for oil and gas fields today is concentrated toward geologic structures or traps into which oil and gas have migrated and been trapped, thus forming a reservoir (Figures 2–10 and 2–11).

Reservoir Properties

People often think of an oil or gas reservoir as a large pool of liquid far beneath the Earth's surface like a subterranean pond. In reality

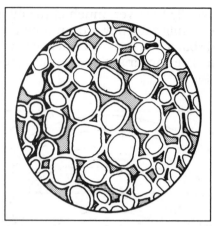

Figure 2-10 Connate water (courtesy Anne McNamara, Petroleum Resources Communication Foundation).

however, the petroleum is entrapped in tiny openings in the rock: the pore spaces. For this to happen, a number of conditions must be present.

(1) There must have been first a source of carbon and hydrogen (plant and animal remains), favorable conditions for their decay and then recombination into the hydrocarbons of petroleum.

(2) There must be porous rock in which the hydrocarbons can accumulate.

(3) The pores must be interconnected so that the fluids can move within the rock; a quality called *permeability*.

(4) Some kind of barrier or closure must be present that will prevent further upward movement of the petroleum and force it to collect in one area.

If any of these four characteristics is not present in an underground formation, then a reservoir cannot exist.

Reservoir Types

Geologic structures of many different sizes, shapes, and types form the reservoirs in which petroleum may accumulate. Sedimentary rock is deposited in layers called *strata*, which are not strong

Modern Petroleum

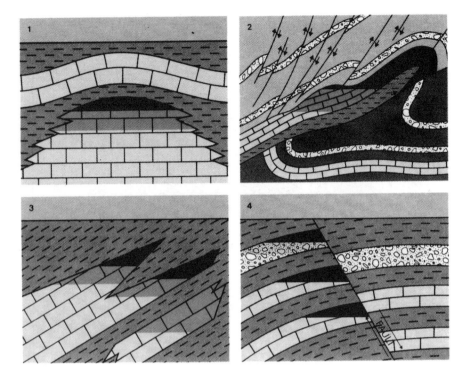

Some other common types of traps for oil and gas.

1. Limestone reef trap of the type found in West Pembina.
2. Reservoirs in folded and faulted strata. Turner Valley's oil was found in traps like this.
3. Stratigraphic trap of the type found in southeast Saskatchewan.
4. Displacement of rock layers along a fault. This type of structure is found in some areas of the Peace River district.

Legend

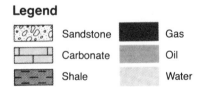

Sandstone		Gas	
Carbonate		Oil	
Shale		Water	

Figure 2–11 Oil and gas traps (courtesy Anne McNamara, Petroleum Resources Communication Foundation).

enough to resist the internal movements of the Earth. Each time a movement occurs, the strata is deformed. The layers may buckle into folds, as in both modern and ancient mountain chains, or they may become small wrinkles or great troughs many mile across. They may also be sharply tilted and then broken off, or twisted and rotated.

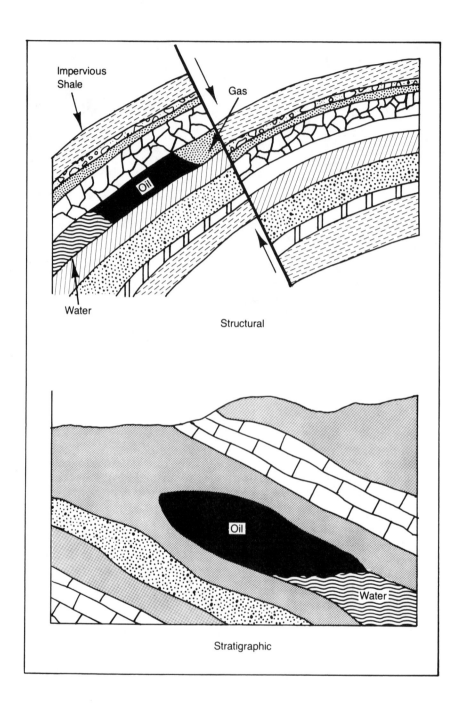

Figure 2-12 Structural and stratigraphic traps.

Modern Petroleum

Generally speaking however, there are two types of impenetrable barriers that can impede the movement of petroleum and thus cause the formation of reservoirs: *structural traps* and *stratigraphic traps* (Figure 2–12).

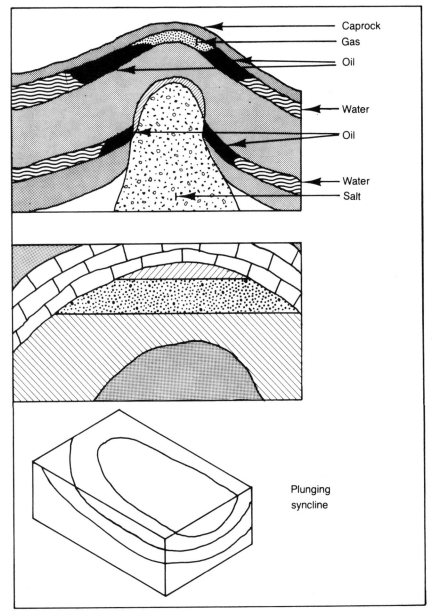

Figure 2–13 A salt dome, an anticline, and a syncline.

The How and Where of Petroleum

Structural Traps

Structural traps are caused by a deformation in the rock layer that contains the hydrocarbons. Two common examples would be *anticlines* and *fault traps*.

Anticlines

An *anticline* is an upward folding in the layers of rock, much like an arch. Anticlines may be symmetrical with similar flanks, or asymmetrical with one flank steeper than the other. The ends of both usually plunge steeply downward. The petroleum migrates to the highest point of the fold where its further escape is blocked by an overlaying barrier of impenetrable rock (Figure 2–13).

A *syncline* is a trough, or downfold. They may be thought of as inverted anticlines.

Dome and plug traps are sharp anticlines that are severely curved at the top. Sometimes accumulations of oil are found in porous rock that has a salt core or plug thrusting up in its center.

Fault Traps

Faults result when the rock on each side of a fracture shifts its position causing the shearing off or offsetting of strata. The actual displacement of a fault may be measured in miles or in inches. *Fault traps* occur when petroleum rock is entrapped by a non-porous layer of rock moving into a position directly opposite the oil-bearing layer. *Normal* and *reverse faults* have vertical movement. *Thrust* and *lateral faults* mainly move horizontally (Figure 2–14).

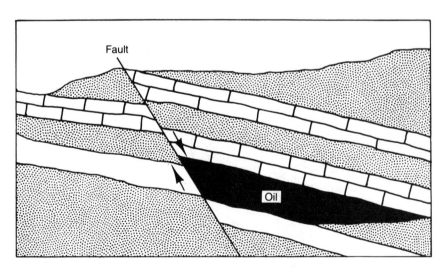

Figure 2–14 A fault is a crack in the Earth along which layers move.

Stratigraphic Traps

Stratigraphic traps result from the reservoir bed being sealed by other beds, or by a change in porosity and permeability (see below) within the reservoir itself. They include *truncation* (a tilted layer of petroleum-bearing rock is cut off by a horizontal, impermeable rock layer), *pinch-out* (a petroleum-bearing formation gradually cut off by an overlying area), or a porous layer surrounded by impermeable rock.

Lens traps are caused by abrupt changes in the amount of connected pore space. This is usually brought about by irregular depositing of sand, shale, or limestone at the time of formation causing porous, oil-bearing rocks to be confined within pockets of non-porous rock.

Combination traps are formed by a combination of faulting, folding, and porosity changes.

Unconformities. The upward movement of the oil cannot continue because a cap has been laid down across the cutoff surfaces of the lower beds.

Joints occur when rock is fractured as a result of Earth movement.

Properties of Reservoir Rocks

In order to evaluate the potential of a reservoir, the petroleum geologist must have the following data: (1) the capacity of the rock to contain fluid, (2) the relative amount of fluid present, and (3) the ability of the fluid to flow through the rock to the well. This last is determined by two factors: *porosity* and *permeability*.

Porosity

Porosity, simply stated, is the capacity of the rock to hold fluids. Or, it is the volume of the non-solid or fluid portion of the reservoir divided by the total volume. Thus porosity is always expressed in percentages. To visualize the concept of porosity, imagine a box full of balls of equal size stacked on top of each other so that only the most outward points of each ball touch the ones above, below, and to the sides. The unfilled spaces in between the balls would be the pore spaces and would represent a porosity of 47.6%, the highest that can be expected.

If the same balls were arranged into layers so that the upper layers nestled into the ones below, the porosity would be reduced to 25.9%. The size of the balls in either case would make no difference as long as they were all the same size. Since in reservoirs the rocks are never all the same size, nor stacked in neat columns, actual porosity may range from 3 to 40% (very rare) with a usual porosity in the area of 20% (Figure 2–15).

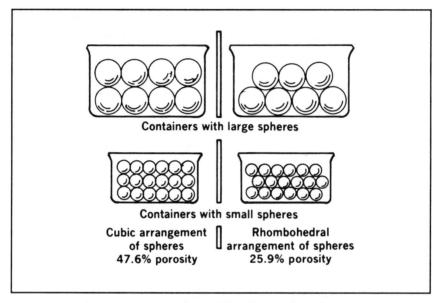

Containers with large spheres

Containers with small spheres

| Cubic arrangement of spheres 47.6% porosity | Rhombohedral arrangement of spheres 25.9% porosity |

Figure 2–15 The arrangement and size of the spheres affects the porosity. Cubic arrangement can have a maximum porosity of 47.6%. A rhombohedral arrangement can yield a porosity of 25.9% (courtesy SPE–AIME).

Porosity of 20% usually occurs only in the "younger" layers near the surface, as porosity usually tends to decrease in the deeper and older layers. This decrease is caused by the weight exerted by the succeeding layers, the effect of time on the rock, and by particles becoming cemented together. This pattern of depth affecting porosity is apparent in shale as well as sandstone, although porosity is generally lower in shales to begin with since it is more compacted, and old shales at great depth have been compressed much more than sandstone at a similar level. Limestones and dolomite do not follow the same depth pattern. They do, however, compress much more than sands.

Processes other than those above create what is known as secondary porosity. This results in much higher permeability in the reservoir because the pores or openings are much larger. It may be the result of ground water dissolving limestone or dolomite which causes larger openings between rocks known as vugs or caverns, or dolomitization, in which the limestone shrinks as it turns into dolomite. Fracturing of the area can also cause secondary porosity.

Modern Petroleum

Pore Space Saturation

If porosity represents the capacity to contain fluids, then *saturation* is the actual amount of fluid present in a given space. If expressed in percentage, a 20% saturation would indicate that one-fifth of the available space contains the fluid being measured, petroleum or water. The extent of petroleum, or hydrocarbon, saturation is one of the factors in determining whether a reservoir is economically worthwhile to develop (Figure 2–16).

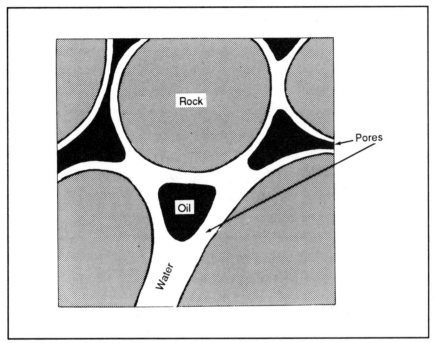

Figure 2–16 Sand grains, coated with a covering of water. Oil occupies the spaces in the larger pores.

Permeability

The permeability of a reservoir is that factor which determines how hard, or how easy, it is for a fluid to flow through the formation. It is not enough for the geologist to know that the oil is present; he must also be able to determine how easy it will be for the oil to flow from the reservoir into the well. This will be based on several factors: the property of the fluid itself, expressed in *viscosity* (thickness; a thin liquid can be pushed though rock more readily than a thick one), the *size* and *shape* of the formation, the *pressure*, and the *flow* (the greater the pressure the greater the flow).

Permeability is usually measured in units called *darcies*, after Henry d'Arcy, the French engineer who, in 1850 found a way to measure the relative permeability of porous rocks. In most reservoirs, the average permeability is less than one darcy, so the usual figures are in thousandths of a darcy or *millidarcies* (md). Permeability for a fine-grained sand may be 5 md or a coarse sand that is highly porous and well-sorted may run to 475 md. However, if the coarse sand happens to be poorly sorted it may run only 10 md (Figure 2–17).

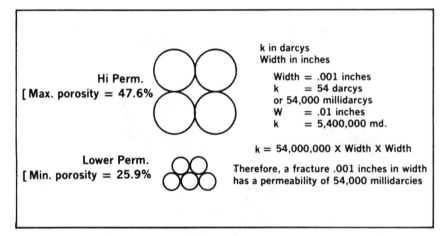

Figure 2–17 Determining permeability measurement (courtesy Dresser Atlas).

Another factor which must be taken into consideration is that at great depths, the weight of the overlaying rock may compact the sand grains closer together. Not only do smaller pores and lower porosity result, there is also a tremendous decrease in permeability. Cementation, which tends to fill the pore spaces, also increases with depth. A reservoir that may be a good producer at one depth may be of no economic value at all at a lower depth if the petroleum cannot flow through the rock to the well.

Reservoir Fluids

A fluid is any substance that will flow, including oil, water, and gas, the three principal fluids found in a reservoir. Oil and water are liquids as well as fluids. Natural gas is considered a fluid even though it is not a liquid.

Water

Since we have said that the reservoir is a sedimentary bed formed from the sea, some of the seawater remains trapped in the reservoir. Geologists call this *connate* (Latin for "born with") *interstitial* (because it is found in the interstices or pore openings) *water*. Thus connate water is that which was present at the formation of the reservoir. From core samples, the amount of connate water can be determined, and these figures are given as a percentage of the volume of the contents of the pore spaces. Besides the connate water found with the oil and gas, virtually all reservoirs have water-filled formations in addition to the petroleum zone.

Bottom water is that which is directly below the oil, and *edge water* is that which surrounds the oil accumulation on the sides. It is this water that provides the water drive, or pressure exerted on the oil in many reservoirs. This is necessary because oil cannot move itself to the surface, so it is the energy furnished by the pressure of the water or gas associated with the oil that drives the oil to the surface.

Oil

Since oil is lighter than water, the oil in a formation is pushed toward the top and the water is pushed downward by the weight of the oil. However, not all of the water is displaced downward. Some of it remains to coat the walls of the pore spaces, and if the interstices are very small (capillaries) the water will remain in them. So water not only occurs below the oil zone but within it as well.

Gas

Natural gas is always present with oil from a reservoir. In fact, gas under pressure provides the prime drive in recovering the oil. Gas occurs as either *free gas*, that is, separate from the oil, or as *solution gas*, which is dissolved in the oil. The relationship between the oil and gas depends on how much the liquid oil is saturated with dissolved gas. Often in the past, and in some remote locations yet today, the reservoir is punctured and the gas is vented to the atmosphere or burned off in flares during attempts to recover the crude oil below.

The gas and oil will remain in solution as long as the temperature remains low and the pressure high. When the oil is pumped to the surface and the pressure is lowered in a separator, the gas will come out of solution. But while it is still in solution in the reservoir it occupies space, and this has to be taken into account when the volume of the oil is calculated.

If there is less gas present than the oil in the reservoir can absorb, then that oil is unsaturated. If the opposite is true, then the oil is supersaturated.

If free gas is present in the formation, it will rise to the top and form a gas cap. This means that the oil below is saturated with gas dissolved in solution. It also means that since saturated oil is lower in *viscosity* (thickness) it will move easily through the formation to the well.

Distribution

As we have seen, gas, oil, and water in a reservoir tend to separate into three layers. In the early stages of development the contact line between oil and water is of primary interest. This is not a sharply defined horizontal line running through the reservoir. Instead, it is an area containing both oil and water that may be as much as 15 ft in thickness which is caused by capillary action. At the upper end, is the area where oil and gas meet. Since there is a greater difference in the specific gravity of oil and gas than there is between the specific gravity of oil and water, the oil does not extend as far into the gas as the water does into the oil.

Fluid Flow

Besides oil being present in sufficient quantities, and water and gas available to help move it to the surface, three other factors must be present in the reservoir to make it economically worthwhile. These are *pressure gradient, gravity,* and *capillary action.*

Pressure gradient means that a difference exists between pressures measured at two different points. If the pressure is lower at the well than at other points in the reservoir, the areas of higher pressure will exert energy that will force the fluids to and up the wellbore resulting in the well flowing. If such pressure is not present in sufficient quantities, then a pump or other artificial means will have to be used to lift the oil to the surface.

If the well is drilled at a low point in the formation, gravity will drive the fluids downhill from the higher points toward the well.

Capillary action can be demonstrated by partially inserting a piece of absorbent paper in a container of colored liquid. You will soon see that some of the liquid has risen into the portion of the paper that extends above the surface. This is the result of capillary action; the liquid has been drawn into tiny openings in the structure of the paper. Oil and water tend to move upward and outward through the pore spaces in the rock the same way.

Drives

Reservoir drive mechanisms are usually termed either *depletion drive* or *waterdrive*. Since oil bearing rock usually occurs as a permeable layer surrounded by other impermeable layers, this enclosure of the reservoir rock is a feature common to any type of drive. Depletion drives are those common to a closed reservoir situation. This is where the oil does not come in contact with water-bearing permeable sands. Since the petroleum is, in effect, isolated in a totally enclosed space, the only energy available to drive it to the surface is from gas in solution with the oil *(solution gas drive)*, or from gas above the oil accumulation in the reservoir *(gas cap drive)* (Figures 2–18 and 2–19).

Waterdrive occurs when water moves in to occupy the space left as petroleum is removed, and the pressure of the water forces the remainder of the petroleum toward the surface (Figure 2–20).

Solution Gas Drive

A typical situation in a solution gas drive reservoir might be one where the pressure on the oil is so great no gas bubbles can form. If a well is drilled and the pressure relieved, the gas will come out of solution and begin to form bubbles. As these bubbles expand, their pressure will force the oil to the well and thence up to the surface. This is analogous to a can of soda pop on a hot summer day. If you shake a can of warm pop violently before snapping the top open, the gas dissolved in the pop, in this case carbon dioxide, will expand rapidly into large bubbles as the pressure is released and spew sticky foam over everything in its path. Solution gas is also called *entrained gas.*

Wells drilled into a solution gas drive reservoir accumulation usually must be pumped at first. The reservoir pressure will show a line of rapid and continuous decline and the gas-to-oil ratio, which will be low at first, will rise to a peak and then begin to drop again. How much of the original oil in the accumulation that can be recovered will vary according to the physical properties of the oil itself and the methods used by the producer.

Gas Cap

If no gas production wells are drilled into a gas cap drive accumu-lation to drain off the gas at the top of the reservoir, the gas will

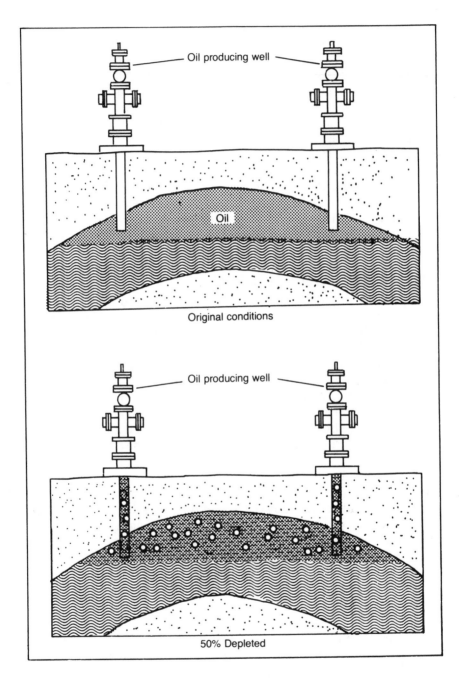

Figure 2-18 Solution-gas drive reservoir.

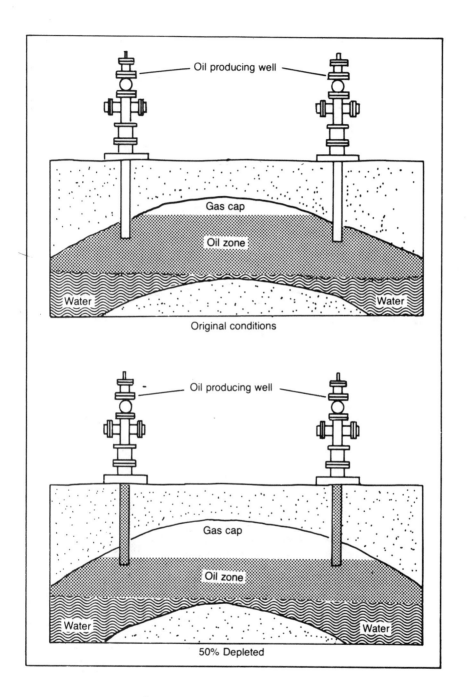

Figure 2-19 Gas-cap drive reservoir.

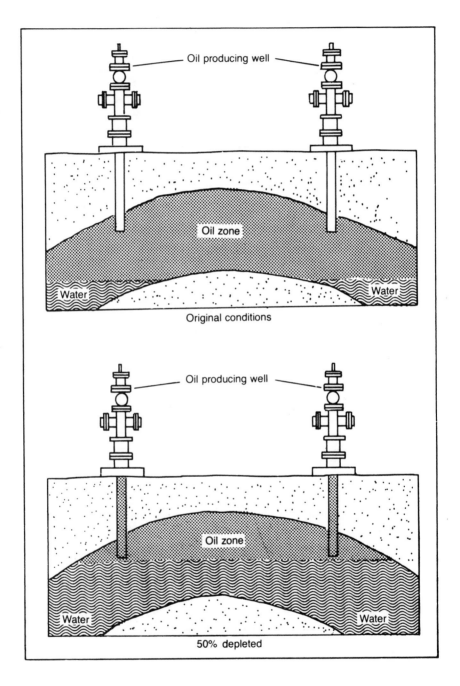

Figure 2-20 Waterdrive reservoir.

continue to expand and push down on the oil. This same energy also pushes the oil up the wells drilled into the oil zone.

Since more energy is present than in other depletion drive reservoirs, production is more stable and does not decline as rapidly. Thus, depending on how large the original gas cap is, there will generally be a long flowing life. While the reservoir pressure will continue to decline, it will be at a slow rate, and the gas-to-oil ratio will continually rise. And, because of the greater energy available for drive, the initial recovery of oil may be twice as much as that from a solution gas drive reservoir.

Water Drive

In neither a gas cap nor a solution gas drive reservoir will there be any appreciable amount of water production unless it comes from wells drilled around the edges of the accumulation. In a water drive reservoir however, water provides the energy for the drive. This may occur as either as a *bottom water drive* or an *edgewater drive* situation. In either case, the water is pressing against the petroleum. As it is withdrawn, the water enters to fill the spaces in the rock.

The edgewater drive occurs usually when the oil accumulation is relatively thin, completely fills the formation, and the formation is surrounded by water. A well drilled on the top of such a formation would reach into the oil zone. However, water would be produced from wells around the periphery.

As the name implies, bottom drive occurs when the oil accumulates in a thicker area underlain by water, which pushes it toward the surface. Edgewater drive is usually considered more efficient than bottom water. Water drive wells are often termed rate-sensitive. That is, oil may be taken from the accumulation faster than the water can flow in to occupy the vacant space left by the oil and thus maintain the reservoir pressure. If this happens, the pressure of the water declines and the well may become a combination: part water drive and part depletion drive. If overproduction occurs, the effect of the water will be so slight as to give the well the characteristics of a depletion drive well.

Generally speaking however, the pressure will remain high as long as the oil removed is replaced with an equal volume of water. And, so long as the pressure remains high, the oil-to-gas ratio will be low since there will be almost no free gas in the reservoir. If a high pressure is maintained, such a well will continue to flow until the oil is exhausted and the well finally produces only water. Some water will be brought to the surface at first, and this amount will gradually increase throughout the lifetime of the well. Since water is more efficient than gas in displacing oil from the rock, anticipated recovery rates are considerably higher for a water drive well. Up to 75% of the oil in place may be recovered as compared with 40% for a gas cap drive and 25 to 30% for solution gas drive.

Combinations of water and gas cap drive may occur naturally and some wells flow by gravity drive. That is where the well is drilled at a point lower than surrounding areas of producing rock and the oil drains downhill toward the well. Such gravity or drainage drive wells offer excellent recovery if the reservoir dips sharply down to where the well is located and the reservoir rock is highly permeable.

Today many methods exist for enhancing the recovery of oil from a reservoir after the natural drive declines. *Enhanced recovery* will be discussed in a later chapter.

Reservoir Economics

Since the primary reason for locating and exploiting reservoirs is economic, some idea of how much oil and gas they contain must be available before any attempts to drill and produce are made. Oil and gas still in the ground are referred to as *reserves. Estimated reserves* are those in reservoirs which have not been tested or produced from. *Proven reserves* are those known to exist in producing reservoirs. By making various tests and measurements, the amount of fluids in place in a reservoir may be determined.

Determining Fluids in Place

The following factors are needed to determine the fluids in place in a given reservoir, as well as the percentage of each that is present.

Acre Feet (ac–ft) = the number of acres the reservoir covers times its thickness in feet.

Barrel (bbl) = 42 U.S. gal or 5.61 ft^3

Acre Foot = 43,560 feet^2 (43,560 ÷ 5.61 = 7,764)

h = Reservoir thickness in feet

A = Reservoir area in acres

ø = Porosity (in percent)

S_o = Oil saturation

S_w = Water saturation

S_g = Gas saturation

To determine barrels of oil in place: 7,764 x h x A x ø x S_o

To determine barrels of water in place: 7,764 x h x A x ø x S_w

To determine cubic feet of gas: 7,764 x h x A x ø x S_g x 5.61

Example:
7,764 X 20 ft X 640 acres X 0.20 porosity X 0.10 oil saturation = 1,987,584 barrels of oil in place.

Note: The total amount of fluids present ($S_o + S_g + S_w$), always equals 100%. The actual percentages of each are determined by testing, such as by taking a core sample. If two of the three are known, then the third can easily be determined. For example, if S_g = 10%, and S_w = 20%, then S_o must equal 70%. (100% minus 30% (10% plus 20%), equals 70%).
It should be remembered that not all these barrels of oil can actually be recovered. To determine recoverable reserves, other factors such as pressure, viscosity, and permeability have to be taken into account. Even then, only a portion of the oil in place can be removed during the initial production phase of the reservoir's life. Producing additional oil during the *secondary* and *tertiary* stages will be discussed in the section on enhanced recovery.

Chapter References

1. U. S. Department of the Interior, U.S. Geological Survey, *Annual Report 1976*, "Plate Tectonics and Man," by Warren Hamilton. (Washington D.C.: Government Printing Office, 1980), p. 3.

Other References

Kenneth E. Anderson, *All About Oil*, (Stillwater: Anderson Petroleum Services, Inc., 1981), p. 2–2.

Bill D. Berger, *Facts About Oil*, (Stillwater: Oklahoma State University, 1975), p. 16.

Thomas Gold, *Power from the Earth: Deep Earth Gas: Energy for the Future*, (London: J.M. Dent & Sons Ltd., 1987), pp. 18–23.

3

Exploration—The Search for Oil

When men first began to seek petroleum, the easiest way to find it was to look for evidence of oil seeps on the Earth's surface. Generally, oil seeps are the result of up-dips or are seepage from along a fracture. Observation of seeps has lead to the discovery of many of the world's great oil fields in the United States, in the Middle East, in Venezuela, and at other places on the globe.

That oil seeps occurred on anticlinal slopes was noted as early as 1842, but not until Drake drilled through rock to bring in his well was it learned that the Oil Creek wells were on anticlines. This was an interesting theory, and it was used in seeking gas accumulations in a limited degree in the United States, and by geologists seeking oil in other countries. But no domestic oil companies seem to have established geological departments until almost the turn of the century.

Shortly thereafter, however, attention began to be given to the role of the geologist and later, the geophysicists, in locating oil accumulations, or, at least places where oil is likely to occur. Since the 1920s there has been a steady increase in the amount and quality of the technology available to oilmen to help them in their search for new supplies of petroleum. These tools and techniques were, of course, unavailable to the early-day wildcatter who often selected drilling sites by intuition, by using dowsing rods, or by "creekology"—searching the bed of a creek for "signs."

During the past decade, rapidly advancing technology, much of it microprocessor-based, has had an explosive impact on the petroleum industry. More new technology and techniques have come about in the past 10 years than in the preceding 100. And, exploration is the segment of the industry that has been most impacted by this overwhelming wealth of new knowledge.

Field equipment is smaller, lighter, more reliable, more accurate, more sensitive, and provides far more detailed data than any that was available at the beginning of the last oil boom in the late 1970s. In between the beginning of the exploration phase and the construction of the final maps that show an oil company where to drill, are new tools and devices that were undreamed of a few years ago.

However, the basic tool in the search for oil remains a knowledge of the Earth itself—how it was formed, its composition, and its present configuration. It is not enough though, to merely become aware of the existence of an oil accumulation at a given location. Before investing what may well be tens of millions of dollars, the operator needs to know if the well will be commercially feasible. Or, simply stated, will he recover his investment and perhaps make a profit? Not every well *spudded* in today is assured of a rich strike. There are still a great many *dusters*, but modern methods lessen the risk. By using state-of-the-art seismic equipment and a powerful supercomputer, Phillips Petroleum hit discoveries on three out of three attempts in northern Texas, and on six out of six in Alberta.

Exploration

The search for oil begins with geologists and geophysicists using their knowledge of the Earth to locate geographic areas that are likely to contain reservoir rock. Once such a "likely" area is found, then more specific tests and investigations are made, and the information from them is used to construct subsurface maps, and today, even three-dimensional models, of what lies under the Earth's surface.

The Geologist

It is the geologist's job to study rocks that are exposed to view and can provide clues to subsurface structures and conditions. Thus outcrops of rocks in river valleys and gorges, in irrigation ditches and mine shafts, and samples taken from water wells and existing oil wells, are all sources of information.

If exploration takes place in a known producing area, geological information is easier to come by. In addition to information gained by examining the surface, detailed subsurface data may be obtained from required government reports, histories of previously drilled wells, and published geological papers.

Geologists study rock cuttings and core samples taken from previous wells to determine the age of the various formations and the environment when it was deposited. Much knowledge is obtained from logs—instruments lowered into wellbores to measure the electrical, radioactive, sonic, thermal, and compositional properties of the rock. This information, together with tests made for fluid content and pressure, enable scientists to determine porosity, permeability, and the ages and sequences of the various rocks, and to look for oil and gas "shows." From all this data, they can suggest where to drill for oil and gas.

Seismology

While the geologist had been the first to play an active role in the search for petroleum, by 1920 the increasing demand for sources of supply made it apparent that merely looking for signs of oil on the Earth's surface was not sufficient, that better methods were needed. So the work of the geologist began to be blended into that of the geophysicist who uses technology, including seismology, to locate and describe subsurface formations. Two of the early methods developed were the torsion bar balance, which measures gravitational forces within the Earth by the amount of torsion, or twisting, they cause on a tightly strung fine wire filament, and the seismograph. With the seismograph, subsurface structures can be deduced by measuring the transit times of sound waves generated by man-made means.

A device called a seismometer to detect underground events was invented in 1841 by David Milne. Then, in 1855, L. Palmiere constructed what he called a *seismograph*, which could both detect and record, at his observatory on Mt. Vesuvius. Its name, taken from the Greek, means literally "earthquake writings," and that was its first, and still one of its major current uses—to plot, record, and study—earthquakes because the *seismograms* it produces are simply that, records of earthquakes. However, in petroleum exploration, instead of waiting for nature to provide an earthquake, geophysicists create their own. Although many times smaller than the real thing, these miniature earthquakes are vital in the search for oil.

Another use for the seismograph was found during World War I when Dr. L. Mintrop invented a portable model that the German Army could use to locate Allied artillery emplacements. Three seismographs were set up in known positions facing an Allied gun. When the gun fired, a record was made of the earth vibrations caused by the report, and from these the exact location of the gun could be calculated.

After the war ended, Dr. Mintrop reversed this process. He set off charges of dynamite at known locations and recorded the vibrations that were produced on portable seismographs. Then he measured the distance and computed the geology using the process described below. His efforts marked the beginning of the modern seismic industry. A company was formed called Seismos, and in 1924 it supplied the first seismic crew brought to North America for use in Texas by Gulf Production Company.

This was shortly after the 1921 Oklahoma City tests conducted by J. Clarence Karcher, William B. Haseman, Irving Derrine, and William C. Kite that according to the inscription on the memorial erected at the site "proved the validity of the reflection seismograph as a useful tool in the search for oil" (Figure 3–1).

Figure 3–1 Plaque erected at the site of the field tests of the reflection seismograph on the grounds of the Belle Isle Library in Oklahoma City (photo by Ken Anderson).

In theory, seismology is really very simple. Because the Earth is composed of layers which vary in density and thickness, then when energy, such as that produced by an explosion, is released on the surface, it travels in all directions. As the energy strikes each of the layers, part of it is reflected back to the surface where it is detected and recorded by the seismograph. The process is comparable to a child bouncing a rubber ball. If the ball strikes a concrete sidewalk it reacts quite differently than it would if it landed in a pile of sand. Seismology is really very similar. A small charge of dynamite is exploded, usually in a shallow borehole. The resulting waves spread out through the ground encountering different strata and formations. Just as with the bouncing ball, each formation reflects the energy waves according to its own "bounce" characteristics. The waves deflect upwards to the surface where they are picked up by *geophones* or *jugs*, sensitive detection devices embedded in the ground at predetermined locations. The geophones are attached to cables which carry their signals to the seismic recording truck. There they are amplified and translated onto a permanent film or magnetic tape from whence they can be used to produce an accurate map of the structures under the surface (Figure 3–2).

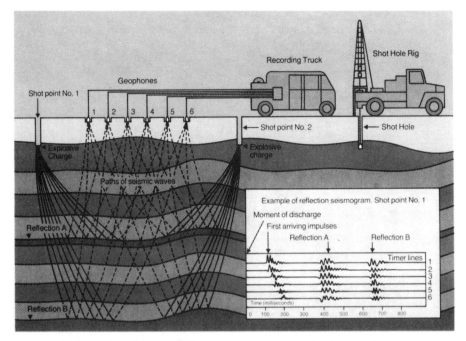

Figure 3–2 Principles of seismic reflection surveying, using shot-hole explosive methods. The explosion at shot point No. 1 creates shock waves which are reflected by the subsurface formations, picked up by the geophones, and recorded in the recording truck (courtesy Anne McNamara, Petroleum Resources Communication Foundation).

The data is gathered over a horizontal distance and compiled to create a vertical cross section of the Earth. By careful examination, the geophysicist is able to ascertain the possibility of the presence of oil and gas (Figure 3–3).

In the Field

In the past, the traditional land-based seismic, or *doodlebug*, crew consisted of the *party chief* who was in overall charge of the crew; the *geologists* or *geophysicists* who decided where the shot would be made, plotted the locations of the various pieces of equipment, and decided on the "pattern" to be used; the *surveyors* who marked the shot hole and geophone locations in the pattern desired, which is often a star or other geometrical configuration; the *drillers* who drilled the shot holes; the *loaders* who made up and loaded the

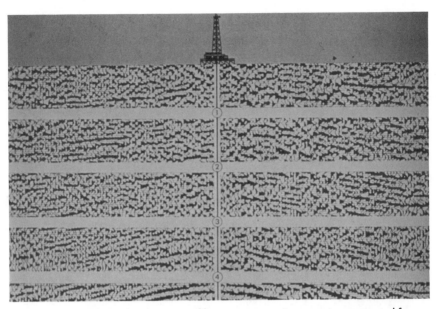

Figure 3–3 Conceptualized version of how a cross-section map is constructed from seismic information.

explosive charges; the *shooters* who connected the charges and fired them on command from the geologist; and finally, the *jug hustlers* who pulled the cables from the cable truck, arranged them in the desired patterns, and attached the geophones. After the shot was fired, they had to "hustle" (run) to pick everything up and "hustle" it to the next location to repeat the process.

Non-Explosive Seismology

In the past 15 years, the use of high explosives by land-based seismic crews has decreased greatly. While some soil and surface conditions still call for the use of dynamite to get accurate data, today much information is garnered by the use of vibrating or weight-dropping machines. One of the pioneers in the field was Conoco, at Ponca City, Oklahoma, the developer of the VIBROSEIS™ system now manufactured by Mertz, also of Ponca City.

VIBROSEIS™ is basically a method of substituting man-made vibrations or waves for those caused by an explosion, much the same as when dynamite was substituted for earthquakes. Specially designed equipment built into either wheeled or tracked vehicles makes contact with the Earth and creates shock waves by either dropping a heavy weight or using a vibrating device to create waves. These penetrate the surface, strike underground formations, and are reflected back to the seismograph in exactly the same manner as explosion-generated waves. At sea, although technically not VI-

54

BROSEIS™, bubbles or bursts of compressed air from devices towed behind boats are used to gather seismic data without the damage to the underwater environment that explosive charges would cause (Figures 3–4 and 3–5).

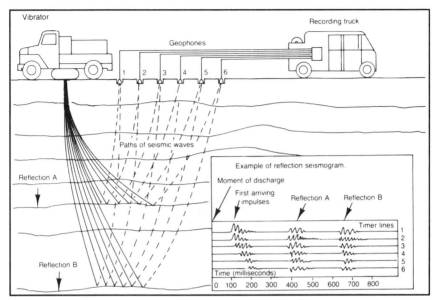

Figure 3–4 When using seismic reflection surveying, the vibrator at point 1 creates shock waves that are reflected by the subsurface formations, picked up by the geophones, and recorded in the recording truck.

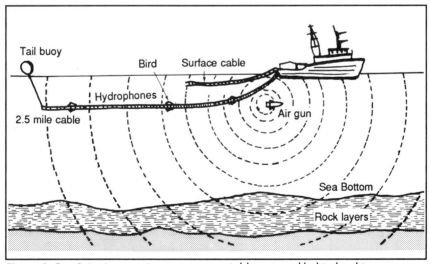

Figure 3–5 Seismic operations at sea use air blasts towed behind a ship to create shock waves that are picked up by hydrophones and transferred to the ship (courtesy Mobile *World*).

Although it cannot be used on every type of soil or terrain, VIBROSEIS™ has many advantages, among which is its ability to be used in populated areas without damage or alteration to the surroundings. It is usually less expensive and less time-consuming than employing drill rigs and shot crews and most obviously, the costs and hazards of transporting and handling high explosives are eliminated (Figure 3–6).

Figure 3–6 Marching in a line, Vibroseis units originate sound waves of varying frequencies at the surface (courtesy Anne McNamara, Petroleum Resources Communication Foundation).

Direct Detection

The above techniques can only furnish information that indicates a probability of the occurrence of a petroleum accumulation at a given site. The only way to confirm the existence of a reservoir is to drill a test well, which can be an expensive and time-consuming project. However, in the mid-1970s scientists at Shell Oil reported a new development called *direct detection of hydrocarbons*, which offered the promise of more accurately pinpointing hydrocarbon accumulations.

Called *bright spots*, the method called for geophysicists to closely study the white bands that often appear on seismic record strips. Previously these had been largely ignored, but it was discovered that they could be interpreted as depicting high levels of energy, and thus revealing areas of hydrocarbon accumulations. The bands were caused by porous rock filled with entrapped natural gas which reflected a stronger seismic echo than water-filled rock.

Modern Petroleum

It was not a panacea, because the bright spots could be misinterpreted, or sometimes they were misleading in that not all bright spots contained hydrocarbons, and not all hydrocarbon deposits showed up as bright spots. However, it did give scientists an important new tool in the search for oil.

Subsurface Mapping

The geologists and geophysicists combined their knowledge of the history of the Earth, its rocks, depositional environments, and geologic characteristics with data obtained from actual surveys. Their information was further enhanced by direct evidence from *core samples* of reservoir rock obtained by drilling into the Earth, or by the data recorded by electronic, sonic, or nuclear *logging devices* lowered into a wellbore.

From all of this it became possible to construct maps of areas beneath the surface. *Contour maps* show geologic structures relative to reference points called *correlation markers*. Contour lines were drawn at regular intervals of depth to add a third dimension, and increasing depths were often illustrated by the use of contrasting colors on the map.

Isopach maps illustrated variation in thickness between correlation markers. Again, color shading was often used in conjunction with the isopach map lines for clarity and contrast. Other maps were constructed to show faults and their intersections with other beds and faults, the porosity and permeability, variations in rock characteristics, and structural arrangements.

All of these, however, still only provided a view from above, much like looking down on the floorplan of a house. To get a clearer perspective, they had to be supplemented with a *cross sectional map* showing the elevation or side view. Cross sections may be made for a large area, or to illustrate just one detail. The various maps were then used together with the cross sections.

Conceptual Models

All of this information was sometimes used to construct *conceptual models* or someone's idea of how a geological area was structured, what it looked like, and where petroleum accumulations were likely to be found. Fault patterns were first shown on conceptual models around 1930, and these proved useful in developing petroleum accumulations around salt domes and faulted anticlines.

By the 1950s, information gained through subsurface studies, plus knowledge of recent sediments, were the basis of conceptual models that provided ideas about reservoir distribution and traps involving both carbonates and sandstone. Most of these included knowledge of the environment of deposition of reservoir layers. This was an important step, because being able to predict pore space saturation and sand trends is important to operations dealing with sandstone reservoirs. It can be extremely disappointing when first developing a reservoir to drill a well near an established producer and fail to find any pay sand. The same can be true later in the development process when a well fails to find the sand present in nearby wells. Missing sands become an even greater problem when it becomes necessary in the life of a producing field to *enhance recovery* through *waterflooding* or other methods of *secondary* or *tertiary recovery.*

Magnetic surveys

Magnetometers are devices that can detect minute fluctuations in the Earth's magnetic field. These fluctuations or "magnetic anomalies" can supply information about buried rock structures. Early magnetometers were bulky, earthbound instruments that had to be wrestled from site to site by hand. This made the making of magnetic surveys arduous and time consuming projects because only a very small area could be covered each day. Finally, models were developed that could be towed behind an airplane and suddenly hundreds of miles of the Earth's surface could be surveyed in a very short time. The development of the helicopter brought even greater flexibility. But such surveys taken at low altitudes could only gather data about nearby geological structures like a mountain or surface fault. Then in 1981, NASA launched Magsat, a satellite capable of surveying magnetic features on a continental scale. While most of the readings came from Earth's main magnetic field, information was also gathered from many faint fields deep within the crust.

Why Is the Study of the Earth's Magnetism Important?

Because of the great depths involved, study of the magnetic fields is one of the few methods available to help man learn about the Earth's mantle, its contents, and its composition. The Magsat data has enabled scientists to construct contour maps using colors to add contrast to the magnetic anomalies which were arranged in order of intensity from low to high showing how subsurface rocks were being magnetized by the Earth's main field. Such data is hoped to reveal more and more information about the tectonic forces that move the continental plates and eventually prove useful in locating areas of buried minerals of interest.

Modern Petroleum

Some anomalies seem to announce the presence of large deposits of iron ore, of *kimberlite pipes*, where diamonds are found. It is interesting to note that these same kimberlite pipes contain deep-source hydrocarbons, a fact which is of interest to the proponents of the inorganic theory of the origins of petroleum.

Other Tools

Other tools developed for use by geophysicists are *gravimeters* which measure variations in the magnitude of the Earth's gravitational field to provide information about the depth and nature of buried rock. *SLAR*—Side-Looking Airborne Radar—can pierce cloud cover or dense jungle foliage to produce picture-like images of the Earth's surface. Today, ground radar, pulled across the surface on sleds can reveal heretofore hidden features. The coming of the space age brought with it *remote sensing*. This involves producing images by using infrared (heat sensitive) means to create photographs or television "pictures" of an area. Carried aboard aircraft or satellites, the sensing devices can gather a wide range of information—crop blights and insect infestations, salt water intrusion, faults, or buried mineral deposits. Today's infrared detection and sensing equipment is so sensitive that one can actually fly over a vacant, paved parking lot, and from high altitude be able to discern which parking space was last occupied from the heat image remaining from the last vehicle to leave (Figure 3–7).

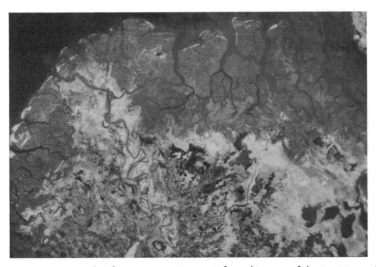

Figure 3–7 An example of remote sensing, an infrared image of the Louisiana delta.

In the past, as newer and newer technology presented more and more raw information to explorationists, the task of gathering, compiling, and arranging it into a usable form grew rapidly. Just physically entering on maps by hand became a long and tedious process that often took hundreds of man-hours by geologists and other earth scientists, cartographers, and mathematicians. In fact, so much data was becoming available this began to be an almost impossible task. Also, despite the skills of those involved, vital information was often not included because the technology could not provide it. Fortunately, the microprocessor, the same device that is now making so much information available, also provides the methodology for making it useful.

A New Era of Technology

Digital Seismic Recording Systems

Today, the use of digital seismic recording equipment has eliminated much of the drudgery, labor costs, and time consumed in gathering, processing, and transmitting high-quality seismic data.

One such system is the OPSEIS® 5586 Portable RF/Wireline Digital Seismic Recording System. Originally developed by Phillips Petroleum of Bartlesville, Oklahoma, in the 1970s, it is now built and marketed by Applied Automation, also of Bartlesville. Today's equipment uses the same microprocessor technology for collecting, editing, and recording seismic data.

OPSEIS® is designed to operate in difficult areas such as rivers, lakes, swamps, mountains, canyons, and jungles. Ideally suited for 3-D operations, it can use any standard energy source: vibrators, dynamite, weight drop, or air guns.

The heart of the system consists of portable *Remote Seismic Units*, also called *Remote Transmitting Units* (RTUs), and a *Central Recording Station* (CRS). The RSUs are connected to the CSU by a single-frequency VHF radio which provides two-way voice and digital communications. Each RSU is able to preamplify, filter, gain range, and digitize the analog data input from its geophones. Each system can accommodate as many as 4,000 stations (Figure 3–8).

Since the RSUs are not normally connected to each other by cumbersome cables, they can easily be moved for a new shot without having to relocate the CRS. However, if natural barriers interfere with the radio link, the RSUs can be connected by a wireline telemetry link. Such versatility allows almost any spread configuration. The exact placement of each RTU, and in fact, the beginning survey points for the entire operation can be achieved through the use of

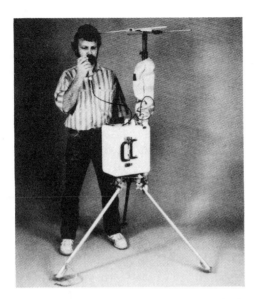

Figure 3–8 OPSEIS 5521 Remote Seismic Unit (courtesy Applied Automation Hartmann/Braun).

satellite positioning equipment. Once the primary and secondary control points are in place, thedolites and electronic measuring equipment can be used to lay out the rest of the sweep (Figure 3–9).

Figure 3–9 OPSEIS 5586 Central Recording Station (courtesy Applied Automation Hartmann/Braun).

Exploration—The Search for Oil 61

Following each seismic event, or series of events, the data from the RSUs are transmitted to the CRS, a portable, or truck-mounted seismic data gathering station. The CRS provides the operational control center and operator interface point for the whole system. It consists of a seismic processing chassis, a desk-top color graphics terminal, tape drive, and digital plotter. The system includes data quality assurance checks, source checks, and diagnostic routines. Additionally, the system writes a tape trailer after every recording trace which documents the test results acquired before and during the shot. The trailer record includes the leakage and continuity readings for that trace as well as the primary battery and supply voltages of the RSUs. These last features can save many thousands of dollars lost in locating leaking or open cables when working in wetlands. Also, the quality checking of data in the field to make sure that all equipment is functioning satisfactorily and furnishing correct data, can prevent extremely expensive and frustrating data loss or distortion, which, in the past, may have gone undetected until after the survey was completed (Figure 3–10).

Since the RSUs are battery powered, periodic charging is necessary to prevent equipment failure or malfunction. The batteries are recharged on a regular and documented basis at mobile charging units located just outside the recording area.

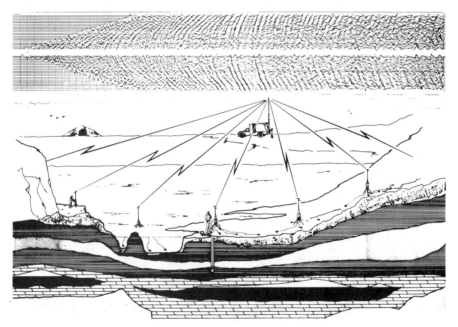

Figure 3–10 OPSEIS 5586 Field Operation (courtesy Applied Automation Hartmann/Braun).

CAEX

One of the biggest breakthroughs in exploration is the use of CAEX or *Computer Assisted Exploration.* This is the use of computer workstations for the interactive processing of the vast amounts of data available today and getting it ready for interpretation. Either *three-dimensional* (3-D), or *two-dimensional* (2-D) systems may be used.

3-D Seismic Imaging

Oilmen say that the odds of finding oil in unexplored territory range from one in five to one in ten. Even when drilling close to where oil has already been discovered, the odds can still be two to one. In order to achieve their perfect string of nine out of nine mentioned earlier, Phillips Petroleum used 3-D Seismic Imaging, the most advanced—and expensive—of the new techniques. This involves recording data from several thousand locations as compared with several hundred with traditional 2-D methods. The 3-D process compiles the data and feeds it into a super computer, in Phillips' case a Cray 1M 2800, which is capable of making 160 million computations in a second. The computer converts the data into a cube-like picture of the underground area under study in place of the old strip charts.

This technique works well in areas where the traps are small and complicated, as in the Hardeman Basin, and where older, less perceptive methods do not work as well. However, the costs of using 3-D must be considered. This includes using more sophisticated equipment, covering a greater land surface area during the sweep, which usually means increased expenses in arranging permission to use the land with the property owners. At current prices, 3-D surveys can cost up to $60,000 per square mile. Therefore, 3-D may be used only after 2-D studies have detected the presence of structures that are likely to contain oil.

2-D CAEX

The 2-D interactive processing of seismic data is an outgrowth of 3-D CAEX not, as one might suppose, the other way around. In 2-D, the workstations try to use as many 3-D attributes as possible, including enhanced color graphic displays as well as horizon and fault mapping. Often, 2-D can complete projects in two-thirds less time than the old "hand" methods. Obviously, it can handle much more incoming data, and it provides images that are far more accurate in detail since the use of color helps to distinguish features that previously may have been overlooked.

GIPSIE

One example of a system that meets the geologists' increasing need to evaluate more and more data in less and less time, produce accurate maps that show underground structures in precise detail, while simultaneously integrating information from other disciplines, is GIPSIE *(Geological Interactive Plotting System for Interpretation and Evaluation)*. GIPSIE, which automates many routine geological tasks, was developed by Shell Oil Company and the computer scientists and geologists of Intergraph Corporation of Huntsville, Alabama.

GIPSIE software actively generates well correlation panels, cross sections, maps, volumes and reserves, and 3-D models. It can also convert 2-D maps to 3-D. Special 3-D functions include both cube and mesh surface generation. Cross sections can be translated into 3-D, and well locations connected with fence diagrams. Intergraph's state-of-the art workstations powered by their CLIPPER micropro-cessors run user-friendly software that offers online help through tutorials and menus. Other hardware and software allows photo-grammetrists to compile digital basemaps from aerial photographs. Optical scanners can input data directly from existing maps, and digital data can be stored on either optical discs or on film.

In short, much of the formerly laborious handwork involved in handling an increasing amount of the analysis and interpretation of geologic and geographic information has been eliminated by micro-processors. Cubic models can be produced that illustrate not only the subsurface features but can also be capped by surface maps that show current land and lease ownership (Figures 3–11 through 3–15).

Figure 3–11 Intergraph's image-processing software is displayed on the new InterPro 3070—the industry's first workstation with 2-megapixel resolution and a generous 27-in. screen (courtesy Intergraph Corporation).

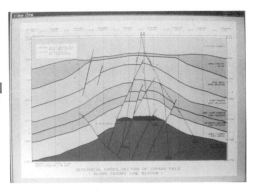

Figure 3–12 Generated with GIPSIE, a geological structural cross section with well paths displayed is created along a seismic line for correlation purposes (courtesy Intergraph Corporation).

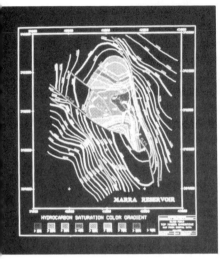

Figure 3–13 This map, generated with GIPSIE, shows hydrocarbon saturation values overlaid on the top of the reservoir structure (courtesy Intergraph Corporation).

Figure 3–14 A true petroleum GIS integrates subsurface interpretations, including porosity and hydrocarbon saturation distribution, with surface features such as lease boundaries. Displayed are the results of a topological query using this data (courtesy Intergraph Corporation).

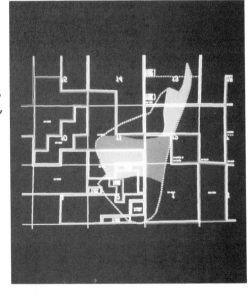

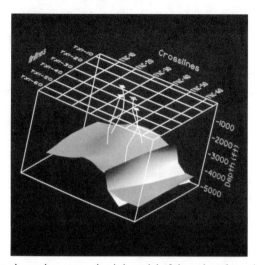

Figure 3–15 This three-dimensional solid model of the subsurface shows the geologists' horizon and fault interpretations based on data provided by deviated boreholes (courtesy Intergraph Corporation).

Chapter References

Bill D. Berger, *Facts About Oil*, (Stillwater: Oklahoma State University, 1975), pp. 18–22.

Focus on Energy, (Calgary: Canadian Petroleum Association, 1979), pp. 1, 2.

Geological Applications, (Huntsville: Intergraph Corporation, 1988), p. 10.

The Geophysical Story: Seismic Exploration for Oil and Gas, (Calgary: The Canadian Petroleum Association, 1979), p. 1, 2.

Thomas Gold, *Power From The Earth: Deep Earth Gas: Energy for the Future:* (London: J.M. Dent & Sons Ltd., 1987), p. 39.

P. Houdry and D. G. Lang, "3-D Land Seismic Acquisitions—A Case History," *The Oil and Gas Journal*, October 3, 1983.

OPSEIS® 5586 System Overview Manual, (Bartlesville: OPSEIS® Publications, 1989).

Lynn Railsback, "High Tech Brings Oil Risk Down," PhilNews, July, 1988, p. 1, 3.

4

Before Drilling Can Begin...

Mapping—Leasing—Permitting

It is not enough to know in what geological structures oil may occur. Before one can begin to drill for oil one must also know exactly where to locate the drill rig, and be able to describe that location in precise legal terms; obtain permission to drill for, and to produce hydrocarbons from the owners of both the surface of the land and the minerals lying beneath the surface (since these are often two separate parties); obtain the proper permits from a myriad of government agencies; and, not least of all, raise sufficient capital to pay for the venture.

Mapping

Land Measurement

The legal description of a piece of land is to a surveyor, landman, or attorney, what a street number, street name, city, state, and ZIP code are to a letter carrier. It describes precisely the exact location of the tract of real estate which it identifies. Therefore correct legal descriptions are vital to the sale, purchase, mortgaging, and leasing of land. And, while simple in concept, most people have no idea of what a legal description is, or how to read or write one correctly.

In order to establish property lines and boundaries, accurately plot proposed well locations, and legally describe land ownership, one must have a system of land measurement that is both highly accurate and relatively easy to work with. The system used in most of the United States meets these criteria. The rectangular survey system used today is the result of a land ordinance passed by Congress in 1785, which had realized that as the nation grew and more land was acquired, an effective means of establishing claims and boundaries was essential. Not included in the survey were the 13 former British colonies plus Maine, Kentucky, Tennessee, and later, Texas, which were already largely settled when they joined the Union. As a result, some areas, Texas included, still measure land by "metes and bounds."

The rectangular survey system called for by the Land Ordinance was confined to public domain that belonged to the federal government. States that were later formed from these public lands are known as public domain states.

The rectangular survey system is based upon the establishment of a *principal meridian* and a *base line*. These principal meridians run in a true north-south direction; the base lines run east and west at right angles to the meridians. The point at which the two lines intersect is called the *initial point* or *starting point*. The geographic locations of meridian and base lines were not fixed by law, therefore the locations of the 34 principal meridians were established by the government surveyors to meet their convenience. Meridians are named, and all legal descriptions of land end, with a notation that names the meridian involved, i.e., "I.M." or, "Indian Meridian" (Figure 4–1).

Townships

The legal description of a tract of land is based on an area known as a *Congressional township*, not be confused with a *Civil* or *named township*. The Congressional township is an area 6 mi square, and contains 36 mi².

Once the principal meridian and base lines are laid out, the surveyor returns to the starting point and lays out lines north and south every 6 mi on either side of the meridian. Then he lays out lines parallel to the base line 6 mi apart north and south of the base. This creates a grid in which the lines parallel to the base are called *township lines* and those parallel to the meridian are called *range lines*. Each square in the grid becomes a township. By determining its location relative to the meridian and base line, each of the resulting townships can now be positively identified and accurately located on a map. For example, a township described as "Township 1 North, Range 3 East," will lie in the first row of townships north of the base line, and in the third column of townships east of the principal meridian. Each township is thus described with reference back to the intersection of the base and meridian lines. It will lie so

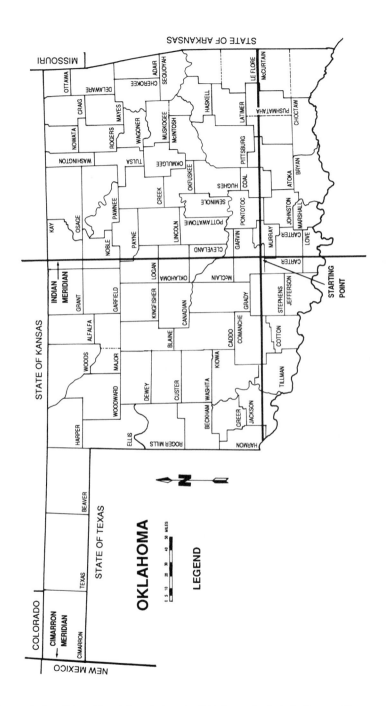

Figure 4-1 All legal land descriptions in Oklahoma include Indian or Cimarron meridian.

Before Drilling Can Begin . . .   69 many squares north or south of the base, and so many squares east or west of the meridian (Figure 4-2).

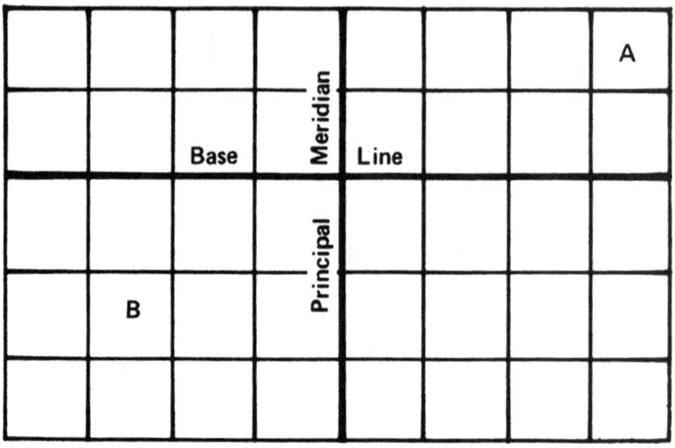

Figure 4-2 Squares 6 mi long on each side form townships on land maps.

Sections

Once the townships are laid out, each of these squares is then divided into 36 one-mile-square *sections* of land. On the resulting map grid, each of these squares is numbered in a serpentine fashion beginning with number one in the upper right hand corner and ending with number 36 in the lower right hand corner (Figure 4-3).

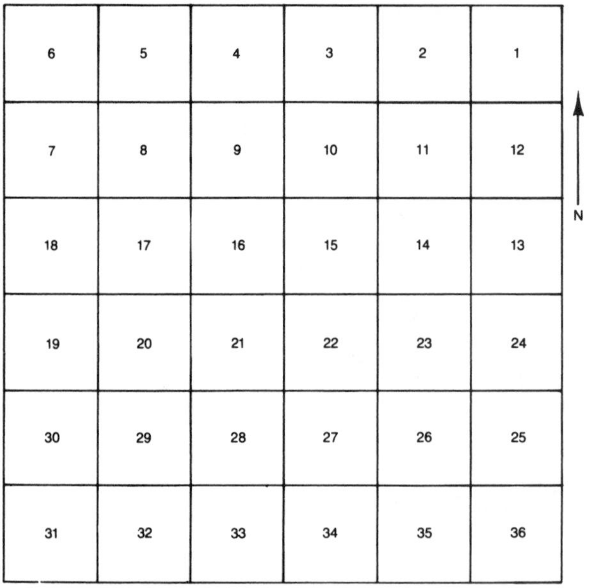

Figure 4-3 Each township is further divided into 36 1-mi-sq. sections of land.

Once the section lines are laid out, then half-mile points are established to divide each section into four *quarters*, which may be further subdivided if necessary.

Mathematically, each *section* should contain *640 ac*, and each *quarter section* should contain one-fourth that area or *160 ac*. However, this is not always the case; the curvature of the Earth affects the convergence of the survey lines. Sometimes surveyors made errors, and in some instances the survey did not match up with earlier ones made before territories became states. Or, in instances where Indian lands were surveyed prior to the establishment of territories or states, lines may not match up. Some abnormalities are due to geographical reasons such as the irregular shores of lakes, or the banks of meandering streams. These are accounted for by the inclusion of odd-sized *correction* or *government lots*. Such size deficiencies usually are confined to sections 1 through 6, 7, 18, 19, 30, and 31. Any shortages will be located along their outer borders (Figure 4–4).

In some states *riparian rights* may affect land ownership and measurement. In Oklahoma for example, all streams that measure two *chains* (132 ft) or more between their high water marks are the

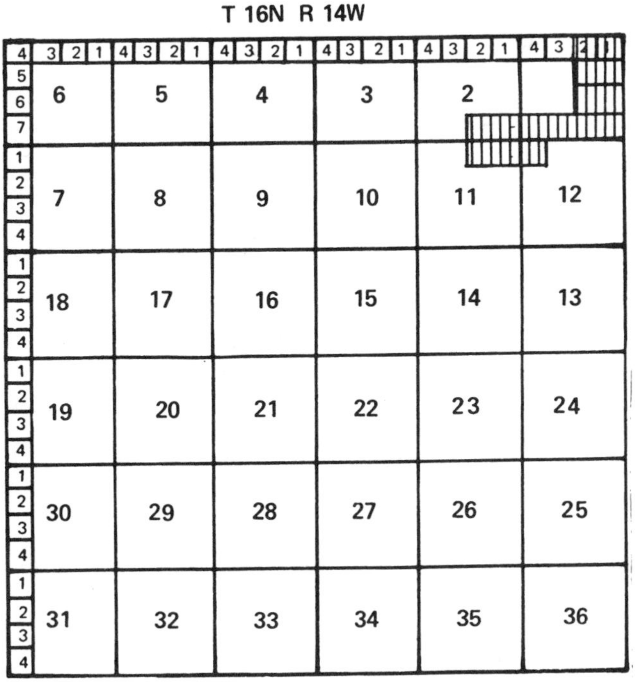

T 16N R 14W

Figure 4–4 Tract smaller than a quarter section is described as part of quarter where it is found.

property of the state. Private ownership ceases at these marks. Therefore, when such a stream cuts into a quarter-quarter section reducing that quarter-quarter to less than 40 ac, it is designated as a lot and assigned a number. While most correction and government lots are close to 40 ac in size, river lots may actually be any size up to 40 ac (Figure 4–5).

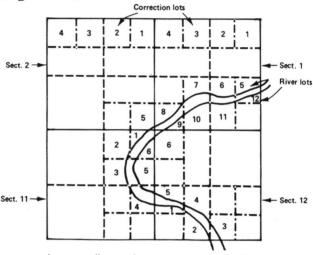

Figure 4–5 Lot numbers usually run downstream to section line then back upstream to line.

A *legal description* that includes a notation such as "Lot 5 and N½ NW¼," indicates that a fractional quarter-quarter (lot) of unknown size is included. The actual acreage of the lot can be determined from the official maps at the county courthouse (Figure 4–6).

Legal Descriptions

While all this may sound a bit confusing, just remember that we are really talking about a checkerboard pattern in which the squares are numbered, then each square itself is evenly divided into smaller squares which are also numbered, and which can be accurately located on the checkerboard by their number and east-west north-south placement. Thus,

One township	equals	36 sections
One section	equals	640 ac
One half section	equals	320 ac
One quarter section	equals	160 ac
One quarter-quarter	equals	40 ac
One lot	may be any size up to 40 ac	

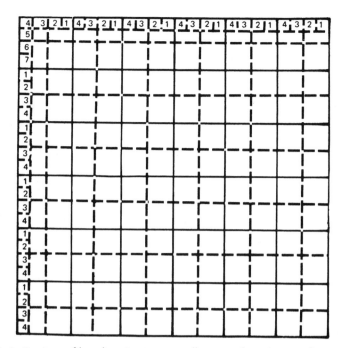

Figure 4–6 Sections of less than 1 mi are usually on north and west sides.

and each of these may be further fractionally divided. For example, a quarter section of 160 ac (notated as ¼), may be divided into half (80 ac or ½), or into quarters (40 ac or ¼), and so on down to the actual amount of land involved. Each of these subdivisions must also be notated as to their geographic placement: NW½, SE¼, NE¼, SW½, etc., (Figure 4–7).

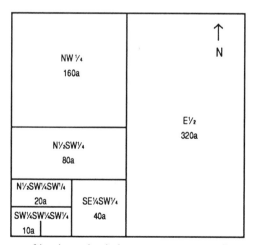

Figure 4–7 Sections of land are divided into quarters or small tracts as shown.

In leasing large tracts of land covering thousands of acres from major land holders such as the government or Indian tribes, often only the townships or sections to be included in the lease need to be named and described. However, most oil and gas leases or land purchases deal with much smaller areas. In such cases the key to writing accurate legal descriptions is usually the quarter section. A tract is usually a quarter section, a part of that quarter, or a quarter plus other land. Therefore, any tract smaller than a quarter section is always described as being a part of the quarter in which it is found. The final letter combination in a description designates in which of the four quarters of section the tract is located. Any time two letters appear in combination, this indicates one-fourth of something.

Legal descriptions enable us to not only determine the location, but also the size of the tract being described if one remembers that a section contains 640 ac.

To Review...
A legal description written as
SW¼ SW¼ W½ Sec 1 Twp 2 South Range 4 East Cimarron Meridian, means
SW¼   the southwestern quarter of the quarter
SW¼   the southwestern quarter of the section
W½   the western half of the section
Section 1 the northeasternmost section in the township
Township 2 South is in the second row south of the base line
Range 4 East is in the fourth column east of the Cimarron Meridian

Mineral Ownership

Having an accurate surface map makes it possible to determine who owns the surface of the land above the minerals, and who owns the minerals beneath the surface. Prospective producers must know this in order to get permission to proceed with their plans. In some areas of the world both the land and the minerals belong to the state. In other places the surface may belong to an individual and the minerals to the state, and in yet other places such as the United States and Canada, individuals may own both land and minerals. Quite often, however, one party may own the surface and another, the minerals. Permission to conduct operations must be obtained from both parties, usually by a *landman*, whose job it is to seek out the owners of both estates and negotiate leases with them on behalf of the intended producer.

Surface Ownership

The surface of the land may be owned by an individual, a group of individuals, in some states by a company or corporation, a church, an Indian tribe, a political subdivision (city, county, state, province, federal government), or any other "person" legally recognized as being competent to own real estate. The owner of the surface has the right to enjoy the use of his property, but the owner of the subsurface mineral estate also has a right to enjoy his property. Usually it is possible to effect an agreement whereby the producer and the mineral owner can drill for and produce oil and gas by paying the surface owner for the use of his property. Such payments can include those for damages occurring to the surface and for loss of crops, fees for the use of water, right of way fees for roads and pipelines, payments for construction of cattleguards, fences, and other specified items. Such arrangements make it possible for the mineral lessee to come on the property, prepare it for drilling according to the terms of the agreement, and extract the minerals from the ground for the period of time called for in the agreement, even though he does not own the land.

If the surface owner also owns the minerals (ownership *in fee*), he is usually just as interested as the producer is in bringing about oil and gas production, and therefore may be more amenable to surface use than an owner with no mineral interest. However, if the surface landowner has already leased his land to another, such as a farmer who has planted a crop or is grazing cattle on it, that lessee's rights must also be protected.

Mineral Ownership

As mentioned above, the mineral owner may also own the surface, or it may be owned by an entirely different person or group of people. Over the years, ownership of minerals has become an exceedingly complex affair. For example, a person might have divided his 100% mineral ownership among four heirs, bequeathing each 25%. They in turn, may have split up their interest. This could eventually result in a large group of owners, each of whom owns just a tiny fraction of the original 100%. Or, an owner may sell his mineral holdings off in fractional amounts. In turn, an owner in fee may sell his surface ownership of the land while retaining the mineral rights. Transfers of mineral ownership are accomplished by a legal instrument called a *mineral deed.*

Government Ownership

Governmental ownership of surface and minerals provides its own set of complexities. On those vast stretches of federal land usually associated with the American West, the government owns both the land and the minerals. If an oil and gas lease on federal land is successful and produces petroleum in marketable quantities, the U.S. Minerals Management Service collects the royalties, but then divides them evenly with the state in which the tract is situated. Joint federal/state mineral income sharing may also occur from offshore leases.

States may also own and lease mineral resources. In Oklahoma, vast tracts of public lands were transferred to the state at the time of statehood. These lands have been held in trust for the public schools. The surface of the arable regions is leased to farmers and stockmen for agricultural use, and the income goes to the schools. Mineral leases are also granted and any resulting royalties also go to the state. In Texas, the university system has traditionally benefitted from oil and gas royalties. Cities and counties (parishes) may also own mineral interests. In Oklahoma City, the major airport is built on a producing oil and gas field which has brought in many millions of dollars in revenue for the Airport Trust Authority.

Mineral operations on federal lands, Indian trust lands, and offshore areas, generally fall under the jurisdiction of the U.S. Department of the Interior and will involve one or more of that department's agencies: the Bureau of Indian Affairs, the Bureau of Land Management, the U.S. Geological Survey, the Minerals Management Service, the National Park Service, or the U.S. Fish and Wildlife Service. Other federal departments that may also have jurisdiction include the Department of Agriculture's U.S. Forest Service if operations are planned in national forests, and the Department of Transportation's U.S. Coast Guard in the case of offshore drilling and production. Leasing federal and offshore lands is a highly complex procedure with many variables depending on tract size, location, and whether or not known producing formations are present. In addition, the rules are often subject to change.

Canada

Petroleum land operations, mineral ownership, leasing, and even mapping vary with location.

The Dominion Land Survey

In the western prairie provinces of Alberta, Saskatchewan, Manitoba, and part of British Columbia, legal descriptions are based upon the Dominion Land Survey (D.L.S.) System. The western provinces are divided by seven north-south meridians. The prime meridian is approximately 12 mi west of Winnipeg, and each of the others lies between 150 and 200 mi apart in a westward progression. Due to the curvature of the Earth they are farthest apart at the border with the United States and closest together at the Northwest Territories boundary. Between the meridians, six 1-mi apart ranges running north and south are established and are numbered from 1 to 26 as they proceed west from the east meridian.

East-west township lines are laid off every 6 mi beginning at the U.S. border. The range and township lines divide the land into townships 6 mi by 6 mi containing a total of 36 mi^2. As in the United States, each 1-mi square is called a section and contains 640 ac. Each section is further divided into 16 Legal Subdivisions (LSD) of 40 ac, the same as a quarter-quarter in the United States.

A DLS legal description may read "15-35-41-4-W3M," which would mean 15th LSD, 35th Section, 41st Township, 4th Range, 3rd Meridian. It should be noted that the section numbering system in Canada is exactly the opposite of that used in the United States.

When using the DLS, one should remember that Section 1 is located at the bottom right of the "checkerboard," and numbering moves in a serpentine fashion to the left.

The National Topographic Survey

The National Topographic Survey System of Mapping is the system used in the part of British Columbia not covered by the DLA and in the rest of Canada.

This system divides the country into Primary Map Units by each four degrees of latitude and eight degrees of longitude. Each of these resulting units is then subdivided into 16 Map Unit Subdivisions, which are identified by letters from A to P. Each of these are then divided into Map Sheets having a scale of 1 in. to 1 mi. Map Sheets are divided into 12 Zones, which are lettered from A to L. These are divided into 100 Units numbered from 1 to 100. Each of these Units can be divided into four Quadrants lettered a, b, c, and d.

Canadian Mineral Ownership

In Canada, as in the United States, surface and minerals may be owned separately. Owners may be individuals, the federal government, or the provincial government. If either surface or minerals are owned by the government, they are said to belong to the Crown. Today, ownership varies widely depending upon the location. In Alberta, one of the richest petroleum producing areas of Canada, 80% of the mineral rights belong to the provincial government. The remainder is divided between the federal government and private owners, including large corporations. Less than 1% is held by "freeholders" who are usually descendants of the original settlers.

The time at which the provinces were settled have affected the number of freeholders having mineral rights. For example, in Manitoba, about 80% of the mineral rights are owned by freeholders. British Columbia, Manitoba, and Saskatchewan all have a higher percentage of freeholder mineral owners than Alberta.

In eastern Canada, which is more densely populated, almost all mineral rights are held by individuals or private enterprises. In southern Ontario, which was the original oil-producing area in North America, minerals are mostly privately owned. However, in northern Ontario, almost all minerals are owned by the provincial government.

Minerals in the Arctic islands and their waters, the Yukon, and Northwest Territories are owned by the Canadian government.

Other Nations

In Latin America and other parts of the world, laws regarding land and mineral ownership vary widely. In some parts of the world today all minerals may belong to the state, or to a state-owned monopoly, or to a ruling family. In others, the private ownership of minerals is very similar to the systems described. However, it should be noted that much of the world measures distances, and therefore land, in metric units.

Indian Minerals

Leasing Indian lands presents its own set of problems. Here the variables may be tribal laws, treaties between the tribes and the federal government, state laws, or any of a dozen or so other factors. One such example is the vast, 1.8 million-ac Osage Reservation in Oklahoma. In one sense no reservation exists because the Osage Nation no longer owns the surface, but they do own the minerals. Therefore all mineral leases have to be made with the Osage Tribe.

Both tribal and individually owned Indian lands may be held *in trust*, which means that the Federal government oversees the leasing or disposal of the land or minerals for the owners. In such cases the Bureau of Indian Affairs supervises the leasing of the land, the conducting of any required environmental surveys, and the Minerals Management Service supervises operations and the collection of royalties from the production of oil and gas. Since there are literally dozens of different classes of Indian surface and mineral ownership, leasing Indian minerals can present some very unique challenges which makes the services of a landman experienced in Indian affairs very necessary.

A knowledge of Indian minerals is important, since in the western portion of the coterminous United States, Indians control 53 million ac of land, much of which is rich in mineral deposits. Some of the largest petroleum reserves in the Western Hemisphere are situated in the domain of the Alaskan Native peoples, and Canadian Indians also own major petroleum-producing areas.

Mineral Leases

A mineral lease is a legal document. It is the instrument used to convey the rights to drill for and produce oil and gas to the lessee, from the mineral owner, the lessor. Leases are written to cover a specific period of time called the *primary term*. This may be any period agreed upon by the two parties involved, but is usually for a number of years such as three, five, or ten. Upon signing the lease, the oil and gas company gains certain subordinate rights. Other rights are reserved for the mineral owner. By signing a lease, the mineral owner can receive:

1. A *bonus* payment, usually based on an agreed-upon amount of dollars per acre, for granting the lease.

2. *Delay rentals*, usually figured at the rate of so many dollars per year per acre if drilling is delayed or a well is shut-in for any length of time.

3. *Royalties*, or payments for the minerals extracted and which are usually expressed as a fraction of the total production. For example, if a well produces 100 barrels of oil per day (BOPD), then the mineral owner would be entitled to a share of the oil in the amount stipulated in the lease. If the lease calls for him to receive a one-eighth share, then he would get 12.5 bbl out of every 100 produced. The lease may specify that payment of royalties may be made either in cash or *in kind*, which means he would receive actual barrels of oil.

4. *Overriding royalties.* Again based on production, this is an additional fractional share. Other parties such as geologists, landmen, investors, and suppliers may also share in overriding royalty income.

5. *Revision of rights.* The mineral owner retains the right to have the ownership of the leased property and minerals revert to him if:

a) The primary term expires and drilling is not underway and minerals are not being produced in marketable quantities. This portion of the lease should be given careful attention by both parties. Does "drilling" mean that a well is almost completed on the expiration date? Or did the operator merely move a rig onto the site the day before the lease was to expire? And what are "marketable quantities"? One barrel? A thousand barrels?

b) The lease may also revert to the mineral owner if the lessee fails to make the agreed-upon annual delay rental payments on the date they are due. This legal requirement as well as the primary term clause came as shock to many federal bank regulatory officials following the Penn Square Bank collapse in 1982. Banks had loaned money on leases and were counting them as part of their assets. When the banks failed, the government seized them. The leases, along with other seized property were placed in vaults to be sorted out and disposed of later. The leases, of course, had time value which the regulators, most of whom were unfamiliar with oil and gas law, were unaware of. Thus they sat locked away until they expired and became worthless except as scrap paper. To avoid forgetting to pay rental payments on time or perhaps toprotect against not having the cash to do so on the due date, cautious operators sometimes pay all the required rental fees in advance upon signing the lease. It then becomes a *fully paid lease.*

c) *Protection of minerals.* The lessee has an obligation to protect the mineral assets of the lessor. For example, if a valid oil and gas lease is in effect and someone on the adjoining property drills a well and begins draining the oil from the formation beneath both pieces of property, the lessee must drill an *offset well* to recover minerals before the "neighbor" drains them all away.

6. *Surface use.* The surface owner may use the surface in any way that he wishes, providing that it does not conflict with the lessee's operations as spelled out in the lease.

State Laws

State laws may also affect mineral ownership and oil and gas production. Louisiana, for example, is a *non-ownership* state. This means that the mineral owner owns only the oil and gas, not any other minerals that may be produced. Texas however, is an *absolute ownership* state. This means that the surface owner owns all of the land that is not mined. The mineral owner owns the oil and gas and the subsoil it is mined from, but not the formation or the surrounding rock. While the surface owner must let the mineral owner have access to his property from above, the surface owner is free to lease the structures for other uses such as underground storage of minerals as long as it does not interfere with the other parties' operations.

Whichever the system, land and minerals remain two separate estates. A surface mortgage may be foreclosed and the owner forced to vacate the premises, but this does not affect the mineral owner and his lessee's operations. Likewise, an oil company might sell its lease rights to another company, but the terms of the lease would still be binding.

What Can the Mineral Owner Expect?

An owner of a quarter-section of minerals (160 ac) who signs a three-year lease with typical provisions might expect the following. The dollar amounts shown are hypothetical and will vary with the price oil is bringing per barrel at the time and also the state of the market in terms of supply and demand.

1. Bonus (On signing) per acre ($25 X 160) = $4,000.

2. Delay rental ($3 per acre per year),
   first year due upon signing   = $ 480.

3. Royalty of ⅜. If well is successful
   and produces 1,000 B/D (barrels per day) at
   $30/bbl, monthly royalty before
   taxes, etc. would be   = $30,337.50

Royalty Interests

Royalty interests differ from mineral interests in that the party does not own the minerals, but only the proceeds or a portion of the proceeds from the minerals produced. An owner of royalty interests has no rights to enter the land or to drill on it.

The Landman's Role

The landman is the field agent who locates mineral owners, verifies their ownership and the legal description of the site where their minerals are located, and negotiates the terms necessary to get a lease signed. Landmen today may well be women, they may be independent businessmen representing sometimes unnamed third parties, or they may be company employees. Often they are attorneys, but in the main they are not. Their success often rests on their personal "PR" strategy, which could perhaps include driving a dusty pickup, wearing faded jeans, and having the ability to carry on a knowledgeable and sympathetic conversation with a farmer in need of some ready cash.

Since their financial well-being depends upon it, landmen use their ability to drive the best bargain possible for themselves or their company. There are no state laws limiting the amount of royalties. The royalty amount is whatever figure the two parties can agree on. For many years one-eighth was a widely accepted figure, but today one-fourth, or even three-eighths is far more common. In Canada royalties of 40% or more may be paid. The wise mineral owner will usually not accept the first offer, but instead shop around and learn what the going terms are in his neighborhood.

While the geologists and engineers use maps to locate areas of possible production, and refer to the production records of any existing wells to determine where and if they want to drill, the landman spends a great deal of his time in the county courthouse researching records and maps of equally important interest. At the courthouse, (or at the BIA Indian agency in the case of Indian-owned minerals), the landman can find out who owns the land, who owns the minerals and how to contact them, and also if the minerals are under lease. And, if they are, he can find out whom they are leased to, the terms of the lease, and when it is due to expire (Figure 4–8).

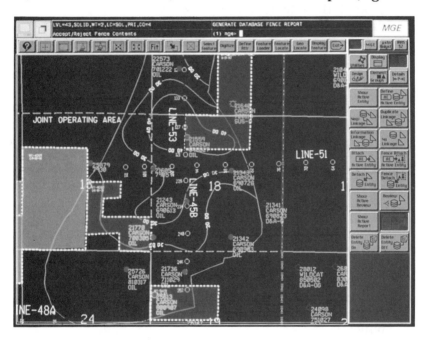

Figure 4–8 An integrated base map, created using PRE/MIER's EPMAP product, includes translated lease, well, and seismic data, linked to a relational database. Queries to the relational database provide the data needed for generating contour maps (courtesy Intergraph Corporation).

Sometimes this can entail quite a bit of detective work as in the case of a mineral estate which has been fractionated among many heirs who may be scattered from northern Maine to southern California. Often there are *defects* in the *title* to the minerals so the landman must know the proper *curative* steps to take to *cure* or correct these defects so that a legal lease may be effected.

Legal Provisions

The landman needs to keep in mind certain legal rights and restrictions of owners. There have been many court battles over the years concerning the rights of landowners, but six major premises hold:

1) *Oil and gas are minerals and are part of the land.* Since they are part of the land, they are considered the property of the landowner.

2) *When the minerals are mined, they become personal property.* In taking the oil or gas from the ground, the mineral owner makes them his personal property. He, in essence, owns the minerals as well as their value.

3) *The owner may either drill for the minerals or transfer his rights to someone else.* Although oil companies are usually selected to come onto a piece of property and produce the oil and gas, the owner, under law, has every right to drill his own well and produce the minerals himself.

4) *The owner can drill a well and drain a reservoir even if it extends under a neighbor's property.* (This is known as *the rule of capture*). If the petroleum reservoir extends beyond the boundaries of an owner's property line, he may continue to drain the reservoir even if he is taking away his neighbor's minerals. Since the minerals flow and there is no way to contain them within a boundary, it is up to the neighbor to drill his own well, in such cases called an *offset well*, in order to reap the profits from his minerals

5) *The owner is liable for any damages he incurs to a common reservoir.* If, through poor drilling or production practices, the owner damages the reservoir because he depletes it too quickly thus losing formation pressure, or he fractures the well so completely that the formation is damaged, he is liable and must compensate the other landowners who share the reservoir for damages to their property.

6) *The owner is liable if he wastes oil and gas.* According to environmental protection regulations, oil and gas and other produced fluids must be conserved. If an owner flares his natural gas, or

allows crude oil or produced salt water to flow over the land, he is liable and must make reparations.

Status of Ownership

In addition to the above, the landman must also consider the status of the owner he is dealing with. If the owner is a minor, the law will often not recognize any kind of contract made with him as legal and binding. The same holds true with an Indian owner who has placed his land in trust with the government of the United States. Those who have been adjudged incompetent may not enter into binding contracts. In such cases legally appointed trustees or others holding valid powers of attorney may sometimes act for the owners. An illiterate person may enter into a contract providing he uses proper witnesses and marks to comply with the law.

There are also other pitfalls which the landman must avoid. If the person living on the land is a life tenant without legal right to inheritance, the landman must deal with both that person and the remainderman—the person who will inherit the rights. If both are not considered, problems may arise in the future. The same goes for owners of future interests, especially if they are as yet unborn.

Partnerships, marital status, and groups of people can form yet other obstacles to obtaining a valid lease. A fiduciary, executor, or administrator of an estate may have the right to sell the land, but not to lease it. For married couples, much depends upon the laws of the state in which they live. Louisiana, which came into the Union retaining the heritage of the Napoleonic Code, has laws that differ in many aspects from those of the states that base their legal codes upon English Common Law. In some states the wife has rights only on her dower property. Other states are community property states, and still others are non-community property states. The landman must be aware which laws apply and if the signatures of both husband and wife are required to make a lease valid.

Oil and Gas Leases

Although you may often hear the term used, there is no such thing as a "standard oil and gas lease." Most leases begin life as a preprinted, fill-in-the-blanks form purchased from a stationer or printer who furnishes blank wills, deeds, mortgages, or promissory notes to the legal trade. There are many types of blank oil and gas lease forms, each designed for a different purpose and each identified by name and number. The "Producers 88" is probably the one most commonly used in the mid-continent area of the United States. It should be noted that the use of such forms is not required. The two parties to the lease may write their own, as long as it meets the legal requirements of a binding and therefore enforceable contract. The blanks may be filled in with whatever figures and terms the two parties agree upon. And anything handwritten on the lease and attested to by both signatories takes precedence over printed matter (Figure 4–9).

There are several sections of a lease, however, that are common to most all oil and gas leases.

1. The *legal names* of the parties to the lease.

2. The exact *legal description* of the property involved.

3. The *granting clause.* This sets out the company's rights and the mineral owner's right by naming the specific purpose of the venture, and the privileges each party is awarded. In some states it is automatically understood that the company has the right to enter and leave the property *(the right of ingress and egress),* the right to set up equipment, drill, and produce, and to reinject (in accordance with law) saltwater into nonproductive formations.

4. *Habendum clause.* This sets out the initial term of the lease. Often, the first part of the clause will set up the time for exploration and drilling and the second part will determine how long production may continue.

5. *Royalty Clause.* This is an agreement to pay the owner a portion of the proceeds of the well, either in the mineral itself or in its equivalent cash value. Royalties are usually expressed in fractions. Minimum acceptable royalty amounts may be set by law, as in the case of U.S. government, Indian, certain Canadian leases, and in other countries

(640 Shut in) (Revised 1963)

OIL AND GAS LEASE

AGREEMENT, Made and entered into this ________________ day of ________________, 19________,

by and between ____

____ Party of the first part, hereinafter called lessor (whether one or more)

and ________________________________ part ________ of the second part, hereinafter called lessee.

WITNESSETH, That the said lessor, for and in consideration of ________________________ DOLLARS, cash in hand paid, receipt of which is hereby acknowledged and of the covenants and agreements hereinafter contained on the part of lessee to be paid, kept and performed, has granted, demised, leased and let and by these presents does grant, demise, lease and let unto the said lessee, for the sole and only purpose of exploring by geophysical and other methods, mining and operating for oil (including but not limited to distillate and condensate), gas (including casinghead gas and helium and all other constituents), and for laying pipe lines, and building tanks, powers, stations and structures thereon, to produce, save and take care of said products, all that certain tract of land, together with

any reversionary rights therein, situated in the County of ________________________ State of Oklahoma, described as follows, to-wit 3

____

____

____

____

____

of Section ________________, Township ________________, Range ________________, and containing ________________ acres, more or less 4

It is agreed that this lease shall remain in force for a term of ________________ years from date (herein called primary term) and as long thereafter as oil or gas, or either of them is produced from said land by the lessee.

In consideration of the premises the said lessee covenants and agrees

1st. To deliver to the credit of lessor free of cost in the pipe line to which it may connect its wells, the one-eighth (1/8) part of all oil (including but not limited to condensate and distillate) produced and saved from the leased premises.

2nd. To pay lessor for gas of whatsoever nature or kind (with all of its constituents) produced and sold or used off the leased premises, or used in the manufacture of
products therefrom, one-eighth (1/8) of the gross proceeds received for the gas sold, used off the premises, or in the manufacture of products therefrom, but in no event more 5
than one-eighth (1/8) of the actual amount received by the lessee, said payments to be made monthly. During any period (whether before or after expiration of the primary
term hereof) when gas is not being so sold or used and the well or wells are shut in and there is no current production of oil or operations on said leased premises sufficient to 6
keep this lease in force, lessee shall pay or tender a royalty of One Dollar ($1.00) per year per net royalty acre retained hereunder, such payment or tender to be made, on or before the anniversary date of this lease next ensuing after the expiration of ninety (90) days from the date such well is shut in and thereafter on the anniversary date of this lease during the period such well is shut in, to the royalty owners or to the royalty owner's credit in the rental depository bank hereinafter designated. When such payment or tender is made it will be considered that gas is being produced within the meaning of the entire lease. Lessor shall have the privilege at his risk and expense of using gas from any well, producing gas only, on the leased premises for stoves and inside lights in the principal dwelling thereon out of any surplus gas not needed for operations hereunder.

3rd. To pay lessor for gas produced from any oil well and used off the premises, or for the manufacture of casing-head gasoline or dry commercial gas, one-eighth (1/8) of the gross proceeds, at the mouth of the well, received by lessee for the gas during the time such gas shall be used, said payments to be made monthly.

If drilling operations or mining operations are not commenced on the leased premises on or before one year from this date, this lease shall then terminate as to both

parties unless Lessee on or before the expiration of said period shall pay or tender to Lessor, or to the credit of Lessor in ________________________

____ bank

at ________________________ or any successor bank, the sum of ________________________

Dollars, ($________________), hereinafter called 'rental' which shall extend for twelve months the time within which drilling operations or mining operations may be commenced. Thereafter, annually, in like manner and upon like payments or tenders the commencement of drilling operations or mining operations may be further deferred for periods of twelve months each during the primary term. Payment or tender of rental may be made by check or draft of Lessee delivered or mailed to the authorized depository bank or Lessor (at address last known to Lessee) on or before such date for payment, and the payment or tender will be deemed made when the check or draft is so delivered or mailed. If said named or successor bank (or any other bank which may, as hereinafter provided have been designated as depository) should fail or liquidate or for any reason refuse or fail to accept rental, Lessee shall not be held in default for failure to make such payment or tender of rental until thirty days after Lessor shall deliver to Lessee a proper recordable instrument naming another bank to receive such payments or tenders. The above named or successor bank or any other bank which may be designated as depository shall be Lessor's agent. Drilling operations or mining operations shall be deemed to be commenced when the first material is placed on the leased premises or when the first work, other than surveying or staking the location, is done thereon which is necessary for such operations.

Should the first well drilled on the above described land, or on acreage pooled therewith, be a dry hole, then, and in that event, if a second well is not commenced on said land, or on acreage pooled therewith, within twelve months from the expiration of the last rental period for which rental has been paid, this lease shall terminate as to both parties, unless the lessee on or before the expiration of said twelve months shall resume the payments of rentals, in the same amount and in the same manner as hereinbefore provided. And it is agreed that upon the resumption of the payment of rentals as above provided, that the provisions, hereof governing the payment of rentals and
the effect thereof, shall continue in force just as though there had been no interruption in the rental payments. If the lessee shall commence to drill a well or commence 7
reworking operations on an existing well within the term of this lease or any extension thereof, or on acreage pooled therewith, the lessee shall have the right to drill such well to completion or complete reworking operations with reasonable diligence and dispatch, and if oil or gas, or either of them, be found in paying quantities, this lease shall continue and be in force with like effect as if such well had been completed within the term of years first mentioned.

Lessee is hereby granted the right at any time and from time to time to unitize the leased premises or any portion or portions thereof, as to all strata or any stratum or strata, with any other lands as to all strata or any stratum or strata, for the production primarily of oil or primarily of gas with or without distillate. However, no unit for the production primarily of oil shall embrace more than 40 acres, or for the production primarily of gas with or without distillate more than 640 acres, provided that if any governmental regulation shall prescribe a spacing pattern for the development of the field or allocate a producing allowable based on acreage per well, then any such unit may embrace as much additional acreage as may be so prescribed or as may be used in such allocation of allowable. Lessee shall file written unit
designations in the county in which the leased premises are located. Operations upon and production from the unit shall be treated as if such operations were upon or such 8
production were from the leased premises whether or not the well or wells are located thereon. The entire acreage within a unit shall be treated for all purposes as if it were covered by and included in this lease except that the royalty on production from the unit shall be as below provided, and except that in calculating the amount of any rentals or shut in gas royalties, only that part of the acreage originally leased and then actually embraced by this lease shall be counted. In respect to production from the unit, Lessee shall pay Lessor, in lieu of other royalties thereon, only such proportion of the royalties stipulated herein as the amount of his acreage placed in the unit, or his royalty interest therein on an acreage basis bears to the total acreage in the unit.

If said lessor owns a less interest in the above described land than the entire and undivided fee simple estate therein, then the royalties and rentals herein provided shall be paid to the lessor only in the proportion which his interest bears to the whole and undivided fee, however, such rental shall be increased at the next succeeding rental anniversary after any reversion occurs to cover the interest so acquired.

Lessee shall have the right to use, free of cost, gas, oil and water produced on said land for its operations thereon, except water from wells of lessor. 9

When requested by the lessor, lessee shall bury his pipe lines below plow depth.

No well shall be drilled nearer than 200 feet to the house or barn now on said premises, without the written consent of the lessor.

Lessee shall pay for all damages caused by its operations ~~to growing crops~~ on said land.

Lessee shall have the right at any time to remove all machinery and fixtures placed on said premises, including the right to draw and remove casing.

If the estate of either party hereto is assigned, and the privilege of assigning in whole or in part is expressly allowed, the covenants hereof shall extend to their heirs, executors, administrators, successors or assigns. However, no change or division in ownership of the lands, rentals or royalties shall enlarge the obligations or diminish the rights of Lessee. No change in the ownership of the land or assignment of rentals or royalties shall be binding on the lessee until after the lessee has been furnished with a written transfer or assignment or a true copy thereof, and it is hereby agreed in the event this lease shall be assigned as to a part or parts of the above described lands and the assignee or assignees of such part or parts shall fail or make default in the payment of the proportionate part of the rentals due from him or them, such default shall not operate to defeat or affect this lease in so far as it covers a part or parts of the lands on which the said lessee or any assignee thereof shall make due payment of said rentals. In case lessee assigns this lease, in whole or in part, lessee shall be relieved of all obligations with respect to the assigned portion or portions arising subsequent to the date of assignment.

All express or implied covenants of this lease shall be subject to all Federal and State Laws, Executive Orders, Rules and Regulations, and this lease shall not be terminated in whole or in part, nor lessee held liable in damages, for failure to comply therewith, if compliance is prevented by, or such failure is the result of any such Law, Order, Rule or Regulation.

This lease shall be effective as to each lessor on execution hereof as to his or her interest and shall be binding on those signing, notwithstanding some of the lessors above named may not join in the execution hereof. The word "Lessor" as used in this lease means the party or parties who execute this lease as Lessor, although not named above.

Lessee may at any time and from time to time surrender this lease as to any part or parts of the leased premises by delivering or mailing a release thereof to lessor, or by placing a release of record in the proper County. After a partial surrender, the rentals specified above shall be proportionately reduced on an acreage basis.

Lessor hereby warrants and agrees to defend the title to the lands herein described, and agrees that the lessee shall have the right at any time to redeem for lessor by payment any mortgages, taxes or other liens on the above described lands, in the event of default of payment by lessor, and be subrogated to the rights of the holder thereof.

____

Figure 4–9 Portion of a typical oil and gas lease.

where minerals are state owned, but the maximums are usually what the market will bear.

6. *Drilling and Delay Rental Clause.* If the company cannot complete the well by the date specified in the habendum clause, it must pay the owner a delayed rental or terminate the lease. This protects the owner from companies who might begin drilling and then abandon the project and move to a more productive field elsewhere.

7. *Pooling Clause.* This allows the company the option to pool or unite their interests with other leases in the area. The companies may share operating costs, and royalties can be shared on an equitable basis. Such an arrangement can provide more efficient recovery and waste less resources.

8. *Special Provisions Clause.* This clause may contain restrictions on drilling too close to existing structures, vegetation, or livestock operations, burial and disposal stipulations, and provisions for crop damage payments. For the oil company, it may provide for access to water, equipment movement rights, and for notification of changes in ownership.

9. *Obligations of Producer Clause.* Once the oil company signs the lease, it has committed itself to explore for, and produce the formation named in the lease (if one is specified), to the best of its ability. Failure to do so is cause for ending the contract since the mineral owner is not receiving the income he deserves from his property.

Leases are generally released because of expiration of the primary term without action by the lessee, failure of the lessee to comply with the other terms of the contract, abandonment or surrender of rights by the lessee, or forfeiture of property rights by the lessor.

Special Leasing Considerations

Special considerations may arise depending upon legal spacing requirements and the type of production desired. Assume someone owns the minerals in an area rich in deep natural gas (20,000+ ft). Such gas is usually given *spacing* by the state regulatory body meaning only one deep well is allowed in every 640 ac. However, this same property may overlie a shallow oil formation where spacing is one well per 40 ac. The mineral owner may be able to *bottom lease* the deep gas zone to one producer and to top lease the shallow oil zone to another producer. Producing mineral zones at different levels may have different owners, a situation which may also affect leasing.

Top leasing has a second meaning in the oil field. It can refer to a lease drawn up by a second company and negotiated to take effect immediately upon the expiration of an existing lease in effect with another company.

It is important for the landman to be aware of the state spacing requirements when he is putting together a lease "package" since, as in the example of a 640-ac spacing, a number of mineral owners might be involved.

Permitting

Once a valid lease has been obtained, the company is free to begin drilling operations once the proper permits have been obtained. The number and type of permits required will vary by location and land and mineral ownership. Each state or province has a regulatory body that oversees oil and gas operations, and each has its own structure of permits. Some wells are drilled inside city limits, and many cities have their own permitting ordinances. The federal government also requires permits to drill on its leases, and companies operating on federal and Indian lands usually must also obtain environmental clearance in the form of an *Environmental Impact Statement* or *Environmental Assessment.* In some cases, at the very least, an *Archeological Clearance* for the drill site must be obtained, and offshore operations must have their own set of highly specialized permitting rules.

Chapter References

David L. Baldwin, *All You Ever Wanted to Know About Leasing Indian Minerals But Were Afraid to Find Out!* (Denver: Baldwin & Associates, 1984), p. 1.

Focus on Energy: Resource Ownership, (Calgary: Canadian Petroleum Association, 1980), p. 1.

Loris A. Parcher and Clint E. Roush, "Legal Land Descriptions in Oklahoma," Extension Fact Sheet No. 9407, (Stillwater: Oklahoma State University Cooperative Extension Service, n.d.), pp. 1–4.

5

Drilling
PART 1

Onshore Drilling

Once the well site has been selected and all the formalities of leasing and permitting have been completed, there are still many steps to be taken before drilling can commence. The end result that the operator is seeking is a well which produces hydrocarbons in commercially acceptable quantities, a well consisting of steel pipe *(casing)* set in cement, which extends from the surface downward into an oil or gas producing formation with all the necessary production equipment in place, and which has been drilled as rapidly and economically as possible (Figure 5–1).

To accomplish this, two things are needed. First, the services of perhaps dozens of highly specialized contractors and subcontractors, and up to 100 or more skilled and experienced workers ranging from heavy-equipment operators and truck drivers to computer operators and geophysicists are required. Secondly, success or failure of the venture may be determined by how much planning and forethought went into the *drilling program,* as the "blueprint" for the well is called.

Since much of the drilling program will be determined by the nature of the formation, planning for the well begins at the bottom of the hole, not at the surface.

If the target is a shallow, low-pressure formation where water saturation could be increased by water-based drilling fluids with the result that permeability might be decreased, then *cable tool* drilling may be the method selected. The elements of cost and time must also be considered in choosing the method of drilling to be used.

If, on the other hand, it is to be a deep hole, or there is a possibility that high downhole pressures may be encountered, then a *rotary rig* would be chosen so that drilling fluids could be used to regulate the pressure.

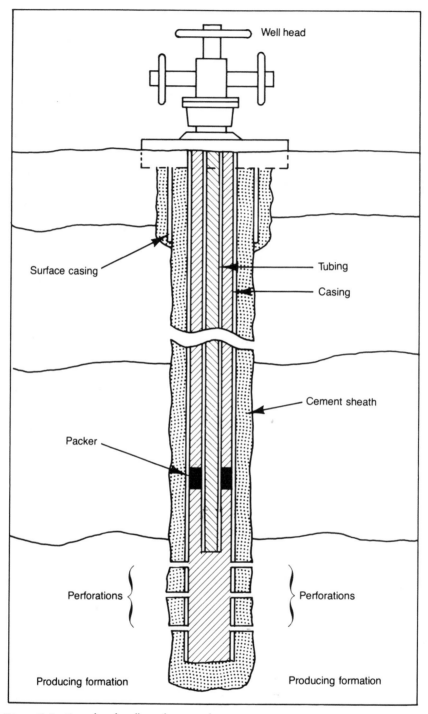

Figure 5-1 Completed well ready to produce.

Thus, determining what type of drilling rig to use will depend on three factors: (a) the character of the producing formation (b) the cost-per-foot and, (c) the nature of the formations that must be drilled through in order to reach the *pay sands*. It is obvious then, that management must not only have knowledge about the extent and nature of the reservoir, but also of what lies between the surface and the reservoir.

The Earliest Wells

Man, for one reason or another, has been digging holes in the Earth's surface since the earliest known times. The first wells were dug mainly for water for drinking and irrigation. These hand-dug wells were usually shallow holes scratched into the Earth that were sometimes lined with rock to prevent their internal collapse.

As man's technology improved, so did his methods of digging. At first, crude tools made of stone or wood were used. Later, digging implements of bronze and iron appeared, and man learned to hand debris up out of the hole in baskets. Then, as wells got deeper, ropes and crude windlasses were used to raise and lower cutting tools and remove litter.

No one knows who was the first to drill instead of dig, but by 600 B.C. the Chinese were using *percussion tools*—the forerunners of cable tools—to dig brine wells. Their rigs were constructed almost completely from bamboo, except for the metal bit on the end of the bamboo line.

Cable Tool Drilling

Percussion, or *cable-tool* drilling, is accomplished by repeatedly raising and dropping a heavy metal bit into the Earth's surface, eventually "punching" a hole downwards. This process is still widely used today for drilling water wells. Depending on the nature of the soil, the bit must periodically be removed and *dressed*, that is sharpened and straightened. And, different types of soil call for different types of bits. Periodically, *bailers*—of which there are various types—must be lowered into the hole to remove the accumulation of rock chips and loose soil so that the bit will have an unobstructed surface to strike.

David and Joseph Ruffner are credited with an important advance in early American drilling. They were trying to drill a brine well near what is now Charleston, West Virginia, but the walls kept collapsing. To support the sides of the hole, they cut sections from a hollow tree and pushed them down as the hole grew deeper. Their "driller" was a man at the bottom of the well with a pickaxe, their "casing" the tree trunk, and their "bailer" was a whiskey barrel cut in half into which the driller would scoop the cuttings.

The Ruffners also improved on another technique called the *springpole*. This consisted of a supple tree trunk or pole, attached to the Earth at one end and laid across a fulcrum. The other end, with a drilling line and bit attached, was positioned directly above the hole. In essence, it worked much like a diving board. Workers would stand in rope loops attached to the free end of the board and kick downward causing the bit to strike the ground. When they released their weight, the bit would spring free. Although unbelievably crude, the method was widely used in the United States, Canada, Romania, and many other parts of the world.

When Drake drilled his well, he adopted the readily available technology of the saltwater well drillers in Pennsylvania: a stationary steam engine that transferred its power by means of ropes, wheels, and a walking beam to the drilling and bailing tools. It used a derrick built on the spot of trimmed tree trunks and the whole affair was likely to be housed in a crude shack covered with rough wooden slabs. Tradition has it that Drake, or more likely, his driller, "Uncle Billy" Smith, achieved success because he adopted the use of steel casing to line his wellbore as it progressed downwards, thus preventing it from collapsing or being flooded with saltwater.

As the industry grew, and drilling proliferated, the "standard" cable-tool drilling rig gradually evolved. This consisted of a wooden *mast* or *derrick* usually built on the spot out of local timber. On top was a *crown block* over which the drilling line ran. Power came from a *steam engine* and was transmitted by belt to the *band wheel*. This was connected by a *pitman arm* to one end of the *walking beam* which raised and lowered the drill bit. Since most of these early rigs were permanent structures, the walking beam was often called upon to operate a pump after the well was completed (Figure 5–2).

The *bullwheel*, the largest pulley on the rig, was used to spool the hemp drilling rope, which had to be long enough to reach the bottom of the hole. It was also used to pull the *tool string* from the hole.

Since the drilling rope ran over just a single pulley in the crown block, it couldn't develop any more pull than it could at the bullwheel. Thus when there was casing to be run, there was simply not enough power available. To increase the lifting ability another spool, called the *calf wheel*, was added, along with a multiple sheave crown block and a multiple sheave traveling block. Not only could this

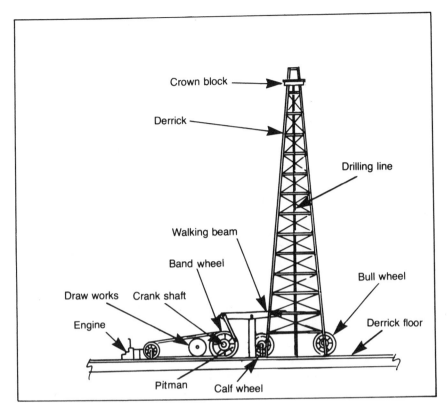

Figure 5-2 Standard cable-tool rig.

arrangement lift more weight with the same engine, it became the forerunner of the draw works, crown block and traveling block of today's modern rotary rigs.

The calf wheel used wire rope to run and pull casing since extra strength was needed to handle the weight of the *drillstring*. Because of abrasion, wire rope was also used for the *sand line*, to which was attached the bailer which removed the cuttings and fluids from the wellbore.

To achieve the weight needed to pound through rock, the drill bit was often connected, by means of threaded *tool joints* to *sinker bars* and *jars*. A circle jack was used to make and break the connections between the tools. This was a large, semicircular toothed rack bolted to the rig floor. One end of a large wrench was connected to a pinion and then jacked along the rack turning the tool held in its jaws. At the same time, the tool that was being attached or detached was held stationary by a second large wrench bolted to the deck. The drillers soon learned that these connections were much easier to make and break, particularly at night or in inclement weather, if the threaded male and female ends were tapered instead of being straight. This arrangement was later adapted to the drillpipe now used in rotary drilling.

The so-called standard rig would have horrified today's safety professionals. Since there were no clutches or transmissions between the engine and the rig, the only way power could be interrupted was by sticking a board between the drilling rope and the bullwheel, or by pulling a cotter pin that allowed the pitman arm to be slipped off the moving band wheel crank. The resulting flying ropes and kicked-back boards caused untold numbers of broken arms and ribs, smashed faces, and cracked skulls. Despite their drawbacks, cable tool rigs were used all over the world from the Baku oilfields of Russia to New Zealand as the demand for oil grew.

By 1900, many oil men were switching to temporary steel structures, a superior derrick that was stronger, safer, and could be used over and over again. Today, cable tool rigs are still used to drill for oil, and are also used extensively for water well drilling. Modern versions, however, are safe, highly mobile, all-steel, truck-mounted rigs equipped with a folding derrick for easy transport. A world depth record for cable tool drilling was set in 1953 when the New York Natural Gas Corporation drilled the E.C. Kesselring No. 1 to 11,145 ft.

Rotary Drilling

After Lucas' success with rotary drilling at Spindletop, many oil-men switched from cable tools, and today, rotary drilling is the method most often used. Today, about 85% or more of all wells are drilled with "conventional" rotary rigs. All deep holes, and many directional, horizontal, and offshore wells are drilled with rotaries. Like cable tool rigs, the rotaries have gone through a period of evolution that have made them safer, more efficient, more powerful, and capable of drilling wells measuring miles, not just feet, into the Earth's surface. The concept is certainly not new; the Egyptians were boring into the ground using a rotary motion in 3000 B.C., and in 1500 Leonardo da Vinci published a detailed set of plans for a drilling machine that had many of the same attributes as today's rigs. By the mid-nineteenth century, French engineers were using diamond-faced bits to drill through rock, and circulating water to remove cuttings.

Components of the Rotary Drilling System

Today's rotary rig is a complex system composed of a number of subsystems that include the *prime movers*, the engines or motors that supply the power, the *hoisting equipment* that raises and lowers the great weights of the drill string, the *rotating equipment* that turns the tools in the hole, and the *circulating equipment* that lubricates the bit, removes cuttings, controls downhole pressure, and sometimes seals the walls of the wellbore (Figure 5–3).

Prime Movers

From the time of Spindletop until World War II, almost all rotaries were powered by steam engines. After the War, most drillers switched to diesel engines as they became available as the wartime shortages ended. While diesel-powered rigs may still be in the majority, there are actually many types of engines and motors in use. Some are diesel-fueled internal combustion reciprocating engines. Others are fueled by natural gas, and common today are diesel or gas powered *recips* or turbines that drive generators that furnish power to electrically driven rigs. Or some rigs may use electricity taken directly from power lines. In the San Juan Basin area of the Navajo Nation, the use of electrically powered rigs has been credited with bringing power lines into remote areas and thus making electricity generally available for the first time.

Many rigs today are in the 1,000–3,000 HP range. Shallow or moderate depth rigs may use 500–1,000 HP for hoisting and circulation, while rigs drilling to 20,000 ft and deeper will be rated well in excess of 3,000 HP. Additional horsepower for auxiliary lighting, water supply, compressed air, and other requirements may be needed.

The super rigs, such as the Parker Drilling Company's Rig 201, the world's largest land-based drilling rig, which was designed to drill to 50,000 ft, need many times the horsepower of an "average" rig.

Power is transferred through chain drives to the draw works and *rotary* and then through belt drives to the pumps. Newer and larger rigs may utilize torque converters or hydraulic couplings with fluid drive in place of chains and belts. Much newer generation rigs, about half of those now in service, derive their power from generator-motors. With these rigs, silicon-controlled rectifiers *(SCR)* electronically control variable speed electric motors which power the operating equipment—rotary, draw works, pumps, and, in the latest equipment, downhole motors and top drives (Figure 5–4).

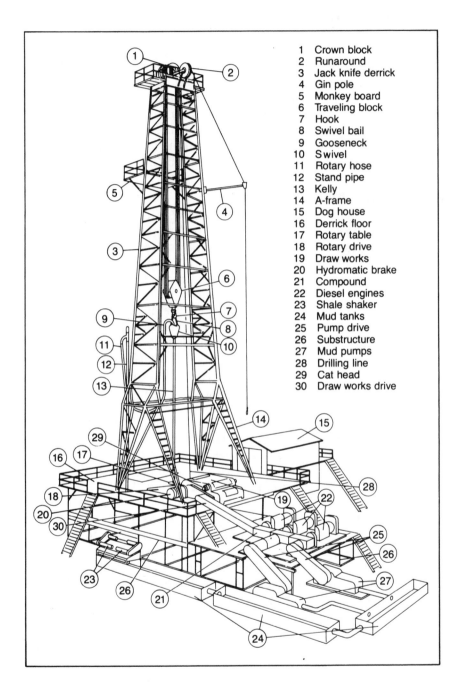

1 Crown block
2 Runaround
3 Jack knife derrick
4 Gin pole
5 Monkey board
6 Traveling block
7 Hook
8 Swivel bail
9 Gooseneck
10 Swivel
11 Rotary hose
12 Stand pipe
13 Kelly
14 A-frame
15 Dog house
16 Derrick floor
17 Rotary table
18 Rotary drive
19 Draw works
20 Hydromatic brake
21 Compound
22 Diesel engines
23 Shale shaker
24 Mud tanks
25 Pump drive
26 Substructure
27 Mud pumps
28 Drilling line
29 Cat head
30 Draw works drive

Figure 5-3 Cross section of a drilling rig (courtesy Canadian Petroleum Association).

Modern Petroleum

Figure 5–4 Electrically driven drawworks installed at ground level are supplemented by electric catworks on the rig floor (courtesy API).

Hoisting Equipment

During drilling, the bits become dull and must be replaced. Since the bit is at the very end of the drillstring, the entire string must be pulled out of the hole in the process called *tripping out*. Each *joint* of pipe is disconnected and stacked upright outside the derrick. When a new bit is attached to the last joint, most often a heavy *drill collar*, the string is reassembled and tripped back into the hole. The hoisting system is the rig component that raises and lowers whatever may go into or out of the well.

Derricks

One of the primary parts of this system is the *derrick*, or *mast*.

These steel "towers" that support many other important components are rated in terms of how much vertical load they can support and also how much horizontal wind force they can withstand. A modern derrick may be able to support as much as 1.5 million lbs and withstand winds of up to 130 MPH when its racks are full of pipe. Some of today's largest rigs—designed for drilling 10 mi or more into the Earth's surface—must be capable of carrying even greater loads.

The height of a mast can also give an indication of what depth well it was designed to drill. Drill pipe comes in joints 30 ft long. If every time a bit had to be changed while drilling a well 20,000 ft deep, it would be hopelessly time-consuming to break that nearly 4-mi-long string down into 30-ft sections. Therefore, rigs designed for deep drilling can handle multiple joints. A small rig may be able to handle a *double* consisting of two joints. A larger rig a *thribble*, which is, as the sound implies, three joints, and a still larger rig a *fourble*. This capability to break the string into 60, 90, or 120 ft sections instead of ones only 30 ft in length saves vast amounts of time and money (Figures 5–5a and 5–5b).

Tripping

This bringing the string out of the hole is known as *tripping out* or *making a trip*, in oil field parlance. In tripping out, the kelly is drilled

Figure 5–5 (a & b) Modern "super rigs."

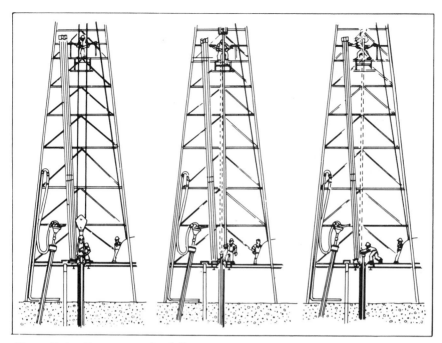

Figure 5–6a Tripping out the drillstring.

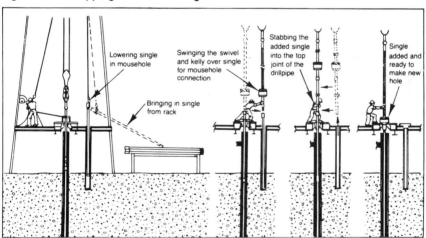

Lowering single in mousehole

Bringing in single from rack

Swinging the swivel and kelly over single for mousehole connection

Stabbing the added single into the top joint of the drillpipe

Single added and ready to make new hole

Figure 5–6b Adding a new joint of pipe.

down before the drillstring is pulled. Then the swivel, kelly, kelly cock, and rotary bushing are broken free and placed in the rat hole out of the way. Next, the elevators on the hook and block assembly are latched around the pipe just below the tool joint box. The pipe is then pulled and broken down into appropriate lengths and stood on the rig floor until it is time to return the string (Figures 5–6a and 5–6b).

Tripping in is, as the name implies, the reverse of the above; the reassembly of the drillstring and its replacement in the wellbore (Figure 5–7).

Draw Works

Deep wells are drilled with long strings of drillpipe and *drill collars* whose combined weight can be measured in hundreds of thousands of pounds. Such weights require tremendous hoisting and braking capability. The modern draw works consists of a revolving drum around which the drilling line is wound, the *catshaft* upon which the *catheads* are mounted, and a series of shafts, clutches, and various drives for changing speed and reversing. The draw works also houses the main brake which is necessary to stop and hold the great weights encountered in tripping, drilling, or *running casing.* When very heavy loads are being lowered, the main brake is assisted by an electric or

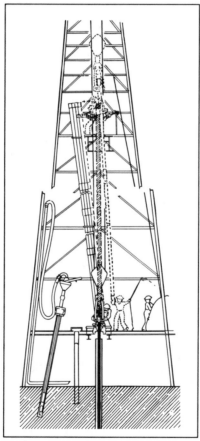

Figure 5–7 Tripping in (courtesy API).

Modern Petroleum

hydraulic auxiliary brake which helps absorb some of the great downward energy created by the mass of the *traveling block, hook assembly*, drillpipe, or other materials being lowered into the hole.

Two types of catheads are used. One is a drum. By wrapping a large rope around it, the crew can perform any number of tasks such as *make-up, break-out*, and general light-duty hoisting. The number of turns taken around the cathead and the amount of tension applied to the tope supplies the friction needed for the pull. The other cathead has a manual or air-actuated quick release friction clutch and drum to which the tong jerk line or spinning chain is attached. It is used for *spinning up* or breaking-out the drillstring during trips and *connections*.

Drilling Lines and Blocks

The *drilling lines* are an integral part of the hoisting system. These are wire ropes, sized form 1 ⅛ in. in diameter up to whatever size is needed to fit the weight requirements of the job at hand. Records of drill line use are carefully maintained. The usage is calculated in ton miles (a 1-T load moved 1 mi). When the safe usage limits of the line have been exceeded, it is discarded, or put to other uses.

The *crown block* is located at the top of the mast, the largest block on the rig, its grooved *sheaves* may be 5 ft or more in diameter. The smaller *traveling block* is located just above and attached to the *hook*, and as the name implies, it moves up and down as the drillstring is raised or lowered.

Both blocks are sized to handle the anticipated loads. The crown block has one more *sheave* than the traveling block. Thus a 10-line string-up would require the use of six sheaves on the crown block and five on the traveling block. As you may recall from high school physics, far greater lifting capacity can be obtained from a rope and pulley system by simply adding more sheaves instead of increasing the power supplied to the system. The traveling block also has a spring unit to absorb some of the shock encountered when handling threaded joints.

At the drillsite, the drilling line is set up on a storage reel located away from the rig and *pipe rack* area. When stringing up, the traveling block is placed on the rig floor, and the free end of the line is threaded over the sheaves in the crown block and through the traveling block a sufficient number of times to lift the anticipated weight. The free end is then attached to the draw works drum which is rotated until it holds at least one solid layer of line before the traveling block is lifted. Without cutting, the other end of the line, called the *dead line*, is attached to an anchor set opposite the two blocks.

The amount of line used can be increased during drilling for efficient use of power or if it should become necessary to run an especially heavy string of casing (Figure 5–8).

Rotating Equipment

The rotating equipment provides the power which makes the string "turn right and make hole" and includes those parts of the rig which begin at the *swivel*, which is attached to the hook on the bottom of the traveling block, and extend down the drillstring to the bit which is in contact with the Earth at the bottom of the hole.

The Rotary

Powered by the prime mover, the rotary is the device that turns the drillstring, and ultimately, the bit. The *rotary drive* transmits power to the *rotary table*, which is set in the rig deck, and through which the drillstring is run (Figure 5–9).

The Drillstring

The swivel not only carries the weight of the string, but it also permits the string to turn freely during drilling. It also provides a rotating pressure seal and entry for circulating the drilling bluid into the top of the string. The *derrickman* can lock the swivel in a particular position to latch the *elevators* when making a trip.

The Kelly

The *kelly* is a hollow square or octagonal steel joint that is attached to the swivel and is *stabbed* into the first joint of the drillstring. By means of large studs on its bottom which fit into corresponding holes on the rotary table, the *kelly bushing* locks into the table, and as the table turns, transmits torque from the rotary to the drillstring. The kelly bushing also allows the kelly to move through it vertically, as the string is lowered or raised. The drilling fluid is pumped through the kelly from the swivel above on to the hollow drill string below. The *kelly cock* is located above the kelly and can be used to shut off back-flow in the event of a *blowout.*

Drillpipe

Drillpipe is heavy steel pipe with a threaded male tool joint at one end and a threaded female tool joint on the other. Manufactured to American Petroleum Institute *(API)* specifications, it is furnished in 30-ft lengths called joints, and during drilling, it connects the bit below with the kelly above. The *drilling contract* between the operator

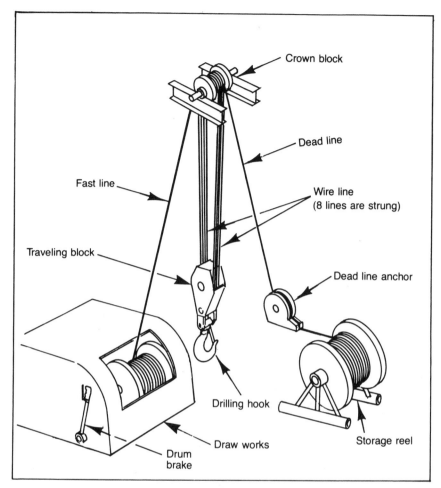

Figure 5–8 Hoisting system for a rotary rig.

and the drilling contractor will specify the API ratings of the pipe to be used. Despite its great strength, it can be surprisingly flexible when *directional* or *horizontal drilling* is necessary. When casing has been set at least partially down the hole and drilling is continuing, rubber protectors will be used to prevent metal-to-metal contact between the pipe and casing.

Drill Collars

Drill collars are much heavier and have thicker walls than drill-pipe. They are also larger in diameter, and are added just above the bit to add weight to the drillstring. Their number and weight will vary depending on the downhole conditions encountered.

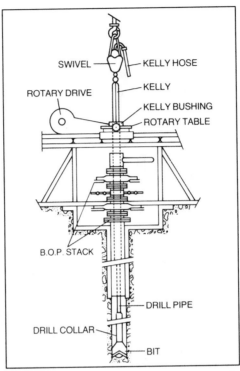

Figure 5-9 The rotary system (courtesy Anne McNamara, Petroleum Resources Communication Foundation).

Bits

At the very bottom of the hole, and at the very end of the drillstring, is the bit, one of the most important parts of the entire drilling process. As with all the other tools used in drilling, bits have undergone a steady evolution over the years. Basically, the job of the bits is to advance the hole by breaking up and dislodging the formation so that the drilling fluid can remove the cuttings. There are literally dozens of types of bits. Each is designed to fit a particular need dictated by the type of soil and rock encountered. When shallow wells are drilled through soft soils and formations, one bit may be used for the entire job. When drilling deep wells through many different types of rock and soil, bits may have to be changed many times, either as they wear out or as soil conditions change.

The decision to change bits is not made easily, because of the time and money involved in tripping out, breaking out, and then reassembling and tripping back in what well may be thousands of feet or drillpipe. However, the alternative may be greatly reduced drilling effectiveness because of a worn-out bit.

Bits are classified by the International Association of Drilling Contractors. There are literally dozens of varieties—each designed for a specific application. Among the many factors affecting bit selection are: the type of formation to be encountered; the use of mud motor, turbine, percussion, or directional drilling; the anticipation of abnormal temperatures; and whether or not cores are desired.

The IADC lists six categories of formations that affect bit selection. These include extremely hard, hard, medium-to-hard, soft-to-medium, soft, and soft sticky. Each formation may have its own classification of bits. These include:

Steel Tooth Rotary Bits. There are at least nine sub-types of these, each differing in shape, bearings, use, etc.
Insert Bits which have tungsten carbide inserts implanted in them.
Polycrystalline Diamond Compact Bits (PDC) have inserts of that material attached to the tungsten carbide inserts mentioned above.
Diamond Bits are implanted with industrial diamonds for use in extremely hard formations.

In addition, there are hybrid bits which combine the features of several of these types; also, variations of each of these types are manufactured as core bits to obtain core samples. And, there are yet other special-use bits such as thermally stable synthetic diamond bits (TSD) which are used in geothermal applications.

Diamond bits are 40 to 50 times harder than the strongest steel and this makes it possible to run them longer between changes. As might be expected, they are correspondingly moe expensive than a steel bit.

Some situations call for the use of downhole drilling motors at the bottom of the string. As the name implies, these are powered motors that drive the bit. Power may come from drilling fluids such as high-pressure natural gas, compressed air, mud, or from electricity.

Included among the many specific designs of bits available are tri-cone roller bits, button bits, tapered bits, fishtail bits, and mill bits. The tri-cones have three inwardly-facing toothed rotating cones mounted on axles with lubricated bearings. As the drillstring turns, so do the rollers, which in so doing, grind their way through the rock (Figures 5–10 through 5–15).

Roller cone bits used in soft formations will have longer "teeth" set in them, and the cones will be skewed to an angle that will allow a scraping instead of a true rolling action. Medium-hard formations require bits with more and shorter teeth and larger bearings. In very

Figure 5-10 Tungsten carbide insert bit (courtesy Dresser Security).

Figure 5-11 Steel tooth bit (courtesy Dresser Security).

Modern Petroleum

Figure 5–12 PDC fixed cutter bit (courtesy Dresser Security).

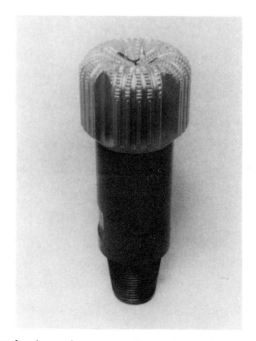

Figure 5–13 TSP fixed cutter bit (courtesy Dresser Security).

Drilling

Figure 5-14 Natural diamond fixed cutter bit (courtesy Dresser Security).

Figure 5-15 PDC fixed cutter bit (courtesy Dresser Security).

hard formations, bits with still shorter teeth or trungsten carbide inserts are used. Openings in the bit direct the flow of the drilling fluid either through the cones, or directly down against the bottom of the hole. Fishtail bits are used to enlarge the hole behind the bit, and mill bits are used to mill or grind away metal debris in the hole such as stuck pipe.

In addition to all these combinations and permutations of bit types, one must also remember that they can be supplied in sizes ranging from 4¾ in. to 48 in. in diameter.

Pound for pound, bits are some of the most expensive items that must be taken into account when designing the drilling program. Not only are they costly to purchase and use, their relatively small size in relation to other rig components and high value makes them easy targets for thieves.

The Circulating System

Drilling Fluids

The drilling fluid is another extremely important component of the rotary drilling system. It serves the following purposes:

1. Lubricates and cools the bit
2. Controls downhole pressures
3. Removes cuttings
4. Coats the wellbore with a *cake*

The type of fluid used depends upon the type of drilling being done and the formations which will be encountered. Most rotary drilling is accomplished with *mud*, which while it may have a clay base, is actually a highly complex mixture custom blended to meet the downhole conditions of the well it is being used for. Mud can also be used to form a cake which coats and protects the wellbore before casing is set.

Mud

The drilling fluid known universally as *mud* is a mixture, custom blended for the formations to be drilled through, and which may contain clays, polymers and chemicals, weighting materials, water or oil, and sometimes lightening materials wuch as air or gas. It is a major contributor to the success of rotary drilling, and is also one of the major items of cost in drilling a well. There is a popular legend that the drilling mud Lucas used at Spindletop was obtained by driving a herd of cattle back and forth over a wet pasture. By contrast, mud engineering today is a highly complex process.

The base material for many muds is bentonite clay. This natural clay, although delivered to the site in dry powder form, tends to swell when mixed with water. This property makes it ideal for drilling use since as it swells, it coats the wellbore with a cake that seals the bore and helps prevent cave-ins, *zones of lost circulation*, and contamination of freshwater *aquifers* until casing has been set.

Mud is also one of the first lines of protection against *blowouts*. During drilling the driller carefully monitors mud weight to make sure it is greater than the upward pressure being exerted by the downhole gas, which, being lighter than air, is trying to escape. Many different types of weighting additives are available, ranging from seed hulls to finely shredded scrap paper.

Proper consistency is also important. The mud must be thin enough to lubricate and cool the bit while at the same time have enough substance to carry the rock cuttings back out of the hole.

It is also extremely important that the mud's chemical composition be compatible with the formations the well will penetrate. For example, if the well is being drilled through known salt formations, then fresh water should not be used in the mix since it would tend to dissolve the salt and possibly cause zones of lost circulation. By the same token, salt water should not be used if there is a likelihood of contaminating freshwater aquifers. In short, the importance of the drilling mud and the circulating system cannot be overly stressed.

Circulating Equipment

At a modern rotary drilling site, bulk storage of the drilling fluid materials, pumps, and mud mixing equipment is placed at the beginning end of the circulation system, and the working mud pits and reserve storage at the opposite end. In between are the drilling equipment, the auxiliary equipment for drilling fluid maintenance, and the well pressure control equipment.

The mud path begins at the mud pit and leads to the mud mixing hopper where the clays and other bagged ingredients are added. From there it goes through the pumps, which force it up the *standpipe* and *kelly hose* to the swivel. It is then pumped down through the kelly and drillstring and out through the bit. Here the mud picks up the cuttings and carries them up the *annulus*, or *annular space*, which is the space between the drillstring and the walls of the wellbore.

When it reaches the surface, it exits the system through the *blowout preventer stack*. The *mud return line* carries it to the *shale shaker* located above the *settling pit.* The reusable portion of the mud filters through the shaker into the pit for recycling. The cuttings (the shale) are carried across the *shale slide* to the reserve pit, which is used for waste. Agitators in the mud pits (actually steel tanks) maintain a uniform mixture of liquids and solids. The system also includes a cone-type desander and desilter for removing fine particles that might not readily settle out, and a vacuum degasser for the quick release of entrained gas (Figure 5–16).

The cuttings are carefully monitored by the mud engineer who sniffs them and examines them under an ultraviolet lamp for evidence of oil and gas *shows.* The cuttings are also evidence of the type of rock the bit is encountering.

Air Drilling

In some situations, however, it is better to use a drilling fluid that is a gas rather than a liquid. Thus, compressed air or engine exhaust gas may be used in place of mud. And in areas of plentiful, high-pressure natural gas, that may be used both as a drilling fluid and to operate downhole drilling motors.

The use of either air or gas is referred to as air drilling, a technique that has been used for more than half a century. In many areas it is both quicker and far cheaper than drilling with mud. The record books are full of instances where shallow wells have been drilled in as little as 24 hours with only two men on the crew. Bit consumption is usually considerably less; in many cases bit usage is reduced as much as 80%. It is particularly useful in crooked formations, and in areas of lost circulation, where thousands of dollars worth of mud can easily be lost. These savings in rig time, drilling mud cost, bit cost, salaries, and time spent tripping in and out of the hole, make air drilling an attractive alternative. In addition to these economic realities, there are any number of engineering and geophysical advantages to be gained.

Since air is the most common low-density fluid, it is easily available in the field and also has the advantage of placing very little outside pressure on the formation being drilled. This causes cuttings to rapidly dissipate as the bit comes into contact with the formation, a phenomena that greatly speeds up penetration.

Air drilling can be accomplished by *dusting*—that is, drilling "dry" without any other substance introduced into the air stream; by *mist* or *foam* drilling—which calls for the introduction of special mud or foaming agents, or by the use of mud which has been *aerated* by the air stream.

Air drilling requires specialized equipment including air compressors for a source of high-pressure air or a source of high-pressure

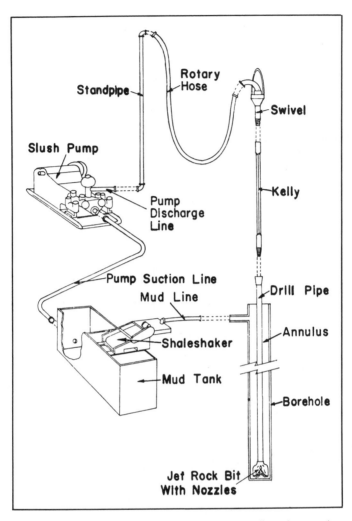

Figure 5-16 Typical circulating system (courtesy PennWell Books, *Workover Well Control*, Adams).

gas. Whichever medium is used, a gas-tight rotating packer must be placed around the kelly. This causes the gas and cuttings to exit the well through the *blooey line*, a long blow line which carries the detritus well away from the rig. Also needed are chemical treatment equipment for prevention of corrosion, which can be severe, mist pumps for injecting fluids and detergent foamers, and bits designed especially for air drilling. The latter may include hammer bits which in some situations may bring about a considerable increase in the rate of penetration.

As air-volume requirements increase rapidly as the depth increases, having sufficient compressor capacity is vital. If fluid-bearing formations are encountered which dampen the cuttings and inhibit their exit, then steps must be taken to dry up the hole. If the problem persists, then a switch to *mist drilling* must be considered. This involves injecting a minimum amount of water mixed with a foaming agent into the air entering the well. The chemical composition of this mixture must be compatible with the formation fluids encountered.

One drawback to misting is that it may cause sloughing, or hole enlargement in some formations. In this case, a thin slurry or mud may be used instead of clear water. This tends to coat the wellbore and reduce the sloughing.

Another technique, *foam drilling*, calls for the use of either salt or fresh water to be mixed with detergent and other chemicals in a foam generator before it is introduced into the air stream. Carefully controlled foam drilling can help maintain a clean hole by removing cuttings that tend to ball up because of excess water in the formation.

If sands containing large amounts of water are encountered, or if lost circulation becomes a problem, then *aerated mud*, or *aerated drilling*, may be used. This involves pumping both air and mud down the standpipe at the same time. This technique can reduce lost circulation and increase penetration rates over those available with "plain" mud.

However, aerated drilling does have its drawbacks including the need for deaerating the mud between cycles, the threat of corrosion destroying the drillpipe in a very short time, and the fact that the hole must be protected by casing.

Air drilling has been extensively used by the Department of Energy in drilling large-diameter holes for underground nuclear tests in Nevada. Air drilling has also seen use in Russia, Canada, Europe, the Middle East, and South America.

Pressure Control Equipment

In addition to mud weight, the basic equipment for well pressure control is in the *blowout preventer (BOP)* assembly which is located on the well casing under the rig deck. The topmost preventer is usually an annular type that can be activated to provide a seal between the inside of the casing and the drillpipe or kelly. In addition, there may be two or three hydraulically operated ram-type preventers that can be quickly closed with great force to seal off the well (Figures 5–17 and 5–18).

There is also provision for a mud return line to the shale shaker, a *kill line* that can be used to pump mud into the hole to restore pressure balance, another line to be used when making a trip, and

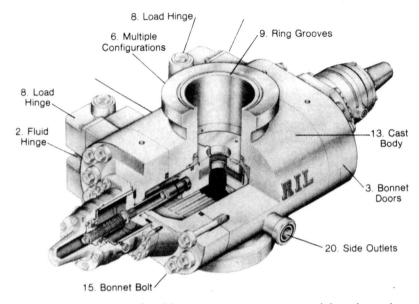

8. Load Hinge

6. Multiple Configurations

9. Ring Grooves

8. Load Hinge

2. Fluid Hinge

13. Cast Body

3. Bonnet Doors

20. Side Outlets

15. Bonnet Bolt

Figure 5–17 Ram-type surface blowout preventer (courtesy Hydril Mechanical Products Division).

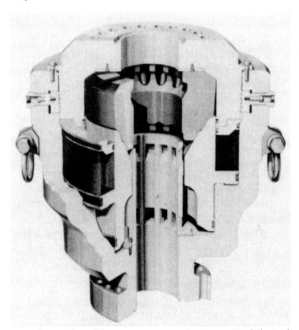

Figure 5–18 Annular surface blowout preventer (courtesy Hydril Mechanical Products Division).

Modern Petroleum

pressure relief lines running to the *chokes* that can be used to relieve pressure in a *blowout* situation.

As the name implies, a choke is a pressure control device. They are used to control pressure in the bore by regulating the flow from the well. Some are automatic, while others may be adjustable or fixed. The gas and drilling gluid coming through the choke go through a *mud-gas separator*, which salvages the reusable mud components and pipes the flammable gas away from the drilling rig to reduce the fire hazard.

As man drills deeper into the Earth, pressures far greater than anything that could have been imagined in the early days of the industry are now commonly encountered. In the Anadarko Basin of western Oklahoma, where natural gas wells are drilled to 30,000 ft, pressures as high as 15,000 psi have been measured at the surface when formations containing high pressure gas are encountered five or more miles down the well. To control these "superwells," entirely new methods and materials have had to be developed.

Preparing To Drill

Today, an immense amount of planning is necessary before actual drilling can begin. First of course, a location likely to be the source of commercially feasible amounts of hydrocarbons must be chosen. Next, the right to exploit the formation must be obtained from the mineral owner via a lease, and also permission for operations must be obtained from the surface owner if both estates do not belong to the same party. Then permits and other needed documents must be obtained from the appropriate regulatory agencies in the jurisdiction where the well is to be drilled, and all environmental laws must be complied with. In the United States, this may mean engaging a specialized contractor to conduct an *environmental assessment* or prepare an *environmental impact statement* before drilling can begin.

Planning the Drilling Program

At the same time, planning for the well itself should be under way. This includes the physical design: total depth, sizes and specifications of casing, cement program, bit program, provisions made for the types of formations to be penetrated as well as the pressure and temperature gradients that might be encountered. The type of rig to be used, the drilling contractor and which type of drilling contract to use must be selected. Also subcontractors from bulldozer services to water suppliers must be lined up.

Each of these steps must be made in its proper sequence in order to insure an orderly drilling program. For example, before the casing and bit program can be formulated the operator must know something of the formations that will be encountered. Will there be areas of abnormally high pressure? Zones of lost circulation? Unusually dense areas where drilling will be slowed? All of these must be taken into consideration before the number of strings of casing needed is decided upon. Also, if the site is located in soil that is so loosely consolidated that it begins to immediately crumble into the hole, then more outer conductor casing than originally planned for must be provided.

Once the length of the casing needed is determined, then its size must be decided upon. The cost of the casing depends on its weight, so since larger diameter casing weighs more, it also costs more. Therefore it is often more economical for the operator to select the smallest casing practical for the job. Also to be considered is outside clearance. Since the cement is placed between the outer wall of the casing and the wall of the wellbore itself, ample room must be provided for enough cement to form a proper bond.

After one string of casing has been set and cemented in place, the next string and bit must pass through it, thus the bits and strings used at lower depths must be successively smaller in diameter than those initially used nearer the surface.

Because more and larger casing increases the weight of the string, the total anticipated load will affect the selection of the rig used to drill the well. The greater the weight, the larger the weight handling capacity the rig must have.

Overall Planning

All of this is more than just an exercise in managerial expertise. Proper judgment in the planning stages can save tens of thousands of dollars in downtime and lost working days. One of the major concerns of both operator and drilling company is coordinating the delivery and availability of all the supplies and services necessary to drill and complete the well at the exact time when they are needed. Logging services must be available when needed. Cementing services must be on hand when needed. Casing should be on hand or readily available. And the logistical problem of providing the proper services and supplies at the proper time can be compounded by both geography and environment. The drillsite may well be in the story North Sea, the frozen Arctic, or a remote corner of a tropical rain forest.

The selection of a drilling company will come early in the overall planning stage. The world of petroleum is a world of contractors and subcontractors. Even integrated oil companies, those companies that are engaged in all phases of oil and gas operations from exploration to the retail selling of gasoline at the corner filling station, seldom perform all of the hundreds of functions necessary to find and produce petroleum themselves. Instead, they rely on the services of many highly specialized contractors to do the work for them under their direction.

Drilling is one of those areas that calls for specialized experts with their own extensive and expensive equipment. In selecting a drilling contractor the operating company must look at a number of factors: the contractor's past record, expertise, and personnel, his equipment, his experience in the location and formations to be drilled, and the dollar amount of his bid. It may well be that the lowest bid may not be the best if the bidder lacks the proper expertise or equipment for the job at hand.

The Drilling Contract

The process will begin with the preparation of bid sheets by the operator. This document will list all the details of the proposed venture: commencement date, depth, formations to be encountered, rig, equipment, and services (mud program, coring, sampling, logging, etc.) to be furnished, BOPs, operational checks to be made, and all the other items that the operator proposes that the driller should do. The drilling company will use the specifications on the bid sheet to prepare a bid, and will submit a contract, usually on one of the standard forms of the *American Petroleum Association (API)*, the *American Association of Drilling Contractors (AADC)*, or the *International Association of Drilling Contractors (IADC)*. Historically, there have been two general types of contracts used. One, the *footage contract*, is based on the driller being paid for the number of feet drilled. The other, the *daywork contract*, is based on time. If the drilling company is to be paid on a day work basis, the size of the crew will be included as will the daily (24 hour) and hourly rates for drilling both including and excluding drillpipe. Also spelled out will be the standby time rate, or amount paid to the contractor for remaining on the job even though no actual drilling may be taking place. Of course, different contract forms are used for offshore and onshore ventures (Figures 5–19 and 5–20).

A third type is the *turnkey contract*. In this case, the drilling contractor assumes total control of the operation and drills the well to the planned total depth for a flat fee.

Site Preparation

Few wells are drilled on ideal terrain; therefore after the site for the proposed well is surveyed and staked, the grounds must be prepared

Published by
Division of Production
American Petroleum Institute
300 Corrigan Tower
Dallas 1, Texas

Model Form 4A2
First Edition
August, 1962

EXHIBIT A
BID SHEET AND WELL SPECIFICATIONS

To:

Gentlemen:

We solicit your bid to drill and complete the hereinafter designated well.

This bid form has been filled in by us to the extent necessary to disclose the manner in which we desire the well to be drilled. If you desire to submit a bid, please complete this instrument in every respect, execute the original and two copies, and return to our office at_____not later than_____a.m. p.m.
_____, 19_____.

Very truly yours,

_____
Operator

By:_____

1. NAME AND LOCATION OF WELL:

Well Name
and Number _____ Parish
County_____ State_____.

Field Name_____ Well location and land description_____

2. COMMENCEMENT DATE:

Contractor agrees to commence actual_____operations at the above location on or before_____,
19_____, or, in the event Operator is to clear and grade location and furnish roadway or other ingress or egress facilities, within
_____days from the date of completion of the clearing and grading and construction of roadway, or such other
ingress or egress facilities, whichever is the latter.

3. DEPTH:

Subject to right of Operator to abandon the well or to have the well completed at a lesser depth, Contractor agrees to
drill the well to a total contract depth of_____feet. Contractor will drill the well on a footage contract basis (See
Section 13a hereof) to_____feet, or the top of the_____formation, or_____
feet of penetration into_____formation, whichever is first reached. Drilling between the footage
contract depth and total contract depth, if any, shall be at day work rates as specified in Section 13 hereof.

At Operator's request Contractor agrees to drill to a depth greater than total contract depth if in Contractor's opinion
equipment at the well site is capable of such drilling. Rates for such drilling shall be negotiated by the parties hereto unless otherwise provided by Section 13 hereof.

4. RIG AND EQUIPMENT TO BE FURNISHED BY CONTRACTOR:

Contractor's Rig No._____
Drawworks:_____
Engines — Number, Make & Models:_____
Slush Pumps — Make, Model & Size:_____
Auxiliary Pump & Power:_____
Derrick or Mast — Make, Size & Capacity:_____
Substructure — Height & Capacity:_____
Drill Pipe — Sizes & Amounts:_____
Drill Collars — Sizes & Numbers:_____

Blow-out Preventers — Power Actuated:

Casing String	BOP Size	API Series	No. and Style	BOP Pressure Tests	
				Frequency	Psi
Surface					
Intermediate					
Production					

Operational checks of BOP Equipment shall be made as follows:

(1)

Figure 5-19 API bid sheet (courtesy API).

Offshore International Contract
Revised June, 1975
0.5M 9-76

International Association of Drilling Contractors
INTERNATIONAL DAYWORK DRILLING CONTRACT — OFFSHORE

THIS AGREEMENT, dated the _____ day of _____ , 19 _____ , is made between:

_____ , a corporation organized under the laws of

_____ , located at _____ ,
and hereinafter called Operator,

and: _____ , a corporation organized under the laws of

_____ , located at _____ ,
and hereinafter called Contractor.

WHEREAS, Operator desires to have offshore wells drilled in the Operating Area and to have performed or carried out all auxiliary operations and services as detailed in the Appendices hereto or as Operator may require; and

WHEREAS, Contractor is willing to furnish the drilling vessel complete with drilling and other equipment, (hereinafter called the "Drilling Unit"), insurances and personnel, all as detailed in the Appendices hereto for the purpose of drilling the said wells and performing the said auxiliary operations and services for Operator.

NOW THEREFORE THIS AGREEMENT WITNESSETH that in consideration of the covenants herein it is agreed as follows:

ARTICLE I — INTERPRETATION

101. Definitions

In this Contract, unless the context otherwise requires:

(a) **"Commencement Date"** means the point in time that the Drilling Unit arrives at the place in or near the Operating Area designated by Operator, or at a mutually agreeable place in or near the Operating Area, or on arrival at the first drilling location, whichever event occurs earliest;

(b) **"Operator's Items"** mean the equipment, material and services which are listed in the Appendices that are to be provided by or at expense of Operator;

(c) **"Contractor's Items"** mean the equipment, material and services which are listed in the Appendices that are to be provided by or at expense of Contractor;

(d) **"Contractor's Personnel"** means the personnel to be provided by Contractor from time to time to conduct operations hereunder as listed in the Appendices;

(e) **"Operating Area"** means those areas of the seabed and subsoil beneath the waters offshore _____
_____ in which Operator may from time to time be entitled to conduct drilling operations;

(f) **"Operations Base"** means the place or places on shore designated as such by Operator from time to time.

(g) **"Affiliated Company"** means a company owning 50% or more of the stock of Operator or Contractor, a company in which Operator or Contractor own 50% or more of its stock, or a company 50% or more of whose stock is owned by the same company that owns 50% or more of the stock of Operator or Contractor.

102. Currency

In this Contract, all amounts expressed in dollars are United States dollar amounts.

103. Conflicts

The Appendices hereto are incorporated herein by reference. If any provision of the Appendices conflicts with a provision in the body hereof, the latter shall prevail.

104. Headings

The paragraph headings shall not be considered in interpreting the text of this Contract.

105. Further Assurances

Each party shall perform the acts and execute and deliver the documents and give the assurances necessary to give effect to the provisions of this Contract.

106. Contractor's Status

Contractor in performing its obligations hereunder shall be an independent contractor.

107. Governing Law

This Contract shall be construed and the relations between the parties determined in accordance with the law of _____ , not including, however, any of its conflicts of law rules which would direct or refer to the laws of another jurisdiction.

ARTICLE II — TERM

201. Effective Date

The parties shall be bound by this Contract when each of them has executed it.

202. Duration

This Contract shall terminate:

(a) immediately if the Drilling Unit becomes an actual constructive total loss;

_____ months after __ __ ____ of termination from Operator but Operator may not give

until __ ' Date, (or if __ __ hen being conducted on

Figure 5–20 Typical IADC daywork contract (courtesy International Association of Drilling Contractors).

for operations. This involves the services of a heavy equipment contractor in preparing access roads capable of supporting the massive weights of the rig components and drill pipe which must be trucked in. A modern rig, such as one of the "Ten-Milers," may require more than 70 large semitrailer trucks to transport it from one location to another. Grading and leveling for the wellsite, including enough room for auxiliary equipment, supply storage, pipe racks, and crew quarters will also have to be done. If permitted, reserve and waste pits will have to be dug, diked, lined with impervious plastic to prevent any hazardous wastes from entering the soil, and in many cases, covered with plastic netting to keep migratory waterfowl from landing in them and suffering harm. In some cases, the "pits" are actually steel tanks that can be used over and over again.

If soil conditions permit, a *cellar* may be dug under the rig deck for the blowout preventers, and a turnaround area for the trucks will have to be prepared. Next, either a well will have to be dug on the site, or a pipeline laid to a nearby source to supply the water which will be needed for the drilling operations.

Personnel

As may be imagined, a great many people are involved in the drilling of a modern, deep well. Consider that a large drilling contractor may have many rigs in operation at any one time and that these may be scattered all around the world.

Typically, such a contractor may establish regional offices, each responsible for all the rigs operating in its geographic area. The divisional or regional manager, his drilling superintendent, and his drilling engineer (who may actually be a petroleum, mechanical, or electrical engineer by training), may be responsible for the day-to-day drilling operations of all the company's rigs working in their assigned area. The actual number of rigs, and the table of organization of personnel, will of course vary with the size of the drilling company, current demand for their services, and company policy (Figure 5–21).

Next in the chain of command, and in charge of operations at the worksite, is the *toolpusher*, an individual of many years' experience who has the knowledge, experience, and common sense necessary to successfully and safely complete a project in which perhaps tens of millions of dollars has been invested. One of the greatest assets a toolpusher can have is his ability to manage his people, to spot and run off the slackers and troublemakers, and to motivate the good workers, who are usually highly independent individuals. In short, he must be able to get the job done correctly, on time, and without any accidents.

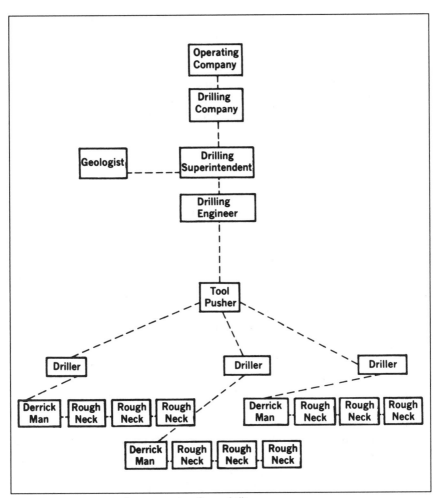

Figure 5–21 Personnel organization for a drilling company.

Next is the *driller*, one of the most important people in the entire operation. A very special kind of person, he must be thoroughly knowledgeable about his rig and its capabilities, the job at hand, and the underground conditions that may be encountered. He must be able to sense trouble before it happens and take immediate corrective action, and cope with any emergency which might arise. From his position on the rig deck, the driller has access to all the rig instrumentation and controls, which now may include computer terminals displaying data from downhole sensors (Chapter 14).

Work on an onshore drilling rig usually goes on 24 hours a day, 7 days a week, and each day is divided into three 8-hour shifts called "tours"—which, in the oil patch, is pronounced *tower*. The driller is the man in charge of his tour. With many drillers it is a point of pride to make more hole than the previous tour, and to set a record for the crew relieving him to beat.

Traditionally, the *derrickman* works high up on the derrick and has the responsibility of racking the pipe as the string is tripped out of the hole, broken down and racked. This can be a very hazardous job, and if warned in time, he can escape to earth by riding down the *geronimo line*. However, there are times when he has no warning and may be caught on the rig should an accident occur.

The *roughnecks* work the rig floor. They make and break pipe connections when tripping in and out and during routine drilling ahead, set the slips, work the tongs, and perform all the heavy, dirty, noisy, arctic cold or desert hot, work that is necessary to keep the rig "making hole."

The Morning Report

One of the time-honored customs of the drilling industry is the morning report. Each morning the toolpusher takes the daily drilling reports filled out by each driller during the previous three tours and assembles the information into the daily morning report (Figure 5–22).

Included in the report will be the depths reached at the end of each tour, how much footage was drilled, drilling fluid record, the assembly used, what formations were penetrated, and all the other highly detailed and specific information concerning operations during the tour.

When the toolpusher has all his data assembled, he phones or otherwise transmits it to the drilling superintendent. Since the superintendent will likely be receiving any number of such reports early in the morning, those wishing to see him soon learn to schedule their calls later in the day. The drilling company will use the information as a basis for the billing they will submit to the operator, and the operator will use his copy as a part of the permanent history of the well.

Spudding In

After the site is cleared, the rig erected, and the drilling lines prepared, the final steps before drilling can commence are *spudding in*, or starting the hole, and then preparing the *mouse hole* and the *rat hole*. The first consists of drilling a usually shallow, large-diameter hole, and lining it with conductor casing set in cement. The depth of the conductor will depend upon the nature of the soil at the drillsite. This surface casing also serves as an attachment point for the blowout preventers.

DAILY DRILLING REPORT
BERGER & ANDERSON Drilling Co.
Oklahoma City, Oklahoma

COMPANY   QR. SECTION   TOWNSHIP   RANGE   COUNTY   STATE   WELL NO.   Report No.   Rig. No.

LEASE   DATE   19

Formations   Coreheads   Bits and   Mud

MORNING TOUR (First)									DAYLIGHT TOUR (Second)									EVENING TOUR (Third)								
From	To	No.	Size	Type	Serial No.	In	Out	Hrs. Run	From	To	No.	Size	Type	Serial No.	In	Out	Hrs. Run	From	To	No.	Size	Type	Serial No.	In	Out	Hrs. Run

Pulled Bit No.   Cond.
Wt.   Vis.   W.L.   Additions   Amount

Derrick Man
Back Up Man
Helper
Driller
Day Work W/DP   WO/DP   Standby
Remarks:

Figure 5–22 Daily drilling report

Drilling   125

The mouse hole and the rat hole are very shallow slanted holes drilled adjacent to the rig by a specialized contractor with a truck-mounted rig. They serve as temporary sheaths or resting places for the kelly and joints of drill pipe when they are out of the hole.

After all these preparations have been made, it is at last time for the rotary to start "turning right and makin' hole," or, as it is more formally known, *routine drilling ahead.*

New Drilling Technology and Methodology

Measurement While Drilling

Over the past 15 years, companies have worked to develop a system to replace the largely unsatisfactory conventional means of measuring actual downhole conditions. The two biggest drawbacks to the old methods were that bottom-hole conditions had to be extrapolated from surface measurements, and that there was a considerable time lag between an occurrence at the bit and its indication at the surface.

Today *measurement while drilling (MWD)*, or borehole telemetry, describes systems that measure factors occurring at the bottom of the hole and immediately transmit that data to the surface. These measurements can be made continuously while circulating or drilling. Thus the need to stop drilling to run separate survey tools is eliminated. By giving the drilling engineer continuous real-time data from the bottom of the hole, both safety and efficiency have been greatly improved.

Systems such as the Smith International Datadril supply directional survey data and other comprehensive information using mud pulse telemetry. The system transmits data to the surface computer for a comparative analysis of the actual versus the proposed directional wellpath. It provides an accurate survey of the wellbore azimuth, hole inclination, sensor temperature, and abnormalities in dip-angle and magnetic field strength, and tool-face updates. When using downhole motors, the system can also be used as a wireless steering tool. If necessary, the downhole electronic package can be retrieved and replaced without having to make a trip.

Horizontal Drilling

Horizontal drilling, heralded today as "causing the greatest change in the industry since the invention of the rotary bit," is the newest and most rapidly growing movement in the drilling industry. Actually, it is an old idea whose time has finally come. The first patent for what would become horizontal drilling tools was issued in 1891. In 1929, Robert E. Lee drilled the first true horizontal well, actually a drainhole for a vertical well, for the Big Lake Oil Company at Texon, Texas. Later, as World War II approached, Lee sold his patents and tools to John Eastman, founder of Eastman Directional Drilling and Oilwell Surveys, forerunner of Eastman Whipstock and now, Eastman Christensen, an acknowledged industry leader in horizontal drilling tools and techniques.

Additional advances were made by John Zublin, the Oilwell Drainhole Drilling Company and many others in the years following the war. But the low crude prices of the time, together with improved perforating and hydraulic fracturing techniques for use on conventional wells put the further development and refinement of horizontal drilling on hold.

Today, however, economic conditions coupled with the burst of new technology, largely based on electronics, make horizontal drilling the most rapidly spreading innovation in the oil patch. In 1988, some 100 wells were drilled horizontally. By 1989, the number had grown to approximately 250 wells, and by 1990 it was estimated that 10% of all wells being drilled in the United States were horizontal. And, horizontal drilling successes in Europe, offshore in the North Sea, in the Arctic, and in many other areas were being reported. Today, the numbers continue to climb with some predicting that as many as 20,000 horizontal wells per year could be drilled by the year 2000, and that in the not-too-distant future the use of horizontal tools and techniques will surpass the use of conventional—and traditional—rotary methods.

Indeed, as the number of successes achieved by horizontal drilling mounts, and its tools and techniques continue their evolution and improvement, horizontal drilling could easily become the method of choice for the future.

What Are Horizontal Wells?

The idea of turning a shaft that is being drilled vertically to a slant, or even to the horizontal, is not new. For years wells have been drilled to purposely deviate from the vertical. *Deviation*, or *slant*, drilling has been used for years in onshore drilling in order to reach pay zones that were located directly under a spot where it was impossible to set up the drill rig.

The use of slant, or deviated, drilling has been one of the major factors in the success of offshore oil and gas production. Offshore gas and oil production platforms are, in most cases, far too expensive to be used to drill just a single well at each location. Instead, by drilling at a slant, 20 or even more wells can be drilled and produced from a single structure. After each well enters the seabed and reaches a predetermined depth, it is made to angle out to its desired target. The big difference in those wells and today's horizontal is that the slant wells may well take 2,000 ft or more to bend from the vertical to the horizontal, whereas today's horizontal may make a 90° turn in a very few feet. Simply stated, horizontal drilling is using state of the art tools to drill a vertical hole, then turning that hole to the horizontal so as to place the producing section of the well laterally into the reservoir. The efficiencies and benefits of draining the reservoir from a horizontal position are readily apparent (Figure 5–23).

Regulatory bodies seem to have a bit more difficulty in defining horizontal drilling. In Oklahoma for example, the State Corporation Commission, which regulates the oil and gas industry, proposed that a horizontal well is one in which "the horizontal component of the completion interval in the geologic formation exceeds the vertical component, and runs laterally at least 150 ft into the formation."

However, the Oklahoma State Tax Commission, which collects gross production taxes, defines a horizontal well as one which is completed from a geologic formation at an angle greater than 70° from the vertical and which penetrates 150 ft into the producing zone. These two definitions can have a definite effect on an operator.

First, wells that qualify as horizontal under the Tax Commission's definition may gain an exemption from gross production taxes for the first two years, or until the costs of the well are recovered, whichever occurs first.

Secondly, the Corporation Commission sets the maximum daily allowables, or production limits for wells. Since horizontal wells can often produce at much higher rates that vertical wells, the commission is proposing greater allowables for them. Also proposed is a bonus allowable based on the number of feet the well extends laterally into the formation.

Why Drill Horizontally?

Why are more and more operators turning to horizontal drilling, particularly since it can cost two to four times as much as a conventional well, and calls for highly specialized tools and expertise? The answer, of course, is greatly increased production, by some estimates five to six times more than that of conventional wells. Some of the benefits of horizontal wells are:

1. A horizontal well can penetrate more than one reservoir, and given the greater production capability of lateral penetration produce six or seven times as much oil or gas as a vertical well.

Modern Petroleum

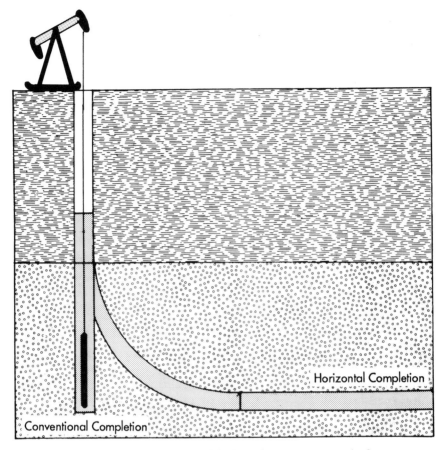

Figure 5-23 Production advantages of the lateral penetration into the formation (Courtesy Eastman Teleco).

2. Salt water production due to coning can be minimized.

3. Far fewer wells are needed to drain a reservoir.

4. The "traditional" primary recovery life of a well can be increased from 25% of the oil in place to 50 to 75%.

5. Later in the life of the well as production drops, it can be converted to a lateral injection well for enhanced recovery use.

6. Hydrocarbons can be produced even while the well is being extended.

7. Horizontal drilling has been credited with vastly increased production in areas long considered to be difficult such as the heavy oil sands in western Canada, the fractured Bakken shale of North Dakota, and the Austin Chalk in Texas.

The Three Categories of Horizontal Wells

Today's horizontal wells may be drilled by more or less traditional rotary methods, by a combination of conventional rotary rigs coupled with new tools and methods such as power top drives, steerable downhole motors, and MWD, or by the use of all new technology. The deciding factor will be whether the well will be short, medium, or long radius. Choosing which category the well will be in calls for even more careful planning than is necessary for a conventional well. The surface location, the target location, the nature of the formation, the nature of the hydrocarbons being sought, and which tools to use must all be considered.

Short-Radius Wells

Short-radius wells are drilled with a 20 to 45 ft radius of curvature. The horizontal section of the well may extend up to 1,500 ft. The tools used may include whipstocks and such new technology as articulated motor systems and compressive service drillpipe (CSDP) (Figure 5–24).

The short-radius well provides easy reentry into an existing well, and can be drilled using a portable workover rig. They are ideal for use in enhanced recovery, in coal-bed methane recovery, in older shallow reservoirs with close spacing, or in tight sands.

Medium Radius Wells

Medium radius wells are those drilled with a 300 to 700 ft radius of curvature. Their horizontal sections may be up to 3,500 ft in

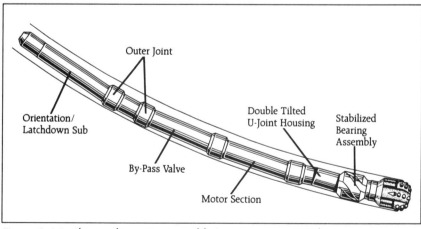

Figure 5–24 Short-radius motor assembly (courtesy Eastman Teleco).

Outer Joint

Orientation/ Latchdown Sub

By-Pass Valve

Motor Section

Double Tilted U-Joint Housing

Stabilized Bearing Assembly

length. They may be drilled with conventional rotaries except that fixed or steerable downhole motors, MWD, power swivels, and top drive motors may be used.

Medium radius wells are usually drilled vertically to reach formations located below 1,000 ft in areas of large spacing, tight gas sands, and to drill in restricted areas where the drillsite must be located well away from the target (Figure 5–25).

Long Radius Wells

Long radius wells are drilled with a radius of curvature of 1,000 to 4,500 ft. They may be drilled using new-generation equipment such as the Navigation Drilling Systems (NDS) which includes custom diamond bits, steerable motor, and MWD. However, conventional rotary drilling tools and techniques may also be used (Figure 5–26).

Long radius wells are generally used to extend out from an offshore platform or to reach an otherwise inaccessible target.

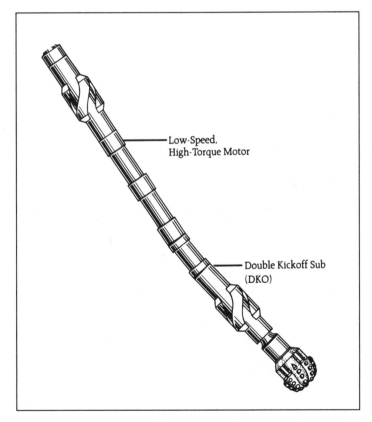

Low-Speed, High-Torque Motor

Double Kickoff Sub (DKO)

Figure 5–25 Medium-range steerable motor assembly (courtesy Eastman Teleco).

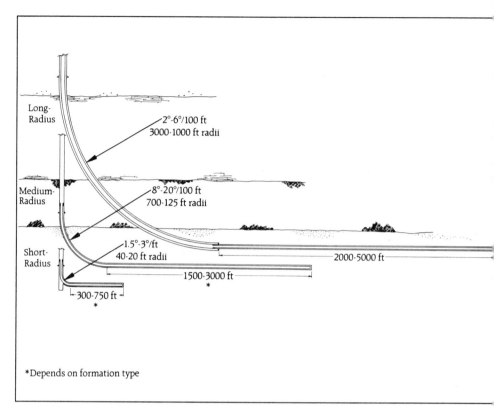

Long-
Radius

2°-6°/100 ft
3000-1000 ft radii

Medium-
Radius

8°-20°/100 ft
700-125 ft radii

Short-
Radius

1.5°-3°/ft
40-20 ft radii

2000-5000 ft

1500-3000 ft
*

300-750 ft
*

*Depends on formation type

Figure 5-26 The three types of lateral completions (courtesy Eastman Teleco).

Automation

Today, automation is rapidly spreading throughout the drilling industry and major changes are taking place both in drilling rigs and in the people who operate them. Tools and techniques developed over a century or more are giving way to equipment and processes designed to meet today's needs.

There are a number of reasons for this. One, of course, is cost. After the price of oil declined in the early to mid-1980s there was a drastic dropoff in the demand for drilling rigs. The drilling companies that remained in business had to slash their costs to the bone in order to survive. As the domestic oil industry in the United States shrank dramatically overnight, great numbers of the trained workforce quit the oil patch in disgust and vowed never to return. As a result, drilling companies now have great difficulty in finding skilled rig hands. Too, today's growing activity in horizontal drilling does not require as many of the old, large rigs, which for today's uses are not as efficient. In fact, many large rigs that were just being completed for the boom that ended in 1982, never saw use. Instead they were either cannibalized for parts to be exported to other countries, or they lie quietly rusting away in dinosaur graveyards along the roads in Texas and Oklahoma.

The need today is for rigs that can produce more oil more safely, more efficiently, and at a lower cost. And that is just what automation can do. One of the greatest boons to automation was the switch to SCR-controlled rigs and their variable-speed electric motors, for today electronics is the name of the game in every phase of the oil

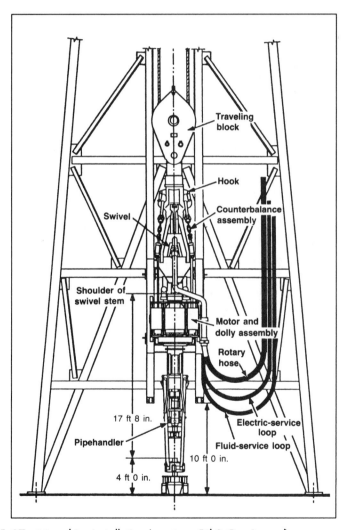

Figure 5–27 A top-drive installation (courtesy *Oil & Gas Journal*).

industry. Another advance lending itself to automation was the development of the *top drive* to replace the traditional rotary drive and table (Figure 5–27).

For today's drilling company the decision is not whether to move to automated drilling, but when and how to do it. The big choice is whether to purchase a complete new rig such as those manufactured by W-N Apache, which offer manual control, remote control, or fully automatic operation. Today, more than 70 of these rigs have been placed in operation. They are designed to be operated by a crew of two or three, and are rated to 6,000 ft, although 20,000-ft models are available.

All of the traditional chores have been automated including pipe handling, positioning, and holding. Automated devices are used for making and breaking connections during trips, which may be why for the past five years there have been no documented injuries on the floors of these rigs. Moving loads have been reduced by 50%, and drilling time is down by 12 to 15%, with consequent savings in the cost of operations.

Other manufacturers are building automated components which can be retrofitted to existing rigs, or automate them one system at a time. These include automated top drive systems, automatic pipe-handling equipment, automatic torque wrenches, and power slips, and the automatic weight-on-bit control systems (Parkomatic) from Parker Drilling.

The crew of a fully automated rig will consist of an experienced driller who fully understands the drilling process, and a technician who is knowledgeable in how the system operates and knows how to analyze and correct any problems that might occur.

Automated drilling, just like horizontal drilling, is another wave of the future that is here today.

Drilling
PART 2

Offshore and Arctic

Offshore Drilling

The objectives and many of the methods of drilling offshore wells are the same as for land-based wells, but the complexities, equipment, personnel, and regulations are all far greater.

One of the earliest patents for an offshore drilling rig was issued to T.F. Rowland in 1869. While it was for use in shallow water, its anchored, four-legged tower was the forerunner of today's platforms. The first actual offshore operation in the United States began in 1886 when a coastal field was discovered off Santa Barbara County, California. The first well was drilled from shore, but by 1890 wells were being drilled into the Pacific from wharves built out into the sea, some as far as 1,200 ft from the beach. More than 200 such wells were drilled, but production was low, and the rigs, placed on pilings, were susceptible to storm damage.

Later, close-to-shore wells were drilled in the Gulf of Mexico along the coasts of Texas and Louisiana. In the 1920s wells again began to be drilled from wharves and piers in California, where, in 1932, the first platform-supported well was drilled at Rincon. And, at the same time offshore wells were being directionally drilled from shore locations at Wilmington and Huntington Beach. By the end of that decade, offshore wells were being drilled off the Louisiana coast in the Gulf.

During World War II, the military borrowed offshore expertise and added new knowledge of their own in offshore construction when the famous Texas Towers were built. These steel platforms, attached to the seabed, much like the ones used for drilling operations, were built to house radar units off the coasts of both the United States and Great Britain to warn of impending enemy air attack. Some of those, off England, abandoned by England after hostilities ceased, became the homes of "pirate" radio stations that competed with the government-owned broadcasting stations in Europe.

In 1947, after the War's end, Kerr-McGee drilled the world's first offshore well located out of sight of land from a Navy surplus landing vessel in the Gulf of Mexico.

Since then the sea has provided a rich harvest of hydrocarbons. Many huge fields have been discovered, and despite the many problems encountered, thousands of successful wells have been drilled. Wherever possible, land-based methods have been used. However, in many cases new techniques and methods in exploration, drilling, production, transportation, and environmental protection have had to be devised.

Exploration

Petroleum exploration goes on continuously on the continental shelves that lie off the world's coasts and in the great inland lakes such as those in Africa. The potential areas are in the sediments laid down millions of years ago, in submerged rift valleys, and other areas that may hold promise for future exploitation. The seismology used is similar to that used on shore, and just as on shore, the use of explosives to generate the needed shock waves has been replaced. At sea, bursts of compressed air are used to lessen the impact on sea dwellers. The sound sources are towed behind seismic vessels. Echoes from the sound pulses they send are recorded by detectors spaced along long electronic cables also towed behind the ship.

Drilling Sites

The biggest difference between onshore and offshore operations is that the actual base of drilling operations is man-made. When exploratory wells are drilled at sea, the procedure is more complicated than it is on land where the surface provides a base for the drilling rig. Therefore, depending on the depth of the water, the climatic conditions of the area, and the costs involved, an operator can choose from several different types of drilling structures.

Special Subsea Equipment

Two items of equipment, not needed on land, have had to be developed to make it possible to drill in water hundreds, or perhaps even thousands of feet deep. In essence, an offshore well is just like an onshore well: a hole drilled into the Earth's surface. But in the marine environment, the drill rig is suspended far above the entrance to the well and, depending on the weather and the type of rig, it well may be pitching, rolling, and otherwise wildly gyrating at the same time it is attempting to drill. Therefore, a mechanism is needed to connect the rig to the well, one that will allow drilling to proceed in an orderly fashion, but at the same time allow for the inevitable motion of the wind and sea.

The first step in drilling an underwater well is to locate the exact site for the wellbore. Today this is often done by the use of advanced LORAN and satellite navigaiton systems. A weighted temporary guide base is lowered to the sea floor. This provides an anchor for the guide lines and a foundation for the permanent guide base that will follow later. Then a large diameter hole is drilled to a depth of about 100 ft. Casing 30 to 36 in. in diameter is used to fabricate a foundation pile to which the permanent guide base or *drilling template* is attached. This is an open steel box-like device with matching holes on its top and bottom. The number of holes will vary with the number of wells anticipated to be drilled at this site. Connected to the drilling platform or ship by cables, it is lowered to the sea floor and anchored in place (Figure 5–28). The foundation pile is then lowered into the hole and cemented into place. Once it

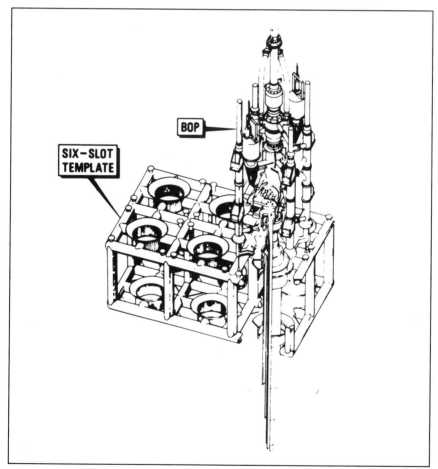

Figure 5–28 Floater-type subsea drilling template (courtesy *Journal of Petroleum Technology*).

is in place, a conductor hole, usually 20 in. in diameter is drilled, cased, and cemented.

Next the subsea blowout preventer stack is attached to the template or guide base. The *marine riser* is attached to it, and routine drilling may then commence.

The marine riser assembly provides a flexible steel pathway from the rig deck to the well opening through which the bit and drillstring can pass. The riser flexibility, which is necessary to overcome the rolling, pitching, and yawing of the drilling platform (Figure 5–29), is provided by slip joints and ball joints at strategic locations (Figure 5–30).

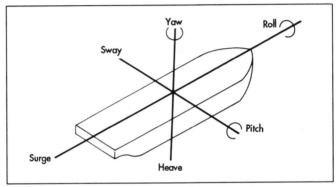

Figure 5–29 Types of motion to which a vessel is subjected.

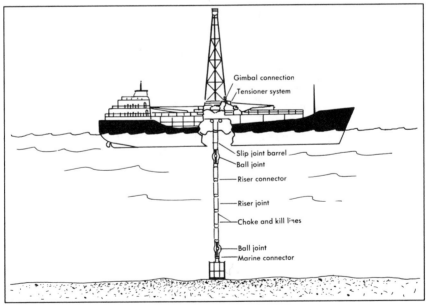

Figure 5–30 Marine riser (courtesy L.M. Harris).

Exploratory Drilling Platforms

Drilling Barges

For inland and shallow water drilling in lakes, bayous, rivers, canals, and estuaries, *drilling barges* provide the ideal drilling platform. Barges differ from ships in that they are not self-propelled, but must be towed to and from the well location by a tug. They are not suitable for deep water or for areas where high waves might occur. In addition to the necessary drilling equipment, barges also house supplies and provide quarters for the crew. Completion and logging services can be provided from other specially equipped barges or workboats (Figure 5–31).

Figure 5–31 Drilling barge.

Jack-Up Rigs

Jack-up rigs can also be towed to the drilling site. They are suitable for use in a large lake where many sites are to be drilled, or in shallow offshore areas. Once the site is reached, the three or four legs are "jacked" down until they rest on the sea bottom, leaving the working platform well above sea level. Since like a submersible rig, it rests on the sea bottom, it is limited to shallow water. Unlike the submersible, however, the working area is farther above the ocean, which makes operations safer (Figure 5–32).

Submersible Rigs

Submersible rigs are among the oldest mobile offshore exploratory rigs. Unlike a large offshore fixed platform, they have the advantage

Figure 5–32 Jack-up rig.

of being able to be moved when the exploratory work is finished. They consist of two hulls. The upper hull provides the work area and the crew's quarters. The air-filled lower hull provides buoyancy when the rig is being towed to and from the site. On site it is flooded, and when it sinks to the ocean floor it provides a stable base, holding the rig in place. Like the jack-up, it is limited to shallow water.

Semisubmersible Rigs

These vessels are the most widely used type of rigs. Some of them can sit on the bottom in shallow water, but they are more frequently used in a partially submerged position, moored in place by massive anchors weighing tens of tons apiece. They are capable of drilling in far deeper water than either the jack-up or submersible since they do not have to rest on the bottom. However, like the other rigs, they are free to move to another location once their mission is completed (Figure 5–33).

Figure 5–33 Semisubmersible rig.

Drillships

Drillships are capable of drilling in waters thousands of feet deep. They are similar in appearance to ordinary ships but with two vital differences. A cutout or hole called the *moonpool* pierces the vessel amidships from the upper deck down through the keel. It is through the moonpool that drilling operations take place. Permanently mounted above the moonpool is a large derrick, just like a shore rig.

While drilling, the ship is kept on exact location by a technique called *dynamic positioning*. Electric motors with propellers called *thrusters* are located under the ship's hull. The exact location of the wellbore is determined by satellite navigation and recorded in the ship's computer. Sensors located on the undersea drilling template continually send information regarding the ship's exact position to the computer. If, because of wind or wave action, the ship moves off location, the computer turns on the appropriate thruster to counteract the movement and keep the ship positioned directly above the wellbore. Like the semisubmersible, the drillship is free to move to another site when its work is completed (Figure 5–34).

Permanent Platforms

Concrete Platforms

Once it is decided to begin exploiting an undersea field, then a switch to permanent platforms, which can be used to drill and

Figure 5–34 Drillship.

workover numerous wells and also serve as production platforms, may be made. This is what occurred in the North Sea when its vast recoverable reserves became apparent. In fact, the North Sea oil fields have been responsible for a whole new generation of offshore drilling and production bases. Not only did these have to be designed with the deep water (up to 500 ft or more) in mind, they also had to be able to withstand the extremely violent weather that occurs there for much of the year. In the North Sea, waves 60 ft high and winds in excess of 90 knots are not uncommon. These factors have resulted in the construction of some of the largest structures ever built by man. And, what is even more amazing is that these structures are usually begun on shore and that their construction continues while they are being towed into place.

These huge structures include the *Condeep*-type used by Shell-Esso. Condeep began in drydock as a series of huge vertical concrete tanks grouped together. As their construction progressed, they were towed into ever-deepening water (Figure 5–35). Once completed, they were placed in position on the seabed (Figure 5–36). During construction and the move, the tanks and their base had been flooded with saltwater to hold them down. Once in place and producing, the seawater could be replaced with crude oil awaiting shipment to shore.

Other concrete North Sea "monsters" are the platforms used by Phillips Petroleum at the Ekofisk Field and those built by Howard Doris, Ltd. for the Ninian Field. These monolithic structures are 455

Figure 5–35 Condeep under construction moonpool.

ft in diameter at the base. Weighing 550,000 T, they rise 776 ft from the seabed to the tip of the derrick. More than 200 ft taller than the Washington Monument, the platform is 100 ft above sea level and the helipad is 100 ft above the platform (Figure 5–37).

Ekofisk Centre, as it is called, became the hub of a vast ocean drilling and production complex that reportedly cost $5 billion. Attached to the Centre are pumping platforms, riser platforms, field terminals, production platforms, quarters platforms, and a hotel. The project serves seven producing fields covering 123 mi². By 1975 it was connected to England by a 225-mi undersea oil pipline capable of delivering 1 million B/D, and by 1977 to Germany by a 275-mi natural gas pipeline 36 in. in diameter capable of transporting 2,000 Mmcf (million cubic feet per day).

By the 1980s it was discovered that such a great volume of hydrocarbons had been removed from the seabed that the under-

Figure 5-36 Condeep concept.

Figure 5-37 Base for Ekofisk-type structure under construction.

ground structures were actually collapsing and the complex was in danger of sinking beneath the sea. A massive injection campaign was undertaken to refill and bolster these structures and halt the sinking.

Steel Platforms

Huge steel permanent platforms have also been used in deep water situations around the world. While some of the North Sea platforms are located in water 475 ft deep, steel platforms have been used in waters 850 ft deep off California, 1,000 ft deep in the Gulf of Mexico, and 1,300 ft in the Bass Strait of Australia for EXXON. These huge structures are built in drydocks and their hollow legs provide buoyancy as they are being towed into place. Once at the site, the legs are gradually flooded and they are tipped into place.

Tension Leg Platforms

The *tension leg platform (TLP)* is yet another structure used for deep water drilling and production. The world's first was designed and

Figure 5–38 TPP concept.

built by Conoco for use in the North Sea's Hutton field in 1984 at a cost of $1.7 billion.

Despite that high cost, TLPs can actually be less expensive than other forms of offshore platforms. Instead of being a solid structure extending all the way from the surface to the sea floor, the TLP is actually a buoyant floating platform tethered by hollow steel tendons to a base embedded on the bottom. The tendons are kept under constant tension and allow no up and down movement; however, there can be a horizontal movement of up to 20 ft (Figure 5–38).

The second TLP was installed in the Jolliet Field in the Gulf of Mexico in 1,760 ft of water 170 mi from shore in 1989. At the time, this set a world production record for water depth. However, engineers claim that TLPs can successfully be used at depths of 8,000 to 10,000 ft. This may open many fields which have previously been inaccessible to man. Jolliet itself is estimated to contain 65 million bbl of oil.

Arctic Operations

Arctic operations, both on land and at sea, present their own unique problems, which, if overcome, can offer the reward of vast amounts of oil. Common to both land and sea efforts are the challenges presented by the extreme low temperatures. Steel and other metal becomes as brittle as glass. Lubricants freeze into unbreakable solids, and engines, if once allowed to stop, may never start again. And, the low temperatures are a constant hazard to personnel.

On Land

There are rich deposits of oil and gas in Alaska, northern Canada, Siberia, and the northern seas. To recover them has called for the development of yet more specialized technology and methodology. Rigs must be covered with shrouds to protect both men and equipment from the bitter cold and high winds. This cost of winterization may easily add a million dollars or more to the cost of an individual well. Specially designed and insulated living quarters for the men, and buildings for the supplies must be furnished (Figure 5–39).

The arctic winter may actually be a better time for drilling than the brief summer. In warm weather, the permafrost would not support the massive weight of the rigs and other equipment, so much of the operations must be carried out when the ground is frozen solid. Rigs

Figure 5-39 Winterized rig (courtesy API).

are not broken down when moved, but are left erect and skidded from location to location.

Offshore

A number of methods have been used to drill in the waters of the far north. One of these is to construct an ice island for use as a platform. Obviously this can only be used when there is no possibility of the ice melting (Figure 5–40).

Another method is to dredge up enough sand and gravel to build an island to use as a drilling platform. This manmade structure is usually circular, broad at the base, and tapers upward to a top just large enough to hold the rig (Figure 5–41). Yet another type of manmade platform was designed like a large, inverted champagne glass. The upside down bowl sat on the sea floor, and the narrow stem extended above the surface and held the rig. The idea here was

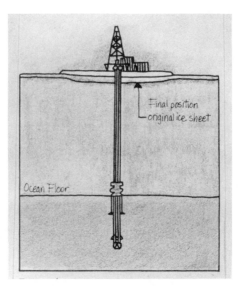

Figure 5–40   Ice island for drilling (courtesy Imperial Oil Ltd.).

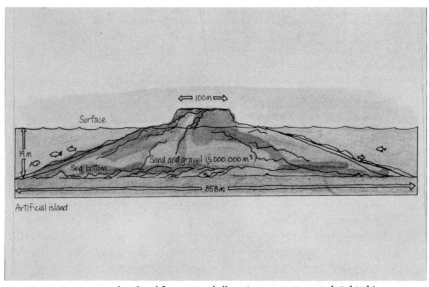

Figure 5–41   Manmade island for arctic drilling (courtesy Imperial Oil Ltd.).

Modern Petroleum

Figure 5–42 Arctic drilling platform concept.

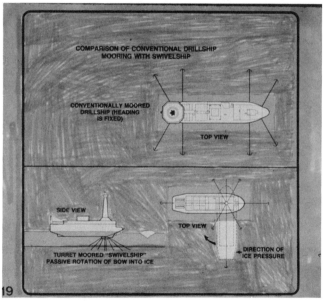

Figure 5–43 Arctic drillship concept.

to present as little resistance as possible to the ice that would surround the rig from all sides (Figure 5–42).

A similar idea was evidenced in a unique arctic drillship. Instead of a moonpool, the drilling apparatus was located in a turret near the bow. Under pressure from the ice, the ship itself was free to rotate around the turret (Figure 5–43).

As mankind's need for hydrocarbons continues to grow and the search for them expands to more and more remote areas of the globe, we can expect the development of technologies not envisioned today.

Chapter References

"Age Old Technology Finds New Application," *On Target*, Vol. 3, No 1., Spring 1990.

John L. Kennedy, *Fundamentals of Drilling*, (Tulsa: Pennwell Books, 1983).

Will L. McNair, "Rig Automation," *Oil & Gas Journal*, April 9 and 16, 1990.

Guntis Moritis, "Horizontal Drilling Scores More Successes," *Oil & Gas Journal*, February 26, 1990.

G. Alan Petzet, "Horizontal Drilling Fanning Out as Technology Advances and Flow Rates Jump," *Oil & Gas Journal*, August 23, 1990.

6
Logging

In Chapter 2 we learned that most of what we know about the interior of the Earth has come to us from secondary evidence gathered from studying rocks and minerals, seismic data, heat flow emanating from the Earth, the Earth's gravity and magnetic fields,and comparisons with other bodies in our solar system. In this chapter we will study the use of well logs, as well as some other methods of formation testing and sampling, to learn more about the subsurface of the Earth. Each time a log is run and analyzed we add to our knowledge of the planet Earth.

The term *log* as used in the petroleum industry refers to a method of testing, or the name of a particular test, or the record made of the test. Thus a log may be a graphic display of the lithology and stratigraphy of the formations penetrated by the well bore. It might also display physical properties such as resistivity, self-potential, and gamma ray intensity or velocity at given depths. And logs may be records, either printed out in long strips or displayed on a computer screen, of formations penetrated, depth records of minerals encountered including gas, water, or oil, the size and length of pipe used, and much more information about a well.

Historically, logging has also come to be known as *wirelining* since the logging instruments were lowered and retrieved by means of a cable—the wireline, spooled on a reel in the rear of the wireline truck operated by the wireline crew who raised and lowered the "wireline tools" and made the "wireline tests."

While that is still certainly true to a large extent, today's "wireline tools" may also be miniature computerized instruments that fit *inside* the drillstring just above the bit and continuously transmit information about downhole conditions by means of telemetry to a computer terminal located at the driller's post on the rig deck.

Today, there are more than 100 different types of logs available. Despite their complexities and miniaturization, today's instruments can withstand downhole pressures and temperatures of 20,000 psi and 400°F.

Why Run Logs?

The primary purpose of running well logs in the oil and gas industry is to determine whether the formation or formations contain profitable commercial reserves. Before actual drilling starts, there have been thousands of dollars spent in securing both seismic information and in buying leases. After drilling has started, costs are usually on a per-foot basis, with the prices increasing along with the depth of the well. Investors want to know as soon as possible when to cease drilling and abandon the project or whether to go on and complete the well. Once a decision has been reached to continue with the venture, enormous costs lie ahead such as running casing, cementing, perforating, testing, and buying and installing tubing and production equipment. The cost of the casing alone is often the greatest single item of expense on a well. Logs help to determine the possibility of a good well, or when to minimize costs on a bad well.

Before the modern art of logging was developed and improved, the only reliable information was obtained from the driller's log, a written diary of each day's events, and records of nearby wells. This information is still very important and useful, but using just these methods still leaves a lot of unknowns. Modern well logging provides information that helps to determine types of formations such as sandstone, limestone, shale, or dolomite, and their thickness as well as the depths of each formation top. Other information that is available includes formation temperature, porosity, permeability, the presence of oil or gas, and much other valuable data.

Logs are used by investors and bankers, engineers, geologists, landmen, suppliers, and practically everyone in the oil and gas industry who has a part in making the decision to either complete or abandon a well. Openhole logs are used to make the initial calculation of reserves in the formation. The reservoir engineer is the one who makes these calculations and updates them periodically from production data, pressure buildup tests, and any other logs that might be run.

One or more of the team assigned to the project should be able to make a calculated "guess" about the outcome of any well before it is drilled. The project geologist, drilling engineer, wellsitter, log analyst, and reservoir engineer should share in this prediction, and hopefully work together harmoniously from project beginning to end.

The project geologist can predict lithologies (information about the rocks in the formation), fluid types, and the possibility of encountering fractures. He or she should also know the area and have information from previously drilled wells.

The drilling engineer should know the approximate depths of each formation and the likelihood of oil or gas occurring in each of them. He or she should also be aware of the possibilities of encountering abnormal pressures.

The wellsite geologist needs to be alert at times for unexpected conditions, and should work closely with the other professional. This evaluation team should observe drilling rates, bit types, and weight indicator readings to gain information about lithology and porosity.

The *mud report* can indicate chemical reactions between the drilling mud and exposed formations, formation breakdowns, or *loss of circulation* (the unwanted leakage of drilling fluid into the formation).

Cuttings

Drill *cuttings* are the first and most obvious source of information obtained from the formation being drilled. Cuttings may reveal the first physical evidence of oil or gas. They can also be an early indicator of increased pore pressure and can reveal formation porosity.

Sidewall Cores

A rotary *sidewall coring tool* is used to cut samples measuring $^{15}/_{16}$ in. X 1¾ in. from the formation and retrieve them. The tool is about 17 ft long and can obtain 30 samples each time it is lowered into the hole.It is operated by wireline and a high-voltage cable in an uncased hole. Drilled sidewall cores offer an opportunity to obtain small physical samples at precise depths. Samples can also be retrieved by use of sidewall core guns.

Conventional Cores

Conventional cores offer a relatively uncontaminated sample of the formation at any desired depth.Because of the large volume sampled, they offer the best rock samples for observing the complete lithologic record. These cores are taken by a special tool that is attached to the drillpipe after it is tripped out and the bit removed. The coring tool replaces the bit at the bottom of the drillstring which is run back into the hole where coring begins, and then tripped out when the desired sample is obtained.

Conventional cores, sidewall cores, and drill cuttings are normally first examined with a low-power binocular microscope which will identify most of the important features of the sample. A *petrographic microscope* may be used to identify some pore-surface and microporosity features. The *scanning electron microscope (SEM)* can magnify from low power up to magnifications of about 2000X, which enables clay mineral textures, microporosity, and pore coatings to be dramatically revealed.

Mud Logging

Traces of oil or gas may be brought up to the surface as the drilling mud is returned during the circulation process. Samples of the returned drilling fluid are captured in a collector, or trap in the return line, and studied. This *mud logging* is useful in determining rates of penetration and lag, for making formation evaluations, and detecting the presence of gas.

Any gases extracted from the drilling mud are usually the first signs that hydrocarbons are present in the formations penetrated. As the bit enters a formation it opens pores that let the fluids they contain mix with the mud. As the fluids and cuttings rise to the surface, pressure drops, which releases more gas from the pore spaces in the cuttings. If the formation pressure is greater than the hydrostatic pressure, and the length of time of this condition lasts, the well will *kick.*

Gas can be detected in various ways with different instruments. One that has been in use for a long time that is simple and inexpensive is the *hot-wire detector.* It can be used continuously in the mud return system.

After a show of gas is detected with the hot-wire detector or other instrument, a sample is taken and analyzed with a *gas chromatograph* which reports the analysis of the gas as percentages of methane, ethane, propane, butanes and pentanes. If the gas saturation of oil in the formation is low, an oil show may not be detected by the gas-in-mud tests.

Logging

Electric Logs

Electric logs are mechanically simple, but they are very difficult to read because of the many environmental effects they are subject to such as temperature, borehole size, salinity, drilling mud and formation composition. In spite of this, the electric log, or E-log, often gives a good indication of formation resistivity and invasion. Many oil and gas deposits have been discovered using these logs. The old E-log has been replaced by more modern and much more easier read logs such as the *induction electric log.* However, it is still of interest because so many copies of E-logs remain in log libraries and are used by geologists to map formations and to put prospects together.

Electric logs are always run in an open, uncased borehole that is filled with water. The instrument, which is lowered into the hole by wireline and insulated electrical cable, usually has two measurements: one short normal spacing of 16 in., and the long normal spacing of 64 in.

Electric current from a constant source is emitted from electrode A and returns to electrode B, which is a long distance away. The current leaves A in an essentially spherical manner. The voltage M is measured with respect to a reference electrode that is at 0 voltage. Because the current at A is constant, any variation in the voltage at M will be due to changes in the *resistivity* of the formation being tested. Electrical current can pass through formations and *mineralized water*. Some earth formations resist the passage of electricity due to low porosity or porosity that contains pure water free of oil or gas. Other formations are good conductors of electricity because of the presence of salt water.

Resistivity is a term that expresses the difficulty an electric current has in flowing through a substance. For example, glass has very high resistivity and is often used as insulation. On the other hand, copper has low resistivity and electric current flows through it easily. Sandstones are essentially glass (silicon oxide) and have high resistivities if they are completely dry and free from conductive materials. Distilled water also has a high resistivity. However, when dissolved salts and other impurities are added, it becomes more conductive. This means that when sandstone contains impure formation water, its resistivity will vary with the amount and salinity of the water. Oil is a good insulator, and when added to the formation water will change the formation's resistivity readings.

When sediments are deposited and compacted, they do not form a solid mass of rock. Spaces exist between the grains that are called voids or pore spaces. These represent the porosity of the formation. Porosity affects the amount of water that the rock can hold, and thus the formation resistivity factor is related to this void space.

Archie's Equation

George Archie pioneered studies in resistivity during the early days of log interpretation research. He ran experiments in which he first measured the resistivity of a 100% water-saturated core and then measured the resistivity as the core was progressively saturated with oil. Archie determined that water saturation is equal to the square root of the 100% water-wet resistivity, R_o, divided by the formation resistivity,

$$R_t: S_w = \sqrt{R_o}/R_o$$

This equation later became known as *Archie's equation*

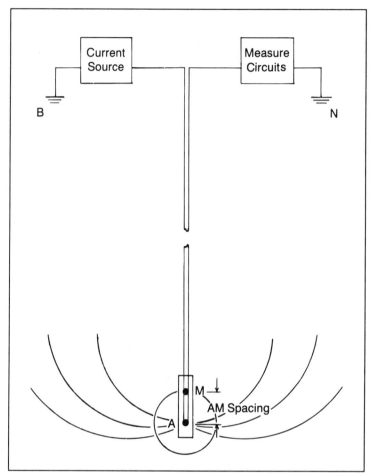

Figure 6–1 Schematic of electric log, normal device. Current flows from electrode A to ground electrode B. Voltage is measured at electrode M with respect to another ground electrode at N (courtesy PennWell Books, *Well Logging for the Nontechnical Person*, Johnson and Pile).

The units of resistivity may be measured with an ohmmeter (Ohm-m) and the readings recorded at the surface (Figure 6–1). Another device, with four electrodes, is called the lateral, which works on the same principle as the normal. The primary difference is that three electrodes are on the *sonde*: A, M, and N. M and N are close together, and the voltage difference between them is measured. The lateral has a deeper radius of investigation.

The electric log usually consists of four measurements, spontaneous potential (SP) and three resistivity measurements with different depths of investigation, short normal (16 in.), medium normal (64 in.), and lateral (18 ft, 8 in.).

Micrologs

Micrologs were developed to locate and define thin permeable zones. The Microlog and the guard-electrode type *microlaterlog*, were widely used for porosity determination until the introduction of the acoustic and radioactivity density logs for that purpose. Technical advances and changing patterns of use have made some of the pioneer equipment obsolete. Today, the induction log has almost completely replaced the older electric log.

Induction Electrical Log

Induction-electrical logs are run with wireline equipment in an open hole. On land, a special truck carrying the logging cable, winch, power source, sonde, recorders, and auxiliary equipment is brought to the well. The downhole instrument, called the *sonde*, is lowered into the well on the logging cable (the *wireline*). As the instrument is moved back up the borehole, electrical signals from the sonde are transmitted through insulated conductors in the logging cable to be recorded by electronic receivers located in the truck. Formerly, these signals were automatically plotted against the depths where they were generated to produce curves on a roll of photographic film. The film was processed, and copies were made on long strips of paper commonly referred to as *logs* (Figure 6–2).

This method is still in use, and millions of these paper logs are still maintained in log libraries and are referred to daily by petroleum and reservoir engineers and company personnel generating prospects. The expertise of these people is still badly needed, since today there is a growing lack of people who can interpret these logs.

Like many other aspects of the petroleum industry, the art and science of logging is rapidly turning to computerization. Today's wireline truck may contain two computer workstations—one for the logging engineer and one for the client's representative.Data is instantly transmitted to the computer and displayed on a color monitor screen. Depths and downhole conditions are graphically and dramatically illustrated in an easy-to-read and understand format that provides a great amount of accurate information on which to base a decision.

Prints are still made, but today they will more than likely be in color and contain computer-generated petrophysical information.Two such systems are the GIPSIE[c] software and LogStation[c] worksta-tions from Intergraph Corporation and the Maxis 500[c] unit from Schlumberger, which is available for both land and offshore use. Such onsite units use advanced computer hardware and software.

Induction-Electrical logs usually record three tracks. Track 1 is for

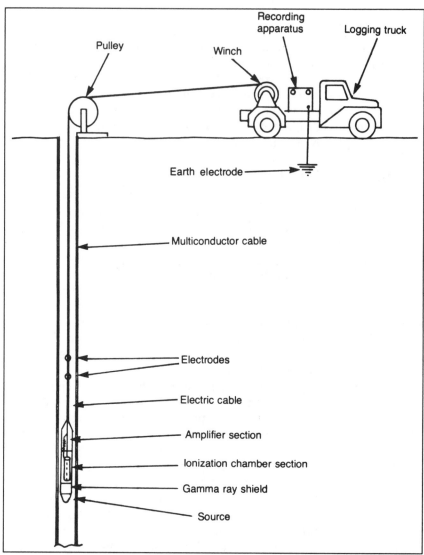

Figure 6–2 A logging truck and equipment.

the spontaneous potential (SP) curve. The *SP curve* is a graph that shows the electric potential, or voltage, that underground formations generate. The SP curve is plotted against depth in the borehole.

Track 2 records resistivity and is plotted against depth.

Track 3 records conductivity and is plotted against depth.

Well depths are shown in the blank space between the SP and the other curves. The normal depth scales are 2 in. = 100 ft for a regular log and 5 in. = 100 ft for a detailed view. For deep holes, a depth scale of 1 in. = 100 ft is used (Figure 6–3).

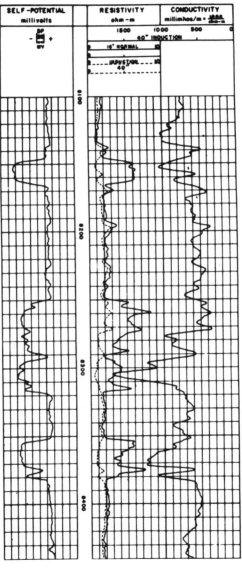

Figure 6-3 An induction log (courtesy Schlumberger).

The *Spontaneous Potential (SP) log* is a record of the naturally occurring voltage, or potential, caused when the drilling mud comes in contact with the formation. This resulting voltage is due to an electrochemical action caused by the differences in salinities of the various fluids. This is similar to the way voltage is produced in a wet-cell battery. Since this voltage occurs naturally, it is termed *spontaneous*. The SP values are very small and they are measured in *millivolts* (MV), which are thousandths of a volt.

Usually the SP line on the log shows a more or less straight line opposite impermeable shales, and will show peaks to the left opposite permeable strata. The shapes and amplitudes of the peaks may be different according to the type of formation. However, there is no definite correspondence between the amplitudes of the lines and how porous or permeable the rocks are. The main uses for the SP curve are to detect permeable beds, that is sand vs. shale formations, and to locate the boundaries between beds. Other uses are to obtain good values for formation water resistivity and to permit correlation of equivalent beds from well to well (Figure 6–4).

The SP deflection is measured with respect to the *shale base line*. That is a reference line which can generally be traced along the extreme positive side of the SP curve, which is generally a straight vertical line. The maximum SP deflections toward the negative side on the log are often the same opposite permeable formations. The sand line is usually traced parallel to the shale line only at the outer edges of the outer extreme negative side of the curve. Field experience has shown that in certain regions there may be shifts of the shale line that should be taken into account.The SP curve is important in geological correlation because the shapes of these curves in different wells for certain geologic horizons will be comparable.

Induction Tool

The *induction tool* was developed to provide a method of logging wells drilled with oil-based muds. The original electric logs used the mud column to conduct the current into the formation. Wells drilled with nonconductive muds, or air-drilled holes (which use no mud) posed a problem. Although the induction tool was developed to meet these needs, it was recognized that the tool worked better than the original electric log in freshwater muds and was easier to read.

The induction tool works by using the interaction of magnetism and electricity, i.e., when an electrical current is sent through a conductor, a magnetic field is created. When the current alternates, the magnetic field will also alternate by reversing poles at the same rate at which the current is alternating. Likewise, a voltage can also

160   Modern Petroleum

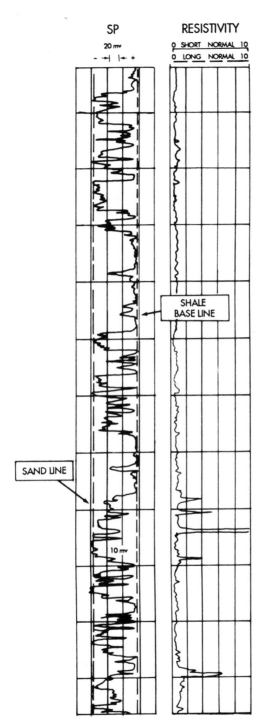

Figure 6–4 An electric log of a well. Note the SP curve on the left and the resistivity curve on the right.

be induced in a stationary conductor by alternating the magnetic field. The tool uses these principles in its operation. The alternating magnetic field created by the transmitter coil on the tool induces eddy currents in the formations. These eddy currents have their own magnetic field which induces a voltage in the receiver coil. Accordingly, the voltage induced in the receiver coil and recorded by the galvanometer on the surface depends upon the conductivity of the surrounding earth.

In addition to the two main coils, induction logging sondes include additional coils whose characteristics and position on the sonde are chosen to minimize as much as possible the influence on the measurement of the mud column and formations above and below the instrument. The recording equipment includes a reciprocal computer, making it possible to record a resistivity curve as well as the conductive curve. Sixteen-inch normal and Spontaneous Potential (SP) recording devices are also run with the *induction electrical survey instrument (I-ES)*.

Acoustic Log

The *acoustic*, or *sonic*, log was originally developed as an aid for interpreting seismic exploration data. However, it proved to be so effective in determining porosity that in many areas it became the standard porosity tool. It measures the depth versus the time it takes for a sonic impulse (a *compressional* or *P-wave*) to travel through the rocks and fluids of a formation. Their speed of travel depends on the type of materials, or mixture of materials, they are passing through.

It is necessary to run an acoustic log in a fluid-filled open hole because the presence of mud in the hole is required in order to provide an acoustic coupling between the tool and the formation.

The acoustic or sonic tool uses sound waves to measure porosity. Sound is energy that travels in the form of a wave (the human ear can detect sound waves traveling between 20 and 20,000 cycles per second (*cps*)), and this wave can travel in several different forms. Compressional waves are the most common form because they are the first wave to arrive, and thus are called P-waves or primary waves.

A compressional wave travels by ever so slightly compressing the material in which it travels. The waves will change speed when the material in which they are traveling changes; this process is known as *refraction.*

A second type of sound wave is the *shear wave* or *S-wave*. It is slower than the P-wave and cannot be transmitted through a fluid. It must be in a solid medium in order to transmit its energy.

The acoustic tool takes advantage of the fact that a sound wave travels at different speeds through different materials and at different speeds through mixtures of different materials. By knowing the sound/speed factor for each of the materials likely to be present in underground formations, it is possible to calculate the amount of each material present as long as there are only two materials involved. If more than two are present, then additional information is needed.

A typical acoustic tool has one transmitter and two receivers. The transmitter is located 3 ft from the first receiver and 5 ft from the second receiver. It emits a strong sound impulse that travels spherically outward in all directions. The mud column and the tool have slower sonic velocities than the formation.

The first sound energy to arrive at the two receivers is the P-wave, which travels through the formation near the borehole. The difference in the times at which the signals reaches the two receivers is divided by the spacing of the receivers. This time, recorded in microseconds per foot, is called sonic interval transit time (t) for the difference in arrival times between the two receivers.

The acoustic tool normally used is very complicated. They have multiple transmitters and receivers to compensate for sonde tilt, washed-out hole, and alteration of the properties of the rock near the borehole caused by the drilling process.

Radioactivity Logs

Electric logs must be run in an open hole to avoid short circuits through the steel casing. This restriction does not apply to *radioactivity logs*, which may be run in either open or cased holes. The gamma ray and the neutron logs are radioactivity logs and will be covered later in this chapter.

Density Logs

A special type of radioactivity log that responds principally to variations in the specific gravity of the formations has been developed for openhole work. The *density logging device* is held against one side of the borehole by a spring arrangement to insure close proximity to the formations. The density- logging sonde is similar to the *gamma ray-neutron instrument*, but has a small window where the gamma rays may enter the detector, which is pressed tightly against the hole wall. It is heavily shielded except for the window on the side adjacent to the formations. In this way the *gamma ray radiation* that reaches the detector is effectively limited to that which penetrates into the rock and is then scattered back in the direction of the detector. Since the absorption and scattering of gamma radiation depends largely upon the amount of matter, and consequently the electrons in its path, the density logging device will respond according to the variations of specific gravity of the rocks. The density log is a good porosity measuring device, particularly for shaley sands. It provides a continuous record of variations in the density of the lithologic column penetrated by the borehole. Most density logging devices simultaneously record a gamma ray log, a density log, and a caliper log, and are thus usually called *G D C Surveys*.

The density log is applied only in open holes because the presence of casing and cement would prevent contact with the rock itself. Porosity correlates with density because the rock matrix must decrease in density as its porosity increases.The density log variations within a uniform sandstone or limestone body must therefore represent increases and decreases of its porosity.

Caliper Log

The *caliper log* measures the diameter of the borehole.Several types of caliper logs are currently in use. One type consists of three or four spring-driven arms which contact the wall of the hole. The instrument is lowered to total depth, and the arms are released either mechanically or electrically. The spring tension against the arms centers the tool in the hole. The arms move in and out to compensate for changes in wellbore diameter. The arm motion is transmitted to a rheostat which insures that any change in the resistance of an electrical circuit is proportional to the borehole diameter. The diameter is recorded by measuring the potential across this resistance. Other caliper tools use similar methods that will that record varying voltages which reflect different borehole diameters (Figure 6–5).

Directional Log

The *directional log* is a record of hole drift, or deviation from the vertical, and the direction of that drift. There is no such thing as a vertically-true drilled oil or gas well. For many reasons, including legal ones, a directional log is necessary to insure that the producing formation is encountered on the proper lease rather than on a neighboring one.

There are a variety of instruments in use such as *drift indicators, single-shot surveys*, and *continuous surveys*. Drift indicator readings are simply observations of hole angle compared to vertical readings taken at intervals as a well is drilled, and are not really

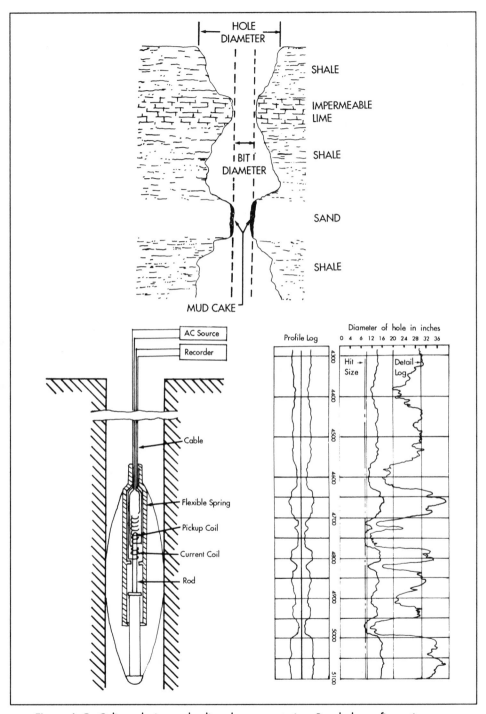

Figure 6-5 Caliper device and caliper log presentation. Borehole configurations differ in various formations.

directional logs. A single-shot directional survey instrument is similar to a drift indicator in that it utilizes a plumb bob and a stylus that is activated by a timing device, but it also includes a compass to obtain a directional reading. Multiple-shot directional surveys are often used to survey the entire uncased portion of a hole, a technique that permits magnetic compass readings to be taken.

As was mentioned in Chapter 5, much of today's drilling is not only purposely directional, but, in fact, horizontal. Many new tools, computer programs, and techniques have been developed to plan the path of the wellbore, calculate its location, and compare its actual location to the target location. Some of these systems and directional packages such as Smith International's Datadril MWD (Measurement While Drilling) system, eliminates the need for wireline equipment. The downhole instrument probe is contained within the drillstring and transmits its information by mud pulse telemetry to the drillsite computer and driller's readout station.

Dipmeter Survey

The *dipmeter* is run to determine the direction and angle of formation dip in relation to the borehole (Figure 6–6).

The instrument for measuring dip uses three or four contact electrodes equally spaced in a plane perpendicular to the wellbore. Each set of electrodes records a separate electric logging curve. By inspecting these curves, it is possible to correlate them, or to locate

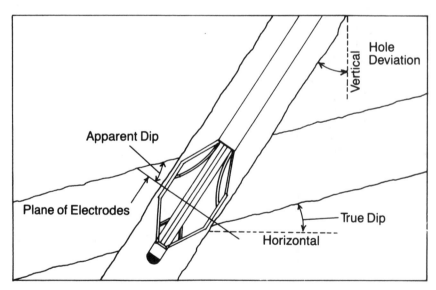

Figure 6–6 Geometry of dip measurement. Relationship between apparent dip and true dip (courtesy Schlumberger).

points that are common to each. If the bedding plane is not the same as the plane of the three electrodes, the curves will be displaced.

One of the major uses of the dipmeter log is to gain an idea of the geological structure from an exploratory well.

Gamma Ray Logging

Natural radiation of unstable elements consists primarily of alpha, beta, and gamma rays. Of these, it is only practical to measure gamma radiation in a borehole. Gamma rays are typical forms of electromagnetic radiation. Some elements naturally emit gamma rays, which are distinctive in both number and energy. Low gamma ray counting rates are related to non-shales and high gamma ray counts are related to shales. The gamma ray log may be run in either cased or open holes.

By measuring gamma ray intensity and plotting this data as a function of depth, a graph is obtained which shows varying formations. It permits the sandstones to be distinguished from clays and shales (Figure 6–7).

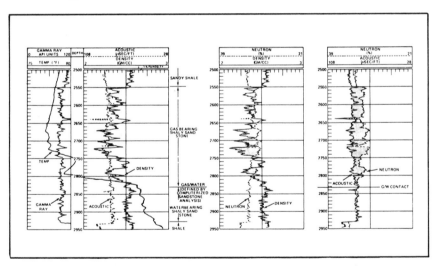

Figure 6-7 Gamma ray and neutron logs (courtesy *Oil & Gas Journal*).

Gamma ray and neutron logs are usually run on the same instrument and the gamma ray curve is recorded on the left side of the chart and the neutron curve on the right. It is not uncommon to run a *collar locator* with the gamma ray and neutron logs. A gamma ray curve will indicate a zone of greater intensity for a limestone that lacks porosity. The gamma ray normally shows medium to high intensity in shale, while the neutron curve records shale at its lowest intensity. The gamma ray curve taken in uncased holes permits evaluation of lithology changes, indicates shale content, and gives accurate depth control when run with a casing collar locator.

Neutron Logging

The *neutron log* is normally a standard counterpart of the gamma ray log. The neutron log is obtainable in both cased and uncased boreholes and is usually recorded simultaneously with the gamma ray log.

Neutrons exist in the nuclei of all elements except hydrogen. They are about the same mass as a hydrogen atom, but have no charge.

Neutron logs are recorded when a instrument, acting as a source of radioactivity by emitting energetic neutrons, is moved up the borehole at the same time as a radiation detector, which is spaced at a fixed distance from the source (Figure 6–8). The neutrons are created chemically from a mixture, such as americium and beryllium, that continuously gives off neutrons. This source bombards the rocks of the formation with a constant flux of neutrons, and the detector records the varying intensity of gamma radiation which results from the bombardment. This creates a curve that is principally influenced by the amount of hydrogen combined in the rocks or by the fluids in the pore spaces.

The downhole chemical neutron source emits a continuous flux of neutrons that are reduced in energy as they migrate spherically away from the source, across the wellbore, and into the formations.

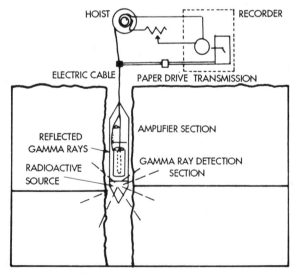

Figure 6-8 Neutron logging device (schematic).

At a very low energy level, the neutrons are eventually absorbed by the nuclei of wellbore and formation constituents. An attached radiation detector senses either the low energy neutrons or the gamma radiation resulting from the slow neutron absorption.

The neutron log basically measures slowing down properties of the neutrons in the formation. The slowing down of fast neutrons is primarily caused by interaction with hydrogen atoms. The hydrogen density inferred by the measurement is then related to porosity. Crude oil and water have essentially the same hydrogen density and they both contain much more hydrogen than rock. Therefore, the behavior of emitted neutrons affords a means of evaluating the fluid content of a formation. If the rocks encountered are dry, the neutrons will not immediately be slowed, and will continue to penetrate the dry rock until they are captured by some element (Figure 6–9).

When logging through casing, the neutron curve may be used to evaluate lithology, relative porosity, formation depths and thickness, and the location of gas-fluid.

Pulsed Neutron Log

The *pulsed neutron logging tool* can be run inside casing to obtain an indication of the presence or absence of hydrocarbons outside the pipe. Tool diameters small enough to run through 2 ⅜ in. tubing are available. A neutron generator irradiates the formation through the casing for a brief time interval, after which the radiation pulse is shut off. A short time later, a gamma ray detector in the tool is opened and a count is made of the radioactivity emanating from the formation. After a brief wait, the process is repeated. There will be less radioactivity measured in the second count because some of the radiation will have decayed. The rate of decay of radioactivity depends on the porosity, the salinity of the water present, and the shale and hydrocarbon content of the formation. Decay time is faster for a saltwater sand than for an oil sand. From the estimates of porosity, water salinity, and available shale content recorded on the log, an estimate of the hydrocarbon content can be made.

This tool can be used at a later date to locate hydrocarbon-bearing formations that may have been previously overlooked, and it can be used years after a well has been in production to locate the oil-water level in the formation.

Chlorine Log

The *chlorine log* is similar to the neutron log in that it is a measurement of the "gamma rays of capture" produced by bombard-

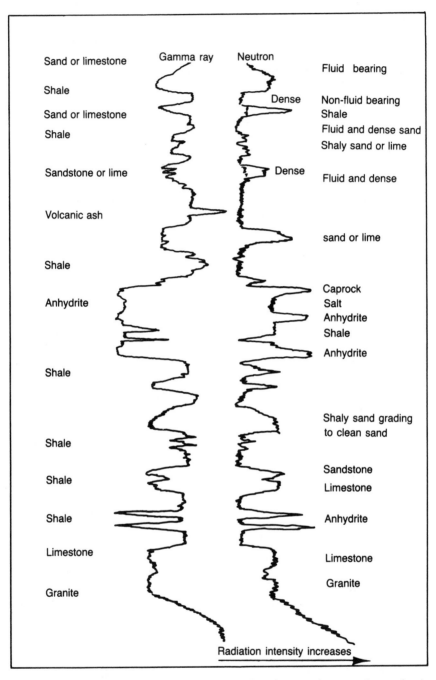

Figure 6–9 Radioactivity log (courtesy PennWell Books, *Petroleum Production for the Nontechnical Person*, Gray).

Modern Petroleum

ing the formation with neutrons. The chlorine logging instrument is the more sensitive to the gamma rays of a certain energy level, that is, those produced by atoms of chlorine. It is a log of the relative amount of chlorine in a formation. Since it is insensitive to porosity variation, it can be used in conjunction with a neutron curve to obtain an indication of the presence of oil.

The instrument used for a chlorine log is similar to the neutron device used for hydrogen detection. Two scintillation-type detectors are placed above the neutron source. The salinity or chlorine-sensitive detector is above the hydrogen-sensitive detector. The neutrons are emitted from the source, slowed down by collisions, and captured by the elements present in the formation. Neutrons are more subject to capture by those elements having a large neutron-capture cross section. The capture cross section of chlorine is approximately one hundred times that of hydrogen , so a small amount of chlorine will have a relative large effect.

The chorine logging instrument produces both hydrogen and chlorine curves. These will fall along the same line opposite oil or fresh water zones on the log strip. The two curves will indicate very low hydrogen density and lower chlorine content opposite gas-filled zones. The two curves will depart appreciably from each other opposite saltwater zones on the log strip, and thus offer a graphic means of distinguishing saltwater from hydrocarbons.

Collar-Locator Log

The *collar locator* is used to locate casing collars for accurate downhole depth measurement (Figure 6–10). It is usually run in conjunction with another cased-hole service, such as a radioactivity log, or when perforating the casing. The depth to the bottom of the well is carefully checked when the electric log is run. Also, the height of the formation of interest from the bottom of the hole is observed, and this distance can be converted to the number of joints of casing required to reach the bottom of the well. Depth control for perforating is almost universally obtained by a radioactivity log taken in conjunction with a casing collar survey (Figure 6–11).

Tracer Log

A *radioactive tracer* is a material such as a liquid, solid, or gas with a high rate of gamma ray emission. Upon injection of a tracer material to any portion of a wellbore, the point of placement or movement may be detected and recorded by a gamma ray instrument. One use might be to determine channeling, or the travel of squeezed cement behind a section of perforated casing, by mixing

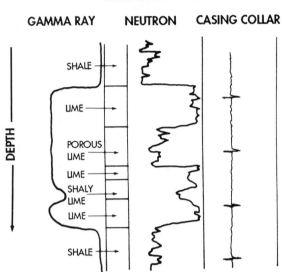

GAMMA RAY   NEUTRON   CASING COLLAR

Figure 6-10   Radioactivity logging curves for cased hole.

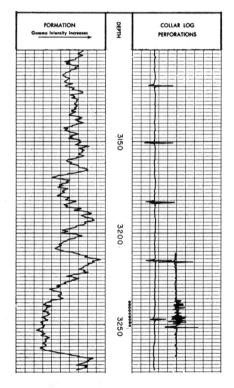

Figure 6–11   Perforation depth-control log (courtesy API).

finely-ground radioactive material with the cement. Radioactive liquids can be used to assess relative permeability because they will enter the porous and permeable zones. Radioactive sand tracers can be used for injection, frac jobs, or in gravel packs.

Cement Bond Log

The *cement bond log* is not a reservoir evaluation log; however, it is a useful tool for well completions. It is an acoustic device used to determine the quality of the cement in the annulus or, the degree to which channels may have developed in the cement behind the casing. Cement bond logs can be used to determine the quality of the cement bond between the casing and the formation. These logs can also be used to determine the location of the top of the cement behind the casing.

Temperature Log

Temperature in the Earth increases with depth at the rate of about 1°F per each 60 ft of depth. This decrease is known as the *thermal gradient.*

Cement generates a considerable amount of heat when setting, and a temperature increase will be found at the level where cement is found behind the casing. A recording *temperature survey tool* is lowered down the casing to detect these higher readings. This same instrument can also record other temperature changes or deviations from normal temperatures. Fluid taken up by the formation through a hole in the casing is continually being replaced by fluid in the casing, and temperatures down to this point of *lost circulation* will tend to be much cooler than the temperature of the formation.

Temperature surveys are often made in cased holes for the purpose of locating the spot where high pressure gas is entering a hole in the casing, since gas flowing through a small opening causes a temperature drop at that point.

Production Logs

Production logs fall into two main categories: those used for injection wells and those used for producing wells. Injection wells normally receive water as a part of a secondary recovery system, or as a means of salt water disposal. The main purpose of a production log is to determine the injection profile. It is also important to check for casing leaks, poor cement jobs, and fluid migration between zones.

The continuous flowmeter is often used where relatively high injection rates of 1,000 B/D or more make it a suitable tool for obtaining an injection profile. The *continuous flowmeter log* (often called a *spinner survey*) is used to determine the contribution of each underground zone to the total injection. This tool is positioned inside the pipe which forms a part of the injection system. Continuous flowmeters may also be used on flowing production wells. In situations where the continuous spinner flowmeter is not suitable, the packer-type flowmeter can be used.

Packer Flowmeter

The *packer flowmeter* uses an inflatable packer bag which assures that all the fluid passes through the measuring devices built into the unit. It must be positioned to measure fluid flow when measuring input volume into an injection well, and positioned differently when it is used to measure the production coming out of a flowing well. More than one tool may be located downhole to measure and record fluids entering the casing at each zone. Recordings at each depth station in the well will show the flow-rate in barrels per day, the average density of fluids passing the metering section, and a watercut index.

Computer Generated Logs

In recent years, well log interpretation has been greatly simplified and improved by the use of computer-generated logs. Wellsite computers now receive the wealth of information that is transmitted to the surface by a downhole telemetry system, and store it on magnetic tape while making corrections for differences in environment. Computers merge data, make complex calculations, and then print out various log formats. The computer's strength is its speed and capability to make repetitious calculations and data presentations. Its weakness is that it cannot think or make judgments, and therefore does only what it is told. Today a computer program simply cannot put together all of the comprehensive formation data available from all the individuals involved in the venture and make true interpretations. Prior knowledge and planning, together with information that a computer can provide, does make well logging easier.

Wellsite computer logs (Figure 6–12), often called *quick-look logs*, and computing center logs (Figure 6–13), are the two most common. The major difference between them is the size of the computer making the calculations. The more powerful equipment can handle more complex software programs and thus make more calculations.

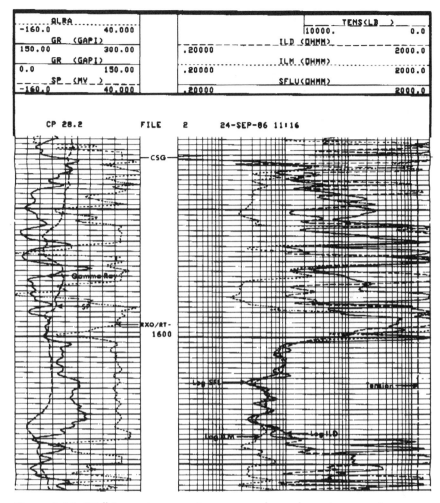

Figure 6–12 Wellsite pre-interpretation pass, used to make environmental corrections and to pick the parameters used in the interpretation pass (courtesy PennWell Books, *Well Logging for the Nontechnical Person*, Johnson and Pile).

Wellsite computer logs indicate water saturation and porosity without the detailed analysis that computer center logs make. Most of them use deep resistivity, neutron and density porosity, gamma ray, SP, and caliper curves to solve for water saturation. The log values are first corrected in a preliminary interpretation pass for environmental effects such as borehole size, temperature, mud weight, and salinity. The corrected porosities from the neutron and

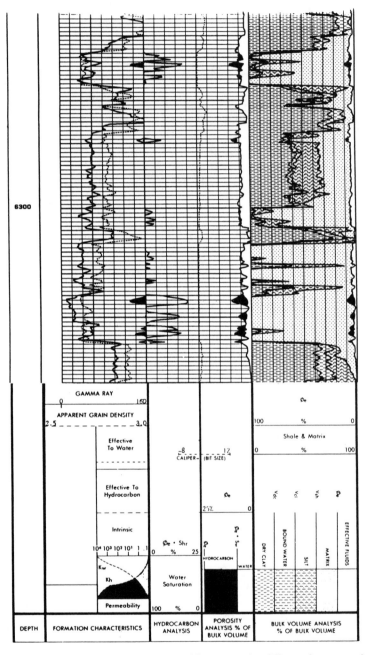

Figure 6–13 Computing center-interpreted log. Note the different format, with the depth track on the left. This is a convention adopted by logging companies to distinguish computing center logs from wellsite logs (courtesy PennWell Books, *Well Logging for the Nontechnical Person*, Johnson and Pile).

density logs are cross-plotted on this pass. The logging engineer then picks the various parameters needed to make the final interpreted log.

One major strength of computer-generated logs is that calculations are made continuously. It helpfully shades the readout so that the amount of hydrocarbons present stands out, or it indicates different rock types by changing the shading or dot pattern between curves.

Wellsite logs come in many different formats to fit the needs of the drilling engineer, geologist, reservoir engineer, or other operating company representative. The geologist needs to know all important interpretations that will aid in pipe-setting decisions, and the production engineer uses these logs to calculate the amount of gas, oil, and water coming from a zone (Figure 6–15).

Computer-generated logs have changed the face of logging forever, and improvements are being made in them almost daily. The capabilities of the newest versions are displayed in a number of ways, including high-quality 3-D maps in color. Today's computerized systems are capable of performing multiple tasks. Thus, at the same time underground information is being acquired, the system can also be transmitting data from wellsite to remote locations, making calibrations and computations, and also playing back data presentations on high-resolution color monitors (Figures 6–14, 6–15, and 6–16).

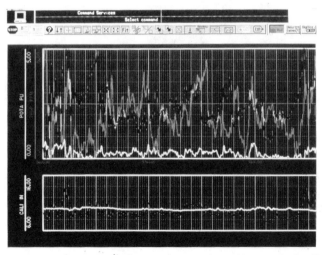

Figure 6–14 LogStation, Intergraphs's workstation-based log analysis software, incorporates sophisticated interactive graphics and object-oriented programming techniques (courtesy Intergraph Corporation).

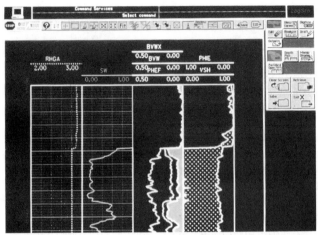

Figure 6–15 LogStation supports various display and presentation options, including color fill and patterning (courtesy Intergraph Corporation).

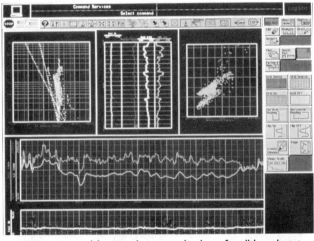

Figure 6–16 LogStation enables simultaneous display of well log data in multiple formats, enhancing an analysts' interpretation (courtesy Intergraph Corporation).

Chapter References

Bill D. Berger, ed., *Principles of Drilling* (Stillwater: Technology Extension, Oklahoma State University, 1978).

Geological Applications, (Huntsville: Intergraph Corporation, 1988), pp. 7, 18.

David E. Johnson and Kathryn E. Pile, *Well Logging for the Nontechnical Person* (Tulsa: PennWell Books, 1988).

7

Completion

Evaluating the Formation

After months of planning, and a great amount of valuable time spent in drilling to the desired depth, the operator has a very expensive deep hole in the earth. Now, he faces the important decision of what to do with it.

If, during the drilling operation, all of the usual drillers' logs, mud logs, core samplings, and other tests indicate that one or more of the formations, including the one at the bottom of the wellbore, shows a possibility of producing oil and natural gas in commercial quantities, the owners will decide to *set pipe (run casing)*.

If the decision is made to go ahead and complete the well, the *production casing* is run by the drilling crew, using the drilling rig for that purpose. Then, depending on the situation, a regular casing crew and other equipment might be brought in.

Casing

The details of the entire *casing* program must be planned before a rig is selected to actually drill the well. Several factors enter into rig selection, including the anticipated casing requirements, casing-bit-size programs, and the nature of the formations to be penetrated.

The selection of casing size should be made before the drilling rig moves on location. The reason is that casing size governs drilling bit size, or sizes. Seldom is a well drilled from top to bottom with only one bit size. The larger diameter hole is drilled first, then it is telescoped down in size as depth increases.

When drilling in a familiar area where formation conditions are known, it is easier to anticipate hole size. Smaller hole size can be drilled in hard rock areas because less casing clearances are required. In soft rock areas where sloughing or caving is common, it is better to plan for larger bit and hole size. Abrupt changes in drilling direction and deviation can be a factor in hole size. Hole size is usually larger when drilling exploratory (*wildcat*) wells than it is when drilling in a known producing field. This allows for the possibility that an extra protective string of casing might be needed.

The nature of formations to be penetrated must be known in order to intelligently plan the casing program. Important considerations are abnormal pressure formation, zones of lost circulation, heaving shales, dense zones where drilling is usually slow, and high formation temperatures. From an analysis of all factors, a casing-bit-size program is formulated. This program is as equally important as planning the actual drilling of the well. In determining casing-bit-size, one of the first considerations is the number of strings of casing which will be required.

Clearance is another consideration in planning casing-bit-sizes. Sufficient clearance should be allowed around the outside of the casing to provide for a satisfactory thickness of cement to form a good bond between the casing and wall of the wellbore. A drilled hole must allow free passage of casing into and down the hole. A typical casing-bit-size program is shown in Table 7–1.

Casing and Bit Programs

Inland

Surface String		Intermediate String		Production String	
Bit size inches	Casing size O.D. inches	Bit size inches	Casing size O.D. inches	Bit size inches	Casing size O.D. inches
17 1/2	13 3/8	12 1/4	9 5/8	8 3/4	7
11	8 5/8	11	8 5/8	7 7/8	5 1/2
13 3/4	10 3/4	9 7/8	7 5/8	6 3/4	4 1/2
12 1/4	9 5/8				

Gulf Coast

15	10 3/4	9 7/8	7	6	5
12 1/4	9 5/8	12 1/4	9 5/8	8 3/4	5 1/2
17 1/2	13 3/8	9 7/8	7 5/8	9 7/8	7
13 3/4	9 5/8	—	—	—	—
20	16	—	—	—	—

Table 7–1 Casing and hole sizes commonly employed.

With cable-tool drilling, several strings of casing were set as the well was drilled. This was because drilling mud was not used. Today, rotary drilling methods can drill more open hole that was ever possible with cable tools.

Kinds of Casing

It is usually necessary to install more than one string of casing because of the many different functions each performs.These various kinds of casing are divided into five classifications:

    (1)   Conductor casing.
    (2)   Surface casing.
    (3)   Intermediate casing.
    (4)   Liner string.
    (5)   Production casing.

Not all drilled wells will use each of the various strings of casing previously listed. A typical relationship of strings of casing is shown in Figure 7–1.

Conductor Casing

It may be necessary to set a short length of *conductor casing* (sometimes called conductor pipe) at the surface to prevent caving around the sides of the hole. This is usually not more than 20 to 50 ft long, and is normally set before the drilling rig moves on site.

Conductor casing is set after the well location has been graded, and mud pits, if any, are dug. It is the pipe that also raises drilling fluid high enough above grade level to return the fluid to the mud pit. It prevents washing out around the rig's base, and it is cemented in after being placed.

Often the hole for conductor casing is drilled with an auger drill mounted on a truck. In swampy areas the pipe is driven down with a pile driver. Then any dirt and mud inside the pipe is drilled out with the rotary drill. Conductor pipe sizes usually measure from 16 to 20 in. in outside diameter.

Offshore conductor pipe is driven with a special pile driver and measures from 30 to 42 in. outside diameter.

Surface Casing

If conductor casing is set, the next string of casing to be set will be *surface casing*. The outside diameter of surface casing will be smaller than the inside diameter of the conductor pipe. In hard rock areas,

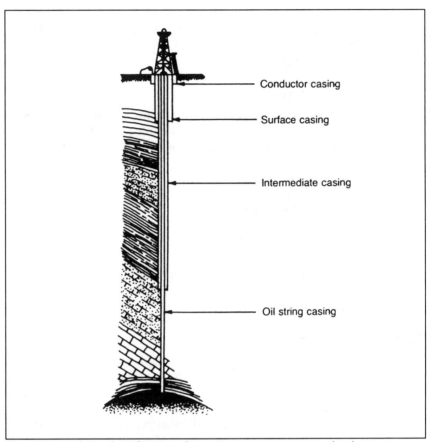

Figure 7-1 Relationship of strings of casing (courtesy McCray and Cole, University of Oklahoma Press, 1959).

a single string of surface pipe is set; usually a few hundred to 2,000 ft in length.

Statutory regulations require setting of surface casing to protect freshwater sands from possible invasion by hydrocarbons or salt water from deeper horizons. These freshwater formations occur near the surface. The surface casing must be properly cemented in after being set. Requirements for how much cement is necessary can vary in different areas, and drilling operations are suspended while waiting on the cement to cure. Surface casing should be set deep enough to reach rock formations that will not break down under the maximum expected mud weight. Usually, there will be a minimum of two strings, the surface casing and production casing, with possibly more in deeper wells. Surface casing furnishes a means of

Modern Petroleum

handling the return flow of drilling mud during operations. It allows attachment of blowout preventers for safety, and protects the drilled hole from damage during drilling.

After each string of casing has been run and cemented, the drill bit must pass through the cemented bottom hole portion in order for drilling to continue.

Intermediate Casing

Use of an intermediate string depends largely on the well depth and on geologic conditions in a specific area. Deep wells may require one or more intermediate strings. When drilling in areas that have abnormal formation pressures, heaving shales, and lost-circulation zones, a string of casing may need be run to minimize such hazards before drilling proceeds to greater depths. Another function is to seal off troublesome zones that could contaminate the drilling fluid, making mud control difficult and expensive. Or a zone could jeopardize drilling progress by causing drillpipe to stick, which would call for expensive hole enlargement, or other tool *fishing* methods.

In deep drilling, there is often reluctance to drill with excessive intervals of open hole, even though no troublesome zones have been encountered. In such cases, an intermediate string may be run as an insurance against possible problems. It also affords greater safety in case of blowouts. However, *intermediate casing* is more a part of the drilling process than the completion process.

Intermediate casing strings are suspended from and sealed at the surface with a *casing hanger*. The bottom portion is cemented by circulating cement down the pipe and out around the open end and up the outside of pipe as necessary.

Liner Strings

Liners are sometimes set in a hole as a protective string, serving the same function as an intermediate string. A liner is usually run only from the bottom of the previous string to the bottom of the open hole. Liners are suspended from a previous string with a hanger. They may be suspended in the well either with or without cementing.

Production Casing

Production casing is sometimes known as the *oil string* or the *long string*. It is the last and deepest string of casing run in the well. Size of the production casing, which provides a conduit from the surface to the producing formation, is determined principally by subsurface considerations, such as:

(1) Subsurface artificial lift equipment required.
(2) Multiple-zone completions requiring several different strings of tubing isolated from each other by packers.
(3) Type of completion method to be used: open hole, perforated casing, screened open hole, or screened perforated casing.
(4) Prospects of deepening the well at a later date.

The size of the production string will usually be the minimum consistent with the demands of the above factors. Ample working room should be provided inside the casing for all the types of production equipment which may be used, plus enough working space for any fishing tools that may be needed to recover lost pieces of equipment. If the well is likely to be deepened later on, the same considerations must be given to these subsequent deeper completions as are given to the present completion interval. Where future deeper completions are considered likely, they will probably be the principal controlling factors. In order to deepen the well, the drilling bit must be run through the previous production string. This means that the bit size will be limited by the size of this casing, which will in turn limit the size of the casing and production equipment for the deeper zone.

Standardization of Casing

Prior to 1920, oil well casing was usually obtained in lengths of about 20 ft, and sizes ranged from 3 to 15 ½ in. inside diameter. At times the inside diameter was reduced when the weight per foot was changed. Threads were usually tapered ⅜ in. on the diameter per foot of length, and numbered 10, 11 ½ or 14 threads per inch. Some larger sizes had eight threads per inch and tapered ¾ in. per foot.

During the 1920s, the API achieved standardization for casing sizes and threads. Outside diameter became the standard reference for size, and grading of the material according to strength was introduced. Thread taper was standardized at ¾ in./ft, with eight threads per inch and the familiar rounded crests and roots were adopted in 1939. API standards are now accepted worldwide and metric equivalents are included in API casing lists. Higher grade material and improved thread forms came as oil and gas well depths increased and pressures became greater (Table 7–2).

Modern Petroleum

Casing is specified by range length, type of construction, coupling type, steel grade, outside diameter, and weight per foot.Detailed specifications are set forth in API Standard 5A and Bulletin 502. API casing is designated by the length range of each joint (Table 7–3). In addition to the API listings, information concerning other modifications and steel grades for special applications is available from *tubular goods* manufacturers.

Casing Measurement

An important part of every casing job is the measurement, or tally, of pipe. Even though most depth determinations of formations, perforating intervals, etc., are by electric log measurements, it is important to have a record of how many feet of casing is run into a well. Each joint could vary slightly in length.

There should be a joint count when the pipe is shipped, and an accurate joint-count must be made when the casing is received. Another count should be made as it is run into the hole, and the final count is made of the remaining balance.

Individual lengths in the string should be numbered in the order in which they are to be run into the wellbore. Different weights or grades should be clearly marked. Each casing joint must be measured, preferably with a steel tape graduated in feet, tenths, and hundredths of a foot (or in metric equivalents). Measurement of each length should then be recorded in a tally sheet. The tape should be read to the nearest hundredth of a foot from the top end of the coupling to the first *scratch* of the run-out for round-thread casing, or between the *shoulders* for extreme-line.

Couplings

API Standards for the most common couplings are shown in Figure 7–2, which depicts generalized details of these connections, showing hand tight and power tight position. By API standard practice, casing couplings are screwed into the pipe power tight, except if specified couplings are made up handset tight. *Handling tight* is to facilitate removal of couplings for cleaning and inspecting the threads, and applying fresh thread compound before using the pipe. Handling-tight is defined as "sufficiently tight so that the coupling may not be removed except by use of a wrench." Sections 6 and 8 of the API Standard 5A give dimensions and other details of the several API thread forms and couplings available.

Casing Grades

There are several grades of steel for casing and tubing that have

Sizes, Weights, Wall Thickness, Grade and Applicable End Finish[1]

1		2	3		Type of End Finish — Grade						
Size: Outside Diameter		Nominal[2,3] Weight Threads and Coupling	Wall Thickness		4	5	6	7	8	9	10
in.	mm	lb./ft.	in.	mm	H-40	J-55 K-55	L-80 C-95	N-80	C-90[4] T-95[4]	P-110	Q-125
4½	114,3	9.50	0.205	5,21	PS	PS	—	—	—	—	—
4½	114,3	10.50	0.224	5,69	—	PSB	—	—	—	—	—
4½	114,3	11.60	0.250	6,35	—	PSLB	PLB	PLB	—	PLB	—
4½	114,3	13.50	0.290	7,37	—	—	PLB	—	PLB	PLB	—
4½	114,3	15.10	0.337	8,56	—	—	—	—	—	PLB	PLB
5	127,0	11.50	0.220	5,59	PS	PS	—	—	—	—	—
5	127,0	13.00	0.253	6,43	—	PSLB	—	—	—	—	—
5	127,0	15.00	0.296	7,52	—	PSLBE	PLBE	PLBE	PLBE	PLBE	PLBE
5	127,0	18.00	0.362	9,19	—	—	PLBE	PLBE	PLBE	PLBE	PLB
5	127,0	21.40	0.437	11,10	—	—	PLB	PLB	PLB	PLB	PLB
5	127,0	23.20	0.478	12,14	—	—	PLB	PLB	PLB	PLB	PLB
5	127,0	24.10	0.500	12,70	—	—	PLB	PLB	PLB	PLB	PLB
5½	139,7	14.00	0.244	6,20	PS	PS	—	—	—	—	—
5½	139,7	15.50	0.275	6,98	—	PSLBE	PLBE	PLBE	PLBE	PLBE	—
5½	139,7	17.00	0.304	7,72	—	PSLBE	PLBE	PLBE	PLBE	PLBE	—
5½	139,7	20.00	0.361	9,17	—	—	PLBE	PLBE	PLBE	PLBE	PLBE
5½	139,7	23.00	0.415	10,54	—	—	—	—	—	—	—
5½	139,7	26.80	0.500	12,70	—	—	—	—	P	—	—
5½	139,7	29.70	0.562	14,27	—	—	—	—	P	—	—
5½	139,7	32.60	0.625	15,86	—	—	—	—	P	—	—
5½	139,7	35.30	0.687	17,45	—	—	—	—	P	—	—
5½	139,7	38.00	0.750	19,05	—	—	—	—	P	—	—
5½	139,7	40.50	0.812	20,62	—	—	—	—	P	—	—
5½	139,7	43.10	0.875	22,23	—	—	—	—	P	—	—

Table 7–2 API casing list.

Size	OD (mm)	Weight	Wall								
6⅝	168.3	20.00	0.288	7.32	PS	PSLB	—	—	—	—	—
6⅝	168.3	24.00	0.352	8.94	—	PSLBE	PLBE	PLBE	PLBE	PLBE	—
6⅝	168.3	28.00	0.417	10.59	—	—	PLBE	PLBE	PLBE	PLBE	—
6⅝	168.3	32.00	0.475	12.06	—	—	PLBE	PLBE	PLBE	PLBE	PLBE
7	177.8	17.00	0.231	5.87	PS	—	—	—	—	—	—
7	177.8	20.00	0.272	6.91	PS	PS	—	—	—	—	—
7	177.8	23.00	0.317	8.05	—	PSLBE	—	—	—	—	—
7	177.8	26.00	0.362	9.19	—	PSLBE	PLBE	PLBE	PLBE	PLBE	—
7	177.8	29.00	0.408	10.36	—	—	PLBE	PLBE	PLBE	PLBE	—
7	177.8	32.00	0.453	11.51	—	—	PLBE	PLBE	PLBE	PLBE	PLBE
7	177.8	35.00	0.498	12.65	—	—	PLBE	PLBE	PLBE	PLBE	PLBE
7	177.8	38.00	0.540	13.72	—	—	PLBE	PLBE	PLBE	PLBE	—
7	177.8	42.70	0.625	15.86	—	—	—	—	P	—	—
7	177.8	46.40	0.687	17.45	—	—	—	—	P	—	—
7	177.8	50.10	0.750	19.05	—	—	—	—	P	—	—
7	177.8	53.60	0.812	20.62	—	—	—	—	P	—	—
7	177.8	57.10	0.875	22.23	—	—	—	—	—	—	—
7⅝	193.7	24.00	0.300	7.62	PS	—	—	—	—	—	—
7⅝	193.7	26.40	0.328	8.33	—	PSLBE	PLBE	PLBE	PLBE	PLBE	—
7⅝	193.7	29.70	0.375	9.52	—	—	PLBE	PLBE	PLBE	PLBE	—
7⅝	193.7	33.70	0.430	10.92	—	—	PLBE	PLBE	PLBE	PLBE	PLBE
7⅝	193.7	39.00	0.500	12.70	—	—	PLB	PLB	PLB	PLB	PLB
7⅝	193.7	42.80	0.562	14.27	—	—	PLB	PLB	PLB	PLB	PLB
7⅝	193.7	45.30	0.595	15.11	—	—	PLB	PLB	PLB	PLB	PLB
7⅝	193.7	47.10	0.625	15.86	—	—	—	—	P	—	—
7⅝	193.7	51.20	0.687	17.45	—	—	—	—	P	—	—
7⅝	193.7	55.30	0.750	19.05	—	—	—	—	—	—	—

Sizes, Weights, Wall Thickness, Grade and Applicable End Finish[1]

1		2	3		4	5	6	7	8	9	10
Size: Outside Diameter		Nominal[2,3] Weight Threads and Coupling	Wall Thickness		Type of End Finish — Grade						
in.	mm	lb./ft.	in.	mm	H-40	J-55 K-55	L-80 C-95	N-80	C-90[4] T-95[4]	P-110	Q-125
7¾	196,9	46.10	0.595	15,11	—	—	P	P	P	P	P
8⅝	219,1	24.00	0.264	6,71	—	PS	—	—	—	—	—
8⅝	219,1	28.00	0.304	7,72	PS	—	—	—	—	—	—
8⅝	219,1	32.00	0.352	8,94	PS	PSLBE	—	—	—	—	—
8⅝	219,1	36.00	0.400	10,16	—	PSLBE	PLBE	PLBE	PLBE	PLBE	—
8⅝	219,1	40.00	0.450	11,43	—	—	PLBE	PLBE	PLBE	PLBE	—
8⅝	219,1	44.00	0.500	12,70	—	—	PLBE	PLBE	PLBE	PLBE	—
8⅝	219,1	49.00	0.557	14,15	—	—	PLBE	PLBE	PLBE	PLBE	PLBE
9⅝	244,5	32.30	0.312	7,92	PS	—	—	—	—	—	—
9⅝	244,5	36.00	0.352	8,94	PS	PSLB	—	—	—	—	—
9⅝	244,5	40.00	0.395	10,03	—	PSLBE	PLBE	PLBE	PLBE	PLBE	—
9⅝	244,5	43.50	0.435	11,05	—	—	PLBE	PLBE	PLBE	PLBE	PLBE
9⅝	244,5	47.00	0.472	11,99	—	—	PLBE	PLBE	PLBE	PLBE	PLBE
9⅝	244,5	53.50	0.545	13,84	—	—	PLBE	PLBE	PLBE	PLBE	—
9⅝	244,5	59.40	0.609	15,47	—	—	—	—	P	—	—
9⅝	244,5	64.90	0.672	17,07	—	—	—	—	P	—	—
9⅝	244,5	70.30	0.734	18,64	—	—	—	—	P	—	—
9⅝	244,5	75.60	0.797	20,24	—	—	—	—	—	—	—
10¾	273,1	32.75	0.279	7,09	PS	—	—	—	—	—	—
10¾	273,1	40.50	0.350	8,89	PS	PSB	—	—	—	—	—
10¾	273,1	45.50	0.400	10,16	—	PSBE	—	—	—	—	—
10¾	273,1	51.00	0.450	11,43	—	PSBE	PSBE	PSBE	PSBE	PSBE	—
10¾	273,1	55.50	0.495	12,57	—	—	PSBE	PSBE	PSBE	PSBE	—
10¾	273,1	60.70	0.545	13,84	—	—	—	—	PSBE	PSBE	PSBE

Table 7-2 API casing list (continued)

Size (in)	OD (mm)	Wt.	Wall (in)	Wall (mm)									
10¾	273,1	65.70	0.595	15,11	—	—	—	—	—	—	PSB	PSB	PSB
10¾	273,1	73.20	0.672	17,07	—	PSB	—	—	—	P	P	—	—
10¾	273,1	79.20	0.734	18,64	—	PSB	—	—	—	P	P	—	—
10¾	273,1	85.30	0.797	20,24	—	PSB	—	—	—	—	—	—	—
11¾	298,5	42.00	0.333	8,46	PS	—	—	—	—	—	—	—	—
11¾	298,5	47.00	0.375	9,52	—	PSB	—	—	—	—	—	—	—
11¾	298,5	54.00	0.435	11,05	—	PSB	PSB	PSB	PSB	PSB	PSB	PSB	PSB
11¾	298,5	60.00	0.489	12,42	—	P	P	P	P	P	P	P	P
11¾	298,5	65.00	0.534	13,56	—	P	P	P	P	P	P	P	P
11¾	298,5	71.00	0.582	14,78	—	—	—	—	—	—	—	—	—
13⅜	339,7	48.00	0.330	8,38	PS	PSB	—	—	—	—	—	—	—
13⅜	339,7	54.50	0.380	9,65	—	PSB	—	—	—	—	—	—	—
13⅜	339,7	61.00	0.430	10,92	—	PSB	PSB	PSB	PSB	PSB	PSB	PSB	—
13⅜	339,7	68.00	0.480	12,19	—	—	PSB	PSB	PSB	PSB	PSB	PSB	PSB
13⅜	339,7	72.00	0.514	13,06	—	—	—	—	—	—	—	—	—
16	406,4	65.00	0.375	9,52	PS	PSB	—	—	—	—	—	—	—
16	406,4	75.00	0.438	11,13	—	PSB	—	—	—	—	—	—	—
16	406,4	84.00	0.495	12,57	—	P	P	P	P	P	P	P	P
16	406,4	109.00	0.656	16,66	—	PSB	—	—	—	—	—	—	—
18⅝	473,1	87.50	0.435	11,05	PS	PSB	—	—	—	—	—	—	—
20	508,0	94.00	0.438	11,13	PSL	PSLB	—	—	—	—	—	—	—
20	508,0	106.50	0.500	12,70	—	PSLB	—	—	—	—	—	—	—
20	508,0	133.00	0.635	16,13	—	PSLB	—	—	—	—	—	—	—

[1] P = Plain End, S = Short Round Thread, L = Long Round Thread, B = Buttress Thread, E = Extreme-line.

[2] Nominal weights, threads and coupling (Col. 2) are shown for the purpose of identification in ordering.

[3] The densities of martensitic chromium steels (L-80 types 9Cr and 13Cr) are different from carbon steels. The weights shown are therefore not accurate for martensitic chromium steels. A weight correction factor of 0.989 may be used.

[4] Grade C-90 and T-95 casing shall be furnished in sizes, weights and wall thicknesses listed above or as shown on the purchase order.

	Range	1	2	3
CASING AND LINERS				
Total range length, incl.		16-25	25-34	34-48
*Range length for 95 percent or more of carload:				
Permissible variation, max.		6	5	6
Permissible length, min.		18	28	36
TUBING				
†Total range length, incl.		20-24	28-32	....
*Range length for 100 percent of carload:				
Permissible variation, max.		2	2	....
Permissible length, min.		20	28	....
PUP JOINTS				
#Lengths — 2, 3, 4, 6, 8, 10 and 12 feet				
Tolerance — ±3 in.				

*Carload tolerances shall not apply to orders of less than a carload. For any carload of pipe, shipped to the final destination without transfer or removal from the car, the tolerance shall apply to each car. For any order consisting of more than a carload and shipped from the manufacturer's facility by rail, but not to the final destination, the carload tolerance shall apply to the total order, but not to the individual carloads.

†By agreement between purchaser and manufacturer the total range length for Range 1 tubing may be 20-28 ft.

#2 ft pup joints may be furnished up to 3 ft long by agreement between purchaser and manufacturer, and lengths other than those listed may be furnished by agreement between purchaser and manufacturer.

Table 7–3 Range Length of API Casing.

been standardized by API. These are designated by both a letter and number (H-40, J-55, P-110, etc.) as listed in Table 7–4. The number represents the minimum yield stress in psi, which is defined as 80% of the average value from test data. Casing of higher strength may be ordered from some steel companies.

Design Considerations

Casing strings are designed to withstand three principal types of loading:

(1) Tensile load: Each joint supports all the weight below it, so the greatest tension occurs at the top joint. Design calculations are made as though the casing were freely hanging in the air.

(2) Collapse pressure: This is the unbalanced external pressure imposed on the casing, such as empty casing with the hydrostatic pressure of the mud column exerted outside. The strongest casing should be placed near the bottom.

(3) Burst pressure: This refers to a condition of unbalanced internal pressure. A common design criterion is to assume that the formation pressure is exerted on the entire length of the string. This condition is approached in gas wells.

Modern Petroleum

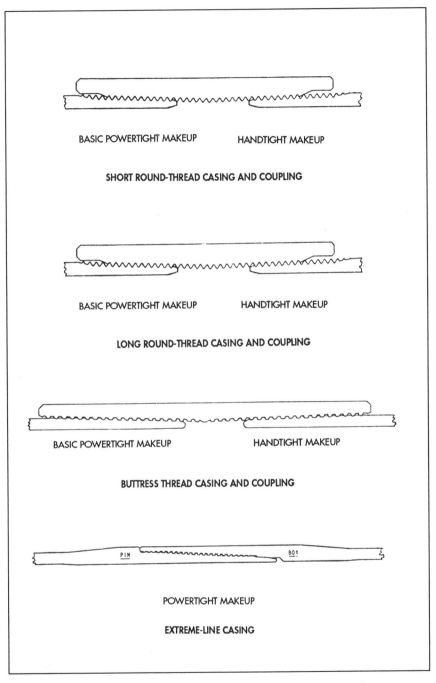

BASIC POWERTIGHT MAKEUP   HANDTIGHT MAKEUP

SHORT ROUND-THREAD CASING AND COUPLING

BASIC POWERTIGHT MAKEUP   HANDTIGHT MAKEUP

LONG ROUND-THREAD CASING AND COUPLING

BASIC POWERTIGHT MAKEUP   HANDTIGHT MAKEUP

BUTTRESS THREAD CASING AND COUPLING

PIN   BOX

POWERTIGHT MAKEUP

EXTREME-LINE CASING

Figure 7-2 API standards for basic powertight and handtight makeup for various casing and couplings.

Special Considerations

Strict adherence to standard design procedures is not feasible in all cases because of unusual well or field conditions. Severely corrosive areas should always received special considerations. High loadings imposed by either operating practices or geologic conditions must be recognized. High collapse pressures may arise during squeeze cementing. This problem may be eliminated by either applying a balancing external casing pressure to the annulus above the retainer, or setting the retainer a substantial distance above the perforations.

In general, however, there is no point in trying to design casing to withstand Earth movements, as the forces involved are too great.

Cementing

The practice of cementing casing was started about 1903 in California, but the modern method of cementing casing off bottom dates from 1920 when Erle Halliburton cemented a well in the Hewitt Field in Oklahoma for W. G. Skelly. Figure 7–3 shows the Halliburton jet cement mixer that is still a basic device for rapid mixing of cement slurry.

In 1930 there was only one kind of cement and no additives. Now there are numerous types listed in Standard 10A, the API specifications for oil well cements. The API lists recommended cement testing practices in their publication RP 10B. Today, approximately 50 additives for oil well cements are available. And, experimentation is continuing on new materials and methods for use with the extremely high temperatures and pressures that are encountered at 30,000 ft and below as greater depths are being drilled. (Table 7–5.)

Bulk-cement handling is the standard procedure. However, it is still customary to specify volume in sacks, where 1 sack = 94 lbs = 1 ft^3 in bulk. The absolute specific gravity of Portland cement is 3.14, so a sack of cement contains only 94/(3.14 X 62.4) = 0.48 net ft^3 of cement, or its porosity is 52%. Slurry volume is computed by assuming the volumes additive. Water-cement ratios are usually expressed in gallons per sack. Example, if 6.0 gal per sack is used, the slurry volume is:

6.0 + 0.48 = 1.28 ft^3 per sack of cement.

A cubic foot (ft^3) of water weighs 62.4 lbs and contains 7.48 gal.

With a few notable exceptions, Portland cement is the principal constituent of most cementing materials. It is the same ordinary cement which has been used by the construction industry for years. Additives have been developed which change the specifications of

TENSILE AND HARDNESS REQUIREMENTS

1	2	3	4		5		6		7		8	9
			Yield Strength				Tensile Strength		Hardness max.*		Specified Wall Thickness, Inches	Allowable Hardness Variation, HRC
			min.		max.		min.					
Group	Grade	Type	psi	MPa	psi	MPa	psi	MPa	HRC	BHN		
1	H-40		40,000	276	80,000	552	60,000	414	...	...		
	J-55		55,000	379	80,000	552	75,000	517	...	...		
	K-55		55,000	379	80,000	552	95,000	655	...	...		
	N-80		80,000	552	110,000	758	100,000	689	...	...		
	L-80	1	80,000	552	95,000	655	95,000	655	23	241		
	L-80	9Cr	80,000	552	95,000	655	95,000	655	23	241		
	L-80	13Cr	80,000	552	95,000	655	95,000	655	23	241		
	C-90	1,2	90,000	620	105,000	724	100,000	690	25.4	255	0.500 or less	3.0
	C-90	1,2	90,000	620	105,000	724	100,000	690	25.4	255	0.501 to 0.749	4.0
	C-90	1,2	90,000	620	105,000	724	100,000	690	25.4	255	0.750 to 0.999	5.0
	C-90	1,2	90,000	620	105,000	724	100,000	690	25.4	255	1.000 and above	6.0
2	C-95	1,2	95,000	655	110,000	758	105,000	724	...	...		
	T-95	1,2	95,000	655	110,000	758	105,000	724	25.4	255	0.500 or less	3.0
	T-95	1,2	95,000	655	110,000	758	105,000	724	25.4	255	0.501 to 0.749	4.0
	T-95	1,2	95,000	655	110,000	758	105,000	724	25.4	255	0.750 to 0.999	5.0
3	P-110		110,000	758	140,000	965	125,000	862	...	...		
4	Q-125	1,2	125,000	860	150,000	1035	135,000	930	...	...	0.500 or less	3.0
	Q-125	1,2	125,000	860	150,000	1035	135,000	930	...	...	0.501 to 0.749	4.0
	Q-125	1,2	125,000	860	150,000	1035	135,000	930	...	...	0.750 and above	5.0

*In case of dispute, laboratory Rockwell C hardness tests shall be used as the referee method.

Table 7-4 API Casing and Liner Casing Tensile Requirements.

1.1 Coverage. This specification covers requirements for manufacturing eight classes of well cements. This includes chemical and physical requirements and physical requirements and physical testing procedures.

A well cement which has been manufactured and supplied according to this specification may be mixed and placed in the field using water ratios or additives at the user's discretion. It is not intended that manufacturing compliance with this specification be based on such field conditions.

1.2 Classes and Grades. Well cement shall be specified in the following Classes (A, B, C, D, E, F, G and H) and Grades (O, MSR and HSR).

Class A: The product obtained by grinding Portland cement clinker, consisting essentially of hydraulic calcium silicates, usually containing one or more of the forms of calcium sulfate as an interground addition. At the option of the manufacturer, processing additions[1] may be used in the manufacture of the cement, provided such materials in the amounts used have been shown to meet the requirements of ASTM C 465. This product is intended for use when special properties are not required. Available only in ordinary (O) Grade (similar to ASTM C 150, Type I).

Class B: The product obtained by grinding Portland cement clinker, consisting essentially of hydraulic calcium silicates, usually containing one or more of the forms of calcium sulfate as an interground addition. At the option of the manufacturer, processing additions[1] may be used in the manufacture of the cement, provided such materials in the amounts used have been shown to meet the requirements of ASTM C 465. This product is intended for use when conditions require moderate or high sulfate-resistance. Available in both moderate (MSR)and high sulfate-resistant (HSR) Grades (similar to ASTM C 150, Type II).

Class C: The product obtained by grinding Portland cement clinker, consisting essentially of hydraulic calcium silicates, usually containing one or more of the forms of calcium sulfate as an interground addition. At the option of the manufacturer, processing additions[1] may be used in the manufacture of the cement, provided such materials in the amounts used have been shown to meet the requirements of ASTM C 465. This product is intended for use when conditions require high early strength. Available in ordinary (O), moderate sulfate-resistant (MSR) and high sulfate-resistant (HSR) Grades (similar to ASTM C 150, Type III).

Class D: The product obtained by grinding Portland cement clinker, consisting essentially of hydraulic calcium silicates, usually containing one or more of the forms of calcium sulfate as an interground addition. At the option of the manufacturer, processing additions[1] may be used in the manufacture of the cement, provided such materials in the amounts used have been shown to meet the requirements of ASTM C 465. Further, at the option of the manufacturer, suitable set-modifying agents[1] may be interground or blended during manufacture. This product is intended for use under conditions of moderately high temperatures and pressures. Available in moderate sulfate-resistant (MSR) and high sulfate-resistant (HSR) Grades.

Table 7–5 Primary cementing.

Class E: The product obtained by grinding Portland cement clinker, consisting essentially of hydraulic calcium silicates, usually containing one or more of the forms of calcium sulfate as an interground addition. At the option of the manufacturer, processing additions[1] may be used in the manufacture of the cement, provided such materials in the amounts used have been shown to meet the requirements of ASTM C 465. Further, at the option of the manufacturer, suitable set-modifying agents[1] may be interground or blended during manufacture. This product is intended for use under conditions of high tempperatures and pressures. Available in moderate sulfate-resistant (MSR) and high sulfate-resistant (HSR) Grades.

Class F: The product obtained by grinding Portland cement clinker, consisting essentially of hydraulic calcium silicates, usually containing one or more of the formsof calcium sulfate as an interground addition. At the option of the manufacturer, processing additions[1] may be used in the manufacture of the cement, provided such materials in the amounts used have been shown to meet the requirements of ASTM C 465. Further, at the option of the manufacturer, suitable set-modifying agents[1] may be interground or blended during manufacture. This product is intended for use under conditions of extremely high temperatures and pressures. Available in moderate sulfate-resistant (MSR) and high sulfate-resistant (HSR) Grades.

Class G: The product obtained by grinding Portland cement clinker, consisting essentially of hydraulic calcium silicates, usually containing one or more of the forms of calcium sulfate as an interground addition. No additives other than calcium sulfate or water, or both, shall be interground or blended with clinker during manufacture of Class G well cement. This product is intended for use as a basic well cement. Available in moderate sulfate-resistant (MSR) and high sulfate-resistant (HSR) Grades.

Class H: The product obtained by grinding Portland cement clinker, consisting essentially of hydraulic calcium silicates, usually containing one or more of the forms of calcium sulfate as an interground addition. No additives other than calcium sulfate or water, or both, shall be interground or blended with clinker during manufacture of Class H well cement. This product is intended for use as a basic well cement. Available in moderate sulfate-resistant (MSR) and high sulfate-resistant (HSR) Grades.

[1] A suitable processing or set-modifying agent shall not prevent a well cement from performing its intended functions.

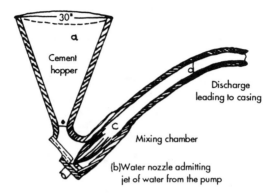

Figure 7-3 Halliburton jet cement mixer (courtesy Halliburton)

Portland cements to adapt them for use in oil and gas industry cementing programs.

Oil well cementing is the process of mixing and displacing cement slurry down the casing and up the annular space behind the casing. In order for it to perform satisfactorily the task allotted it, certain requirements must be met:

(1) The cement slurry must be capable of being placed in the desired position by means of pumping equipment at the surface.

(2) After being placed, it must develop sufficient strength within a reasonably short time, so waiting-on-cement (WOC) is minimized.

(3) The cement must provide a positive seal between the casing and the formation.

(4) Sufficient strength must be developed in the cement to avoid mechanical failure.

(5) The cement must be chemically inert to any formations or fluids with which it may come into contact.

(6) The cement must be stable enough that it will not deteriorate, decompose, or otherwise lose its qualities of strength for the duration of the well life.

(7) The cement must be sufficiently impermeable so that fluids cannot flow through it when it has set.

Primary Cementing

Primary cementing pertains to the initial cementing jobs performed in conjunction with setting the various casing strings. It also affords additional support for the casing, and helps retard corrosion.

The typical procedure for a single-stage primary cement job is the conventional two-plug method, in which neat cement (no sand or gravel) is introduced into a hopper where it is thoroughly mixed with water by a high velocity jet mixer. The resulting slurry is then pumped down the casing between two rubber plugs with wiping fins, which are placed in the system at the proper time via a cementing head. When the bottom plug reaches the float collar, it stops. Pressure builds up which quickly ruptures the plug's diaphragm and allows the slurry to continue. The top plug, however, has a solid core, so that when it seats in the float collar, the surface pressure builds up sharply, thereby signaling the pump operator that the job is complete. The position of the top plug may also be checked either by metering the displacing fluid, or with a wireline measurement. The casing below the float collar is left full of cement and can be drilled out if necessary.

Numerous variations of primary cementing techniques are in use. One or both plugs are sometimes omitted. There is a wide diversity of opinion on cementing practices, so studies and tests will be made to decide which is best for a particular well. Figure 7–4 shows a conventional cementing job with some auxiliary equipment being used.

Multistage Cementing

Multistage cementing techniques are used for cementing two or more separate sections behind a casing string. Excessive pressures caused by placing these long columns of slurry may cause fluid losses to incompetent formations, or the cement slurry, in its unusually long travel length, may become contaminated. To eliminate these possibilities, stage-cementing equipment has been developed. In this technique the desired amount of slurry is placed around the lower part of the casing in the conventional fashion. Then, by use of a shut-off plug and a stage-cementing collar, the cement slurry is delivered through ports in the stage-cementing collar and out into the annular area, by-passing the lower part of the casing where the slurry has already been placed. Several isolated zones can also be cemented in one operation by placing the shut-off plugs and stage-cementing collars at desired places.

Secondary Cementing

Secondary cementing is done after primary, most often because of some functional failure of the primary cement job. It may be used to *squeeze cement* through perforations, to plug a dry hole, or it may be used to plug an open zone. Secondary cementing normally

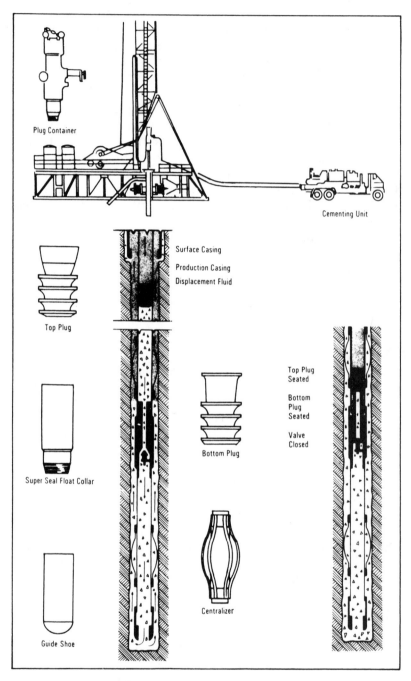

Plug Container

Cementing Unit

Surface Casing

Production Casing

Displacement Fluid

Top Plug

Super Seal Float Collar

Bottom Plug

Top Plug
Seated

Bottom
Plug
Seated

Valve
Closed

Centralizer

Guide Shoe

Figure 7-4 Diagram of a casing cementing job (courtesy Halliburton).

Modern Petroleum

involves the injection under pressure of a slurry into a porous formation. The pumping rate is slow enough to allow for dehydration and/or initial setting of the slurry. This effect is indicated by increased pump pressure, and injection is continued until the desired final or squeeze pressure is attained. Secondary cementing is used to:

(1) Supplement or repair a primary job.
(2) Repair defective casing or improperly placed perforations.
(3) Abandon a permanently nonproductive or depleted zone.
(4) Reduce the danger of lost circulation in a open hole while drilling deeper.
(5) Isolate a zone before perforating.
(6) Fracture the formation.

Cement Additives

The slurry density depends on the amount of mixing water and additives in the cement and the amount of slurry contamination from drilling mud or other foreign material. Slurries range in density from 10.8 to 22 PPG (pounds per gallon). Additives can reduce slurry density as well as increase slurry volume. Some of the additives are retarders and some are accelerators that alter cement setting times. Other additives reduce water loss and help prevent premature dehydration.

Conductor and surface casing cements have lower temperatures and require an accelerator to promote setting of the cement. Cement retarders help extend the cement's pumpability for deep wells. The well's bottom-hole temperature is the primary factor that governs the use of additional retarders because the chemical reaction between cement and water is accelerated as temperature increases. This reduces the cement's thickening time and pumpability.

Heavyweight additives are used when abnormally high pressures are expected. Lightweight additives are used to reduce slurry density. Fluid-loss additives are widely used in squeeze cementing and in high-column cementing such as deep liner cementing.

Friction-reducing additives promote turbulent flow at low displacement rates to ensure better flushing of mud from the annulus.

Salt-saturated cements were developed for cementing through salt zones because fresh water does not bond properly to salt formations. The water from the cement slurry dissolves or leaks away at the interface, which prevents an effective bond. Salt slurries also help protect shale sections sensitive to fresh water.

Pozzolans are siliceous materials which will react with lime to form a cementitious material. As such, they may be used as an additive to ordinary cement or prepared as a lime-pozzolan blend without Portland-type cement. It is a satisfactory deep-well cement.

Diacel cement systems have a large range of densities and thickening times, which gives them a wide scope of applicability.

Latex-Cement is a special cement composed of latex, cement, a surface active agent, and water. It is useful in such special applications as plug-back jobs for water exclusion.

Diesel-Oil-Cements (DOC) are mixtures of Portland cement, diesel oil (or kerosene), and a chemical dispersant. This is useful in well repair work to seal off water-bearing strata. It does not set until brought into contact with water and has an unlimited pumping time.

Oil-In Water Emulsion Cements have low water loss, low density cements of adequate strength and thickening time. They have been prepared from kerosene, water, cement, and 2% to 4% bentonite. Calcium lignosulfonate is used as emulsifying agent and retarder.

Resin cements are in proper combination of synthetic resins, water, and Portland cements that are often used to provide an improved formation-cement bond in certain remedial operations.

Gypsum cements are special mixtures that have high early strength and easily controlled setting times. Gypsum is the basic ingredient. Their principal use is to provide temporary plugs during testing and remedial work.

Calcium chloride is sometimes added to water to decrease freezing time and to accelerate setting.

Nut shells, cellophane flakes, etc. are added when necessary to prevent lost circulation.

Casing Accessories

In order to achieve the desired objectives in cementing, special equipment and accessories have been designed to aid the job, such as *guide shoes, float collars, centralizers* and *scratchers.*

The guide shoe, which protects the lower end of the pipe from damage, is placed on the bottom of the casing. Heavy and with a rounded nose, the guide shoe has a cement collar inside the bottom to aid in bonding the cement to the casing string. It also usually contains a back-pressure valve arrangement to permit circulation from the inside of the casing to the outside. Reverse circulation from the outside of the casing to the inside is not possible as the back pressure valve will seat, preventing fluids from entering the casing from the outside.

Float collars are multipurpose devices that are equipped with a back-pressure valve. In many cases where the casing is set through the producing formation, the lower end of the casing will be in a water-bearing formation. This water may enter the casing, necessitating remedial cementing operations, but if the lowest joint of casing is left filled with cement, the hazards are reduced. If the cement has become slightly contaminated with the displacing fluid, the contaminated portion will probably be left in this last joint of casing. This collar, in addition to performing its regular duty of attaching two joints of casing, has a back-pressure valve, similar to the one in the guide shoe, which prevents circulation back into the casing. The internal diameter of the float collar is reduced to provide a positive seat for the cementing plugs.

Centralizers and scratchers are attached to the casing for specific purposes to aid in the cementing process. There must be a predetermined minimum thickness of cement throughout the cemented interval to insure a good cement job. The borehole will not be perfectly straight and the casing will be in contact with the hole in some places, unless provisions are made to correct this problem. Casing centralizers are flexible springs attached to the casing to insure the centering of casing in the hole and provide a more uniformly cemented casing string.

The scratchers are mechanical devices with wire figures attached to the casing along with the centralizers. They remove mud cake from the wellbore and provide a better surface for the cement to bond to.

As covered in the previous chapter, the engineer runs a cement bond log to test the cement for quality and bonding. A temperature log may be used to locate the top of the cement, or it can be estimated based on the size of the hole drilled and the outside diameter of the casing. Some wells may have cement all the way from the bottom of the casing to the surface.

Zone Completions

Formation evaluation methods are also useful in defining certain individual characteristics of the zone which dictate the completion method best suited (Figure 7-5). Various schemes are used to classify zone completions, and some overlapping occurs. There are six main categories of zone completions:

(1) Open Hole Completions
(2) Conventional Perforated Completions
(3) Sand Exclusion Types
(4) Permanent-Type Completion
(5) Multiple Zone Completions
(6) Drainhole Completions

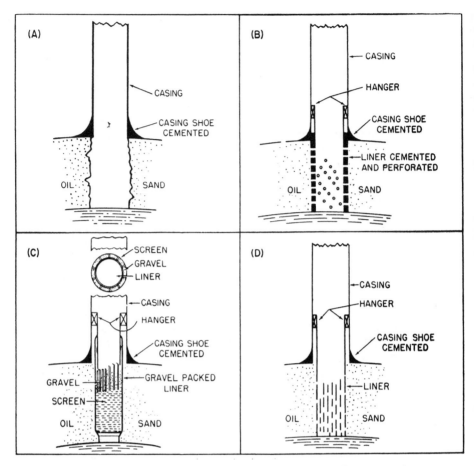

Figure 7–5 Open-hole, perforated, and sand exclusion types of zone completions.

Open-Hole Completions

The term *open-hole completions* refers to those in which the production casing is set on top of the productive zone, leaving the productive interval as an open borehole without pipe to protect it. This is also known as a *barefoot completion.* This method of completion is only applicable to highly competent formations which will not slough or cave. Rotary tools are used until the production casing is set, at which time the rotary rig moves out and a cable tool rig is brought in. The cable tool uses a sand line and bailer to bail out the mud, and then attaches a percussion drilling tool to drill the desired depth. This method permits testing the zone as it is drilled, eliminates formation damage by drilling mud and cement, and cautiously deepens the hole to avoid drilling into water.

Another advantage is having the cable tool rig on location to swab the well if that should become necessary. Swabbing is sometimes used to open up the formation so that oil and gas flow more freely. The swab is a tool that is run inside the casing, and has rubber rings to make close contact with the casing. As it is pulled up the casing by a steel cable, a vacuum is formed below it, which in turn helps open the formation. When the tool is run back down the casing, a valve opens, which allows it to be lowered with ease.

Open-hole completed wells are usually stimulated by *hydraulic* or *acid fracturing.*

Conventional Perforated Completions

Conventional perforated completions are restricted to wells in which the production casing is set through the zone, cemented and subsequently perforated at the desired interval. These completions are very common and are feasible in all formations except those in which sand exclusions are a problem. The major factor involved is the perforating process.

Bullet Perforating equipment was developed in the early 1930s and is still used sparingly. Later, *jet perforating*, or the shaped charge principle, was adopted and has taken the lead in perforating.

The bullet perforator is essentially a multi-barreled firearm designed to be lowered into the casing and electrically fired from the surface at the desired interval. Penetration of the casing, cement, and formation is accomplished by high velocity projection or bullets. A number of bullet types are available for particular purposes.

Today, jet perforating is the accepted standard in the oil and gas industry. All jet perforators have the same ignition system; an electric detonating cap, which is connected to the primer cord which runs the entire length of the gun and makes contact with each jet charge. The primer cord ignites the booster in the charge, which in turn ignites the main charge and produces the jet. Two basic types are retrievable and expendable.

The *expendable jet charge* was designed to allow more explosives to be run in a fixed diameter. The enclosures are made from die-cast aluminum which disintegrates upon detonation.

A *retrievable gun* is composed of a cylindrical steel carrier that resembles a piece of pipe with the charges facing the perimeter of the carrier. The carrier ranges in size from 1 ⅜ to 5 in. OD. It is designed to provide a fixed and precise perforating pattern, to protect the charges from wellbore fluids, and to absorb the major portion of the shock from detonation. Following detonation, it contains the debris from the charge casing and alignment sleeves, thus preventing plugging of chokes, valves, and flowlines.

All perforating guns should be shot from the bottom up. It is better to perforate with the pressure inside the wellbore lower than the rock formation pressure.

The typical shaped charge is a simple device in terms of construction, as shown in Figure 7–6. It usually consists of only four basic components: a conical metallic liner, primer, main high explosive charge, and case or container.

Though simple in construction, the shaped charge in operation is a highly complex, high- speed reacting device. This stems, in part, from the character of explosive used in the charge. *RDX (Cyclonite)* is typically used. It is similar in explosive characteristics to TNT and nitroglycerin.

Once the primer is initiated, it in turn initiates detonation of the main charge. A detonation wave sweeps through the charge at about 30,000 ft/sec generating pressures on the liner of 2 to 4 million psi.

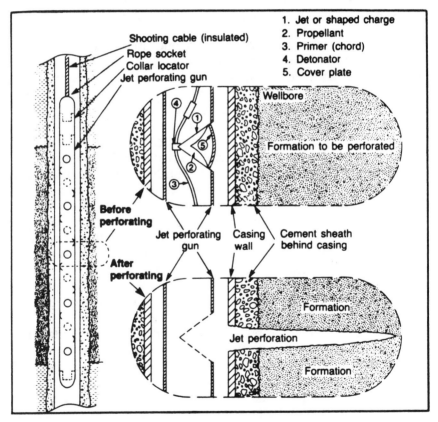

Figure 7–6 Schematic of jet perforating procedure (courtesy PennWell Books, *Petroleum Production for the Nontechnical Person,* Gray).

Modern Petroleum

The result is that the liner collapsing forms the jet and the slug. Modern charges have eliminated the slug by utilizing particulated liners fabricated from powdered materials. Such liners still yield a mass of nonjet material that follows the jet. This results in the slug material being more uniformly distributed in its dissociated form along the perforated hole, rather than creating a solid slug at some point.

Improvements are being made as more tests are run, which will result in some changes in the future.

For penetration depths of ½ to 1 well diameter, little increase in productivity is gained from increasing perforating densities above four ½ in. holes per foot. For example, if the well diameter is 9 in. or less, you gain no advantage if you shoot more than four ½- in. holes per foot of pipe being perforated in the production zone.

Sand Exclusion Completions

The completion of a well in an unconsolidated sand is more complicated than open-hole and conventional perforated completions because of the additional problem of excluding sand. Sand production, if unchecked, can cause erosion of equipment and retard production to an uneconomical level. At very low production rates, very little sand may be produced, while at high rates large quantities will be carried along in the production stream. This is especially true where wells are being pumped.

Screening seems to be the most common method of excluding sand, and there are two general techniques used:

(1) Use of slotted or screen liners.
(2) Packing the hole with aggregates such as gravel.

The principle behind these two methods is that the opening through which the fluids must flow must be properly sized to cause the formation sand to form a stable bridge and thereby be blocked.

The formation sand should be analyzed for grain size from samples of the formation. Special core barrels are used for sampling purposes. From the sample, it is possible to obtain the maximum opening screen size which will exclude the sand and allow fluid flow.

The slotted or screen liner is usually run on tubing and hung inside the casing opposite the formation. This may be done in either an open-hole or a cased hole.

Gravel packing can be used in various ways in either open-hole of perforated intervals the same as slotted or screen liners. Gravel packing is perhaps the most widely used method of sand control. It involves the use of a selected size gravel and screen. The normal procedure is to place the screen in the casing or wellbore of an open-hole completion opposite the production interval and then pump sized gravel between the screen and the formation. The slots in the screen may be larger than for the other method and are generally only slightly smaller that the gravel. The formation sand bridges with the pores of the gravel pack, and the gravel is prevented from entering.

Plastic consolidation types of sand control are effective in certain formations. The main objective of plastic sand consolidation is to increase the strength of the formation around the wellbore so that grains of sand are not dislodged by the drag forces of flowing fluid at production rates. It is applicable in small diameter casing and suitable for through-tubing application. These plastic consolidation methods can be used in abnormally high pressure wells, and works effectively in fine sands difficult to control with a gravel packing system. The technique is suitable for multiple zones.

One advantage of plastic consolidation sand control is that a notable reduction in permeability can exist.

Permanent-Type Completions

A *permanent-type completion* is a type in which the tubing is run and cementing, perforating, swabbing, and other completion work is performed with small diameter tools. The wellhead is assembled only once in the life of the well and artificial lift equipment will be limited as to type. There is a great savings in cost with this type that makes it attractive in some areas, especially where sand exclusion and high pressure flowing wells are produced.

Multiple Zone Completions

A *multiple zone completion* allows simultaneous production of two or more separate formations in a single casing without commingling the fluids. It is useful for the purpose of reservoir control and cuts costs. Two-zone completions are more common, although triple and quadruple zone completions are used in some areas.

One objection to this practice is the complexity of the downhole equipment necessary to achieve and maintain segregation. The problem becomes more complicated when one or more of the zones require artificial lifting equipment.

The application of multiple zone completions receives favorable consideration when two or more marginal zones may be produced, saving the cost of separate wells. Another consideration is well spacing, or offset clauses in a lease that may be fulfilled with this system.

Figure 7–7 shows one method of multiple completion in which two zones are being produced and a triple completion with three zones being produced. There are several configurations that may be employed, such as using the annulus of the casing as a means to produce a flowing well, as well as using the tubing.

Multiple completions could not be possible without some equipment capable of separating the tubing and segregating one zone from the other, or others. Packers are the specialized equipment used in multiple completions. They are a rubber packing element or packer that can be expanded against the casing and around the tubing. Rubber elements with varying hardness and composition are used. The most commonly used rubber is nitrile with a 70 durometer hardness. High bottomhole temperatures will require that a greater hardness be used.

Packers must be able to withstand high differential pressure and still maintain a good seal, and the tubing must be free to move when temperature changes occur.

There are two general packer classifications: the permanent and the retrievable. The permanent type is held in place by opposing slips and can be set with wireline or tubing methods. The retrievable packers can be weight set, mechanical set, or hydraulic set. There are four major components including the slip assembly and outer seal which secures and seals the packer/casing interface. The packer bore and the tubing seal assembly provide the tubing-packer pressure seal while also allowing tubing expansion or contraction to occur.

Packers are manufactured for other purposes as well as for multiple completions.

Drainhole Completions

A form of horizontal or near-horizontal drilling is the major method of *drainhole completion.* Basically it is a form of controlled directional drilling, in which the prospective formation is reached with the drill bit in a horizontal position. Special equipment is required to start the drilling to make an arc at a desired depth then continue in a long, short, or medium radius until the zone is reached. Swivel joints are used behind the drill bit to aid in the process. This horizontal drilling results in a long productive interval in the well.

Coning

In some reservoirs, the oil zone occurs with an overlying gas zone or an underlying water zone, or both. In these cases it is desirable to complete the well in such a manner so that only the oil zone will produce. Production of large quantities of gas per barrel of oil is

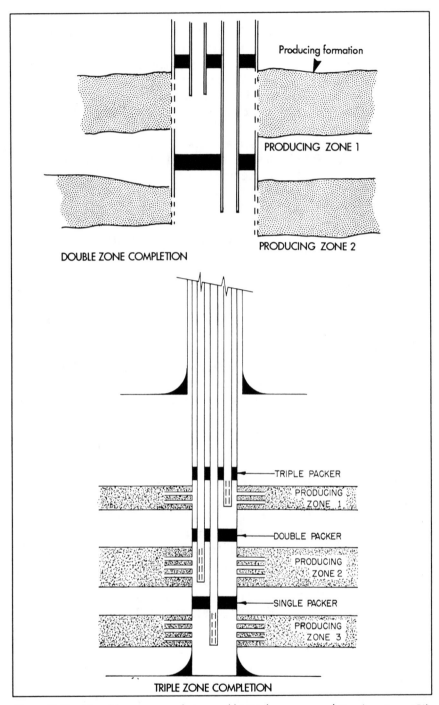

Producing formation

PRODUCING ZONE 1

PRODUCING ZONE 2

DOUBLE ZONE COMPLETION

TRIPLE PACKER

PRODUCING ZONE 1

DOUBLE PACKER

PRODUCING ZONE 2

SINGLE PACKER

PRODUCING ZONE 3

TRIPLE ZONE COMPLETION

Figure 7-7 a) Double zone completion and b) Triple zone completion (courtesy API).

undesirable because it wastes reservoir driving energy. Production of water is undesirable for a multitude of reasons.

In order to produce the well, a pressure gradient must be established between the well and its drainage radius that extends both vertically and horizontally and acts on the water as well as the oil. Consequently, both fluids tend to flow toward the wellbore. If the producing rate is too high, *coning* may occur. Since coning is primarily a rate-sensitive problem, the obvious solution requires a reduction in production rates.

Water or gas exclusion problems other than coning, are sometimes due to irregular permeability and saturation distributions. In reservoirs composed of alternate and separate layers, exclusion problems may be solved by squeeze cementing and reperforating.

Well Treatment

Very often it is necessary to treat a formation to improve the natural drainage pattern, or to remove barriers with the formation which prevent easy passage of fluids into the wellbore. These processes are known as *well-stimulation treatments*. Stimulation treatments are classified as *acidizing* or *fracturing*. Often both processes are used in combination because they frequently help each other.

Acid was used before 1900 to enhance productivity but did not come into widespread use until the 1930s. Acidizing involves the injection of acid into an acid-soluble formation where its dissolving action enlarges existing voids and increases the permeability. The acid commonly used is 15% hydrochloric (by weight) which reacts with limestone or other carbonates.

Only carbonate rocks are normally susceptible to acid treatment; however, some sands do have sufficient calcareous content to justify acidization. Many additives are used in the acid, including inhibitors to retard corrosion of casing and tubing.

Hydraulic Fracturing

Fracturing is a process involving the injection of a fracturing fluid and a propping agent into the formation under sufficient pressure to open (crack) existing fractures or create new ones. These are extended some distance around the wellbore by continued high pressure injected after the initial breakdown has occurred. The propping agent, a carefully sized silica sand, glass beads, epoxy, or other material, is pumped into the fractures of formation and holds them open as the pumping pressure ceases. This process may be used on practically all reservoir rocks and can be combined with acid treatments in limestone areas.

Produced crude oil from the same lease is often used as the fracture fluid because native oil is less prone to damage the permeability of the formation (see Figures 7–8 and 7–9).

Tubing

Tubing is used as the flow string in the well, and its selection is based on the same considerations given the selection of casing. The

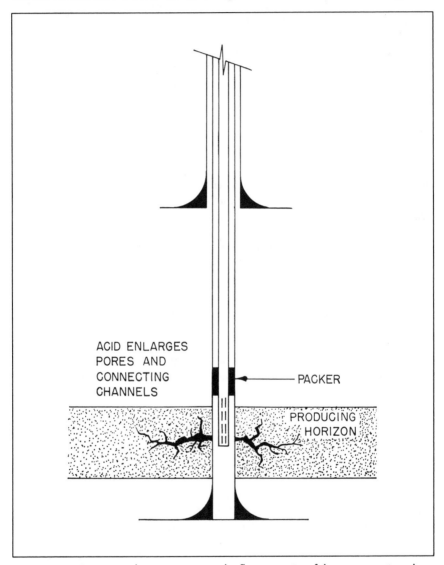

ACID ENLARGES
PORES AND
CONNECTING
CHANNELS

PACKER

PRODUCING
HORIZON

Figure 7–8 Crevice acidizing to increase the flow capacity of the pay zone into the well bore (courtesy API).

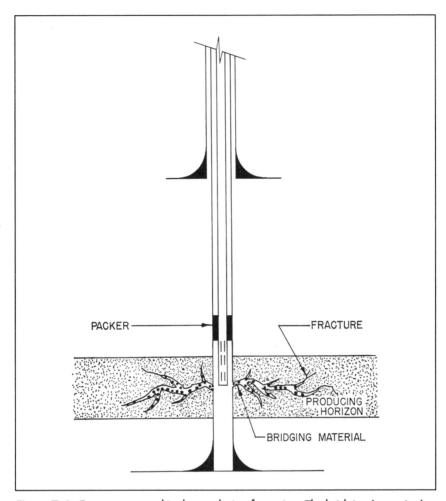

Figure 7-9 Fractures opened in the producing formation. The bridging (propping) materials are placed in the fractures to keep them open (courtesy API).

design factors applied to tubing strings are practically the same as those used for casing. Specifications for casing, tubing, and drillpipe are included in API Spec 5A. Although regular or non-upset tubing is available, the majority of tubing used has the externally upset end. Tubular goods that are *upset* are made thicker in the area of the threads in order to compensate for the metal cut away in making threads. The thickening is on the outside, and is known as exterior-upset tubing. Sizes from 1.050 in. to 4½ in. OD are available. The common sizes are 2 ⅜ and 2 ⅞ in. OD (Table 7-6).

Sizes, Weights, Wall Thickness, Grade and Applicable End Finish[1]

1		2	3	4	5	
		Nominal Weights[2,3]				
Size: Outside Diameter		Non-Upset T & C	External Upset T & C	Intregal Joint	Wall Thickness	
in.	mm	lb./ft.	lb./ft.	lb/ft.	in.	mm
1.050	26,7	1.14	1.20	—	0.113	2,87
1.050	26.7	1.48	1.54	—	0.154	3,91
1.315	33,4	1.70	1.80	1.72	0.133	3,38
1.315	33,4	2.19	2.24	—	0.179	4,55
1.660	42,2	—	—	2.10	0.125	3,18
1.660	42,2	2.30	2.40	2.33	0.140	3,56
1.660	42,2	3.03	3.07	—	0.191	4,85
1.900	48,3	—	—	2.40	0.125	3,18
1.900	48,3	2.75	2.90	2.76	0.145	3,68
1.900	48,3	3.65	3.73	—	0.200	5,08
1.900	48,3	4.42	—	—	0.250	6,35
1.900	48,3	5.15	—	—	0.300	7,62
2.063	52,4	—	—	3.25	0.156	3,96
2.063	52,4	—	—	—	0.225	5,72
2 3/8	60,3	4.00	—	—	0.167	4,24
2 3/8	60,3	4.60	4.70	—	0.190	4,83
2 3/8	60,3	5.80	5.95	—	0.254	6,45
2 3/8	60,3	6.60	—	—	0.295	7,49
2 3/8	60,3	7.35	7.45	—	0.336	8,53
2 7/8	73,0	6.40	6.50	—	0.217	5,51
2 7/8	73,0	7.80	7.90	—	0.276	7,01
2 7/8	73,0	8.60	8.70	—	0.308	7,82
2 7/8	73,0	9.35	9.45	—	0.340	8,64
2 7/8	73,0	10.50	—	—	0.392	9,96
2 7/8	73,0	11.50	—	—	0.440	11,18
3 1/2	88,9	7.70	—	—	0.216	5,49
3 1/2	88,9	9.20	9.30	—	0.254	6,45
3 1/2	88,9	10.20	—	—	0.289	7,34
3 1/2	88,9	12.70	12.95	—	0.375	9,52
3 1/2	88,9	14.30	—	—	0.430	10,92
3 1/2	88,9	15.50	—	—	0.476	12,09
3 1/2	88,9	17.00	—	—	0.530	13,46
4	101,6	9.50	—	—	0.226	5,74
4	101,6	—	11.00	—	0.262	6,65
4	101,6	13.20	—	—	0.330	8,38
4	101,6	16.10	—	—	0.415	10,54
4	101,6	18.90	—	—	0.500	12,70
4	101,6	22.20	—	—	0.610	15,49
4 1/2	114,3	12.60	12.75	—	0.271	6,88
4 1/2	114,3	15.20	—	—	0.337	8,56
4 1/2	114,3	17.00	—	—	0.380	9,65
4 1/2	114,3	18.90	—	—	0.430	10,92
4 1/2	114,3	21.50	—	—	0.500	12,70
4 1/2	114,3	23.70	—	—	0.560	14,22
4 1/2	114,3	26.10	—	—	0.630	16,00

[1] P = Plain End, N = Non-upset T & C, U = External Upset T & C, I = Integral Joint.
[2] Nominal weights, threads and coupling (Col. 2, 3, 4) are shown for the purpose of identification in ordering.
[3] The densities of martensitic chromium steels (L-80 types 9Cr and 13Cr) are different from carbon steels. The weights shown are therefore not accurate for martensitic chromium steels. A weight correction factor of 0.989 may be used.

Table 7–6 Tubing made to API specifications.

6	7	8	9	10	11	12
			Type of End Finish[4]			
H-40	J-55	L-80	N-80	C-90[5]	T-95[5]	P-110
PNU	PNU	PNU	PNU	PNU	PNU	—
PU	PU	PNU	PU	PNU	PNU	PU
PNUI	PNUI	PNUI	PNUI	PNUI	PNUI	—
PU	PU	PNU	PU	PNU	PNU	PU
PI	PI	—	—	—	—	—
PNUI	PNUI	PNUI	PNUI	PNUI	PNUI	—
PU	PU	PNU	PU	PNU	PNU	PU
PI	PI	—	—	—	—	—
PNUI	PNUI	PNUI	PNUI	PNUI	PNUI	—
PU	PU	PNU	PU	PNU	PNU	PU
—	—	PN	—	PN	PN	—
—	—	PN	—	PN	PN	—
PI	PI	PI	PI	PI	PI	—
P	P	P	P	P	P	P
PN	PN	PN	PN	PN	PN	—
PNU	PNU	PNU	PNU	PNU	PNU	PNU
—	—	PNU	PNU	PNU	PNU	PNU
—	—	PN	—	PN	PN	—
—	—	PNU	—	PNU	PNU	—
PNU	PNU	PNU	PNU	PNU	PNU	PNU
—	—	PNU	PNU	PNU	PNU	PNU
—	—	PNU	PNU	PNU	PNU	PNU
—	—	PNU	—	PNU	PNU	—
—	—	PN	—	PN	PN	—
—	—	PN	—	PN	PN	—
PN	PN	PN	PN	PN	PN	—
PNU	PNU	PNU	PNU	PNU	PNU	PNU
PN	PN	PN	PN	PN	PN	—
—	—	PNU	PNU	PNU	PNU	PNU
—	—	PN	—	PN	PN	—
—	—	PN	—	PN	PN	—
—	—	PN	—	PN	PN	—
PN	PN	PN	PN	PN	PN	—
PU	PU	PU	PU	PU	PU	—
—	—	PN	—	PN	PN	—
—	—	PN	—	PN	PN	—
—	—	PN	—	PN	PN	—
—	—	PN	—	PN	PN	—
PNU	PNU	PNU	PNU	PNU	PNU	—
—	—	PN	—	PN	PN	—
—	—	PN	—	PN	PN	—
—	—	PN	—	PN	PN	—
—	—	PN	—	PN	PN	—
—	—	PN	—	PN	PN	—
—	—	PN	—	PN	PN	—

[4] Non-upset tubing is available with regular couplings or special bevel couplings. External-upset tubing is available with regular, special-bevel, or special clearance couplings.
[5] Grade C-90 and T-95 tubing shall be furnished in sizes, weights, and wall thicknesses as listed above, or as shown on the purchase order.

API Non-upset

API Integral Joint

API External Upset

Low-fluid-level pumping tubing must support its own weight, plus the weight of the liquid inside. And variations of load in pumping wells make the use of upset joints attractive because the greater metal area is subject to lower stress and has increased resistance to fatigue.

During the life of the well the tubing will be used as part of the system to control the fluid of a flowing producer. Later, the same tubing may be used to support a pump and sucker rods when artificial lift is necessary. Sizes are sometimes tapered in deep pumping wells to accommodate larger sucker rods in the top section. Size tapering (using heavier and larger tubing at the top) also allows greater tension setting depths.

Sucker Rods

Sucker rods are solid high-grade steel rods that are run inside the production casing to connect the down-hole pump to the pumping unit on the surface. Even though they are part of the down-hole equipment, they are covered in the following chapter with rod pumping equipment.

The Wellhead

The *wellhead* is the equipment used to maintain surface control of the well. It forms a seal to prevent well fluids from blowing or leaking at the surface. The kind of wellhead equipment to be used is determined by well conditions. High-pressure wells require equipment to withstand pressures greater than 20,000 psi and low-pressure wells require only simple or ordinary equipment. Wellhead equipment generally consists of the *casing head, tubing head,* and *christmas tree.*

Casing Head

During the drilling operation it is necessary to install heavy fittings at the surface as each string of casing is run into the hole. The casing is attached to the casing head and it in turn is supported by the previously installed casing; for example, the surface casing will support the production casing (Figure 7–10).

Each part of the casing head provides a gripping device to hold the weight of the casing. It has a sealing device between the casing strings to prevent flow of fluids.

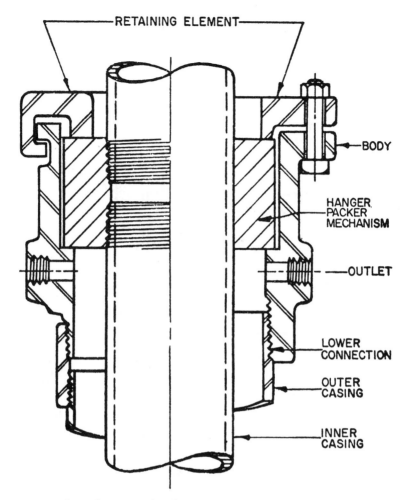

Figure 7-10 Independent casing head (courtesy API).

The casing head is used during drilling and work-over operations as an anchor for pressure-control equipment, such as *blow-out preventers.*

Tubing Head
The *tubing head* is used to support the tubing string and seal off pressures between the casing and tubing. It provides connections at the surface with which fluids can be controlled. For pumping wells, the stuffing box attaches to the top of the tubing, and for flowing

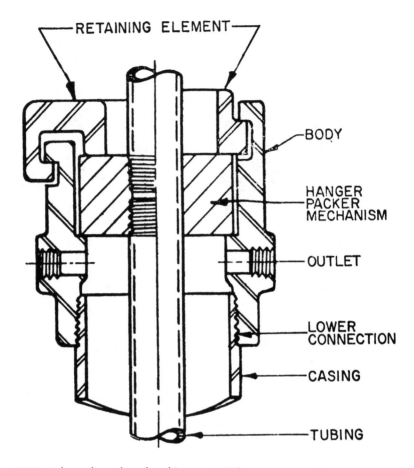

RETAINING ELEMENT

BODY

HANGER
PACKER
MECHANISM

OUTLET

LOWER
CONNECTION

CASING

TUBING

Figure 7–11 Independent tubing head (courtesy API).

wells, valves and gauges are attached. It must be easily assembled
and disassembled to make well-servicing operations simple (Figure
7–11). Figure 7–12 shows a typical wellhead assembly with both
casing and tubing heads attached.

The Christmas Tree

The christmas tree is a group of valves that control the flow of fluids
from the well. It is called a christmas tree because of its shape and

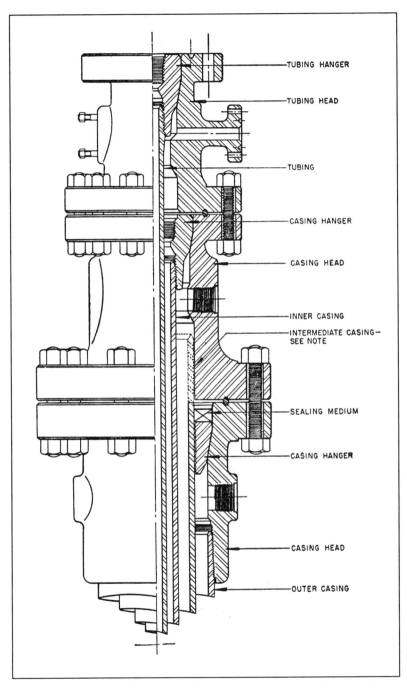

TUBING HANGER

TUBING HEAD

TUBING

CASING HANGER

CASING HEAD

INNER CASING

INTERMEDIATE CASING—
SEE NOTE

SEALING MEDIUM

CASING HANGER

CASING HEAD

OUTER CASING

Figure 7–12 A typical wellhead assembly with both casing and tubing heads attached (courtesy API).

Completion   *217* the large number of fittings branching out above the wellhead. The master valve is placed at the bottom of the assembly and is normally the last valve to be operated when emergency situations arise.

Wells expected to have high pressures, or corrosive gases are normally equipped with special heavy valves and control equipment above the casing head or tubing head before such wells are completed. Many variations in arrangement of wellhead and christmas tree assemblies are available to meet the needs of any type application.

Christmas tree assemblies for flowing wells differ from those used for a pumping well, primarily because of the stuffing box required for the polished rod, from which the sucker rods are hung, to pass through.

Chapter References

Neal Adams, *Workover Well Control*, (Tulsa: PennWell Books, 1981).

Preston L. Moore, *Drilling Practices Manual*, 2nd. Ed. (Tulsa:PennWell Books, 1986).

8
Production

Millions of years ago processes were underway beneath the surface of our earth to produce products that are vital to the lifestyle that we enjoy today. Just how these processes produced the crude oil, natural gas and condensate is theorized by scientists and engineers but without their complete agreement on any one method. Another bit of controversy is how petroleum fluids came to be in the formations in which we find them.

Ever since we have learned that petroleum fluids are held in reserve in formations beneath the surface of our Earth, we have spent countless hours and untold amounts of money to locate the productive zones. The search continues, and yet more time and money will be spent to locate still more prospective locations in which to drill.

After experiencing many failures in locating a productive formation, when success does come, the next very important step is to determine how to bring this valuable fluid to the surface. There are various methods available, and the one selected should take into account the depth and type of formation, the gas-to-oil ratio, the viscosity of the crude oil, and the economics of the entire project. Once the well has been completed with casing set, cemented, perforated, and stimulated, if necessary, the well is ready to be equipped for production. During the early days of the oil industry the most common thing to do with a well that was not flowing of its own accord was to install a rod pumping unit and hope it would prove to be the proper size.

Well Testing

In producing gas and oil, more and more importance is being placed upon the *most efficient recovery (MER)* performance of the producing wells. Efficient recovery takes proper engineering and planning along with the right equipment. Controlling the rate of production is one of the very important factors for efficient recovery. Experience has shown that excessive production rates can result in premature declines in reservoir pressure that results in the trapping or bypassing of fluids in the formation. Other damages might be premature release of dissolved gas, dissipation of gas and water, and irregular movement of the displacement forces. In general, excessive production rates may reduce the ultimate primary recovery of oil. Generally, some sort of test must be made in order to determine the best performance of the well.

Potential Test

Recovery from many formations is directly dependent on the rate of production. The most frequently conducted well test is the *potential test*, which is a measurement of the maximum amount of oil and gas a well will produce in a 24-hour period under certain fixed conditions. The produced oil is measured in an automatically controlled production and test unit or by wire-line measurement, once the oil is out of the formation and in the lease tank.

The potential test is normally made on each newly completed well, and then again periodically during its production life. The MER for the well is defined as the greatest rate that can be sustained during a predetermined length of time without damaging the reservoir. This data, along with a flow test, gives the total pressure drop of the well. This can help to determine the reserves in that particular reservoir. All of this information is valuable in determining the feasibility of drilling offset wells.

Potential tests will be different for oil wells and natural gas wells, but the purpose of the tests are basically the same: to help establish proper production practices.

Bottom-Hole Pressure Test

Bottom-hole pressure tests measure the reservoir pressure of the well at a specific depth in the producing interval. This test measures the pressure of the zone in which the well is completed. In making this test, a specially designed pressure gauge is lowered into the well by means of a wireline. The pressure at the selected depth is recorded by the gauge, and then the gauge is pulled to the surface and removed from the well.

There are several variations of this type of test, such as the *flowing bottom-hole* pressure test, which is a measurement taken while the well continues to flow. A *shut-in bottom-hole pressure test* is a measurement taken after the well has been shut in (closed) for a specific length of time. These tests also give information about the fluid levels in the well. Other bottom-hole pressure tests furnish valuable information about the decline or depletion of the zone that the well is producing from.

Productive Tests

Productive tests are made on both oil and gas wells and include the potential tests and the bottom-hole pressure test. This reading determines the effects of different flow rates on the pressure within the producing zone of the well and thereby establishes certain physical characteristics of the reservoir. In this manner, the maximum potential rate of flow can be calculated without risking possible damage to the well.

In the procedure, the closed-in bottom-hole pressure of the well is first measured. Then the well is opened and produced at several stabilized rates of flow. At each rate of flow, the flowing bottom-hole pressure is measured. These data provide an estimate of the maximum flow expected from the well.

Fluid-Level Determination

Fluid-level determination is a test most commonly performed on oil wells that will not flow and must be made to produce by pumping or by other means of artificial lift. To help select the proper equipment, the standing fluid level in the well is determined. First, a small explosion or other acoustic signal is created at the wellhead. The sound is deflected by tubing couplings. By counting these deflections (echoes), the fluid level in the well can be determined.

Bottom-Hole Temperature Survey

A *bottom-hole temperature survey* is normally made along with the bottom-hole pressure test. It determines the temperature of the well at the bottom of the hole or at some point above the bottom. First, a specially designed recording thermometer is lowered into the well on a wireline with the pressure gauges. After the thermometer is extracted, the temperature of the well at the desired depth is read from the instrument. These data and the bottom-hole pressure calculations are used to solve problems about the nature of the oil or gas. Temperature tests using different thermometers are sometimes helpful in locating leaks in the pipe above the producing zone. They are also used to determine whether gas-lift valves are operating, the location of top and bottom cement in newly cemented wells, and the injection interval in injection wells.

Flowing Wells

Only a small percent of the oil wells completed flow freely of their own accord. Gas wells are produced by pressure flowing through the formation. Some oil wells may flow naturally due to a driving force during the early stages of their productive life, but at some point before depleting they will require an external energy source. *Flowing oil wells*, like gas wells, only require a well-head assembly, or christmas tree, at the surface for well control (Figure 8–1).

Figure 8–1 Barton API 6A gate valves on a wellhead assembly used at the surface for well control of flowing wells (courtesy Barton Valve Company, Inc.).

Artificial Lifting Systems

The majority of producing oil wells require some means of *artificial lifting system* to bring the hydrocarbons to the surface. The method chosen depends on the depth of the well, the nature of the sands, and if it is a single completion or multiple completion. A multiple completion is a process by which it is possible to produce from different pay zones through the same well bore. This method uses packers and two or more tubing strings, to obtain the greatest amount of fluid with the minimum use of casing.

Surface Lifts

The oil wells drilled in the Drake era required some means of bringing the oil from its natural level in the wellbore up to the wellhead because the fluid did not flow. Thus lifting methods have developed concurrently with drilling methods ever since that point in history.

Drake and Uncle Billy Smith devised a *bailer* system which was both crude and slow, but sufficed for their production of 30 B/D. Other oil wells producing during that time used similar methods of bringing their production to the surface. It was almost immediately apparent that a better way was necessary. Today, lifting methods fall into two general categories: surface and subsurface.

Rod Pumping

Ten years after Drake's well was drilled, the rod pump system was in wide use in the new fields. Actually, all this involved was leaving the standard cable-tool drilling rig on the well site letting the walking beam furnish the up and down stroke for the pump. In the past, rod-activated pumps had been used on the salt brine wells, so the idea was not really new. Wooden *sucker rods* with wrought-iron fittings were used at first. The walking beam gave the reciprocating movement, which moved the rod up and down, activating the pump. This principle, although very old, has evolved into what has become known as the *standard pumping rig*. Like other oil-field equipment, it has been improved over the years. The rod pumping system is probably the most well-known and widely used artificial lift technique in the world.

A familiar sight in oil fields around the world is the *horse head* bobbing up and down on the *conventional beam pumping* unit (Figures 8–2a and 8–2b). This method of bringing oil to the surface (or a variation of it) accounts for approximately 80% of the artificial lifting of oil. Originally powered by steam, the pump may also use an internal-combustion engine or an electric motor. Those that are electrically powered require speed reduction between the motor and the pitman crank. This is usually accomplished by a combination of V-belt drive and gear reducer. A speed reduction ratio of 30 to 1 is necessary to operate the unit at 20 strokes per minute (spm) with a prime mover speed of 600 revolutions per minute (rpm). With one end of the pitman connected to the crank and the other end to the walking beam, the rotation is transferred to an up-and-down movement that is necessary to operate the sucker rods. A set of weights are attached to one end of the walking beam or to the crank

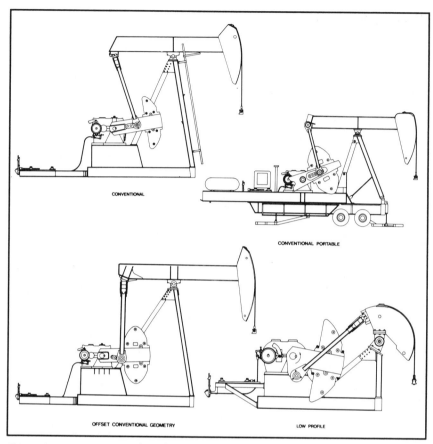

Figure 8–2a Conventional, and variations of conventional, beam pumping units (courtesy Lufkin Industries, Inc.).

that counterbalances the weight of the rods and fluid in the tubing. These weights are carefully positioned to help the motor lift the rods and fluid on the upstroke and increase efficiency. The sucker rod string is lifted by means of a cable looped over the horse head and connected to the carrier bar to support the polished-rod clamp.

Stuffing Box

The polished rod works up and down through a *stuffing box* which contains packing for sealing off the pressure inside the tubing to prevent leakage of liquid and gas outside the polished rod exit. Stuffing boxes are installed inside the upper side of the simple christmas tree that is assembled for use with a rod pumping unit. It consists of flexible material or packing housed in a fitting (box) which

Figure 8–2b This double exposure illustrates the action of a conventional beam pumping unit (© Gary Gibson, Yale, Oklahoma).

provides a means of compressing the packing. The stuffing-box packing must be replaced when it comes worn and loses its seal. The *polished rod* is connected to the sucker rods which in turn are connected to the downhole pump.

Sucker Rods

The heart of the standard pumping unit is the sucker rod, which does the actual job of lifting the oil to the surface. They are solid, high-grade steel rods that are run inside of the producing tubing string to connect a subsurface pump to the pumping unit. Sucker rods are made in various diameters from ½ to 1⅛ in., and are usually 25 to 30 ft long. Those less than 25 feet are called *pony rods*, or *sub rods*. Sucker rods having threaded pins on each end are known as the double-pin type. Those using a threaded box on one end are called box-and-pin type and need no coupling. Even though it is a rod, it acts much like a flexible spring and operates under great stress. Sucker rods can be easily damaged by improper handling. Bends, nicks, or dents can lead to metal fatigue and early failure. If the pump sticks, the rod may be stretched beyond its elastic limits and break.

Plunger Pumps

At the bottom of the sucker rod string is a piston or *plunger pump* submerged in the fluid of the well. They are cylindrical pumps equipped with plungers and valves. As the sucker rods are drawn up by the beam unit at the surface, the pump opens and allows the fluid to enter the pump chamber. As the pump closes, the fluid inside is forced up through a valve into the tubing, with another valve closing behind it to prevent escape back into the formation. Then the process is repeated. Slowly but steadily, the fluid is raised to the surface.

Plunger pumps may be generally classified according to whether they are connected to the tubing, which requires the pulling of the tubing to remove the assembly. These are called *tubing-type pumps.* The other type, in which the plunger and barrel assembly form a unit that fits inside the tubing and may be installed or withdrawn by the rods, is called a *rod pump.*

The operation of tubing-type and rod pumps are the same. The basic parts include the outer shell or working barrel, the plunger, a *standing valve*, and a *traveling valve* (Figure 8–3).

Air Balanced Beam Pumping Unit

Variations of the standard unit have included the *air balanced beam pumping unit* (Figure 8–4). This uses air pressure against a piston in a cylinder to counterbalance the weight of the fluid being lifted instead of a weight attached to the beam. This system allows for changing the counter-balance by merely adjusting the air pressure in the cylinder and it also eliminates much of the mass of the unit and its foundation. Otherwise, it is a rod pumping unit that can be installed as a replacement for any of the other units that use sucker rods and plunger pumps.

Another type of rod pumping unit has the crank mounted on the front and utilizes an upward thrust instead of a downward motion to raise the sucker rods. It is like the air-balanced beam pumping unit without the air cylinder. The gear box and crank are so arranged that the beam has a slow upward stroke and a fast downstroke.

Cable-Operated Longstroke Pump

The *cable-operated long stroke pumping unit* is another variation of the rod pumping unit. This unit, instead of having the standard walking beam, is mounted on a 50-ft tower erected over the well and has a 34-ft stroke, much longer than any standard unit. Since there are fewer cycles, the unit is supposed to lift more oil with fewer rod

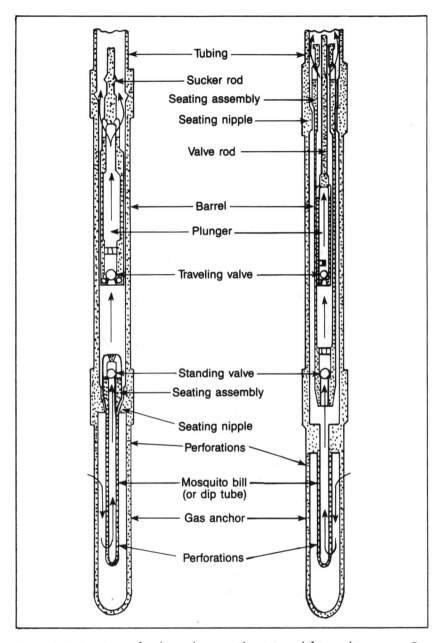

Figure 8–3 Two types of sucker rod pumps. The pump at left is a tubing pump. On the right is a stationary barrel, top hold-down rod type. The rod-type pump and barrel can be removed from the well without removing the tubing (courtesy PennWell Books, *Petroleum Production for the Nontechnical Person*, Gray).

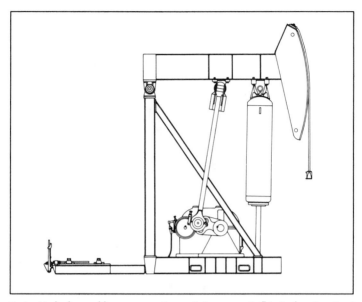

Figure 8–4 Air-balanced beam pumping unit (courtesy Lufkin Industries, Inc.).

failures. Because the stroke is longer and speed more uniform, a more viscous fluid can be lifted and the downhole efficiency is increased (Figure 8–5).

Other Rod Pumping Units

Often, new types of pumping units being put on the market often that are introduced as a new generation of energy-saving pumping units. Some use a winch-and-cable system that operates like a lightweight draw works to raise and lower the rod string in the well. One unit uses only 3 spm as compared to 12 to 15 spm on a well of similar depth. Another advantage cited is that the long, slow stroke of the unit extends rod string life and improves the efficiency of the bottom-hole pump, which results in reduced maintenance.

Energy is saved because the electric motor turns off on the downstroke. When the motor stops, the unit rolls to a stop and a cam starts the unit rolling in the opposite direction, then the motor restarts. Thus the motor only runs five-eighths of the time instead of full time. However, the surge electrical energy required to start the motor each time should be taken into account when figuring efficiency.

When long strokes are used with any of the rod pumping units, it is important to match the plunger pump with each unit.

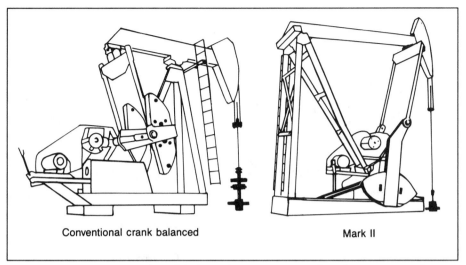

| Conventional crank balanced | Mark II |

Figure 8–5 Conventional crank balanced pumping unit and Mark II (courtesy Lufkin Industries, Inc.).

Subsurface Lifts

An *electrically-powered submersible pump* consists essentially of a centrifugal pump with the shaft directly connected to an electric motor. The entire unit is cylindrical and is sized to fit inside the well casing. It is connected to the tubing and has an insulated cable attached to the outside of the tubing. The submersible equipment and cable are lowered into the well as the tubing is being run in the hole. The cable is attached to the electrical fittings of a control box on the surface. The pressure created by the rotation of the pump's impellers then forces the fluid to the surface through the tubing.

The electrical submersible pumps often used where it is necessary to pump large amounts of fluids. In cases where the water-to-oil ratio is high, these pumps are ideal. A typical submersible pump may lift from 250 to 26,000 B/D, depending on the size of the casing and the depth of the well (Figure 8–6).

When first introduced, there were problems from failure of the insulated power cable. After new and better insulation was developed for this type of service, reliability was improved. Today the electrical submersible pump is used throughout the world in petroleum production and is used in other industries as well.

Subsurface Hydraulic Pumping

Subsurface hydraulic pumping is another method of pumping oil from the formation using a bottom-hole pump without sucker rods.

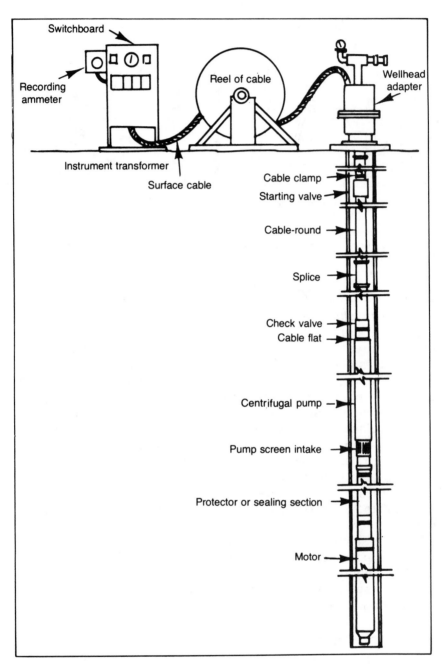

Figure 8–6 Centrifugal pumping unit.

Unlike the electrical submersible pump, hydraulic power and piping is used instead of electricity and wiring. The subsurface hydraulic pumping system uses a bottom-hole production unit consisting of a hydraulic motor and a pump directly connected to it. Power at the surface is supplied from a standard engine-driven pump (Figure 8–7).

One system uses two strings of tubing placed along side one another. The power oil is pumped down in one and the produced oil and the power oil is pumped to the surface through the other. Another system has a small string of tubing inside the other. Clean crude oil is drawn from the top of a settling tank at the surface and a high-pressure pump forces it down through the larger tubing to become the power oil. The power oil goes to the engine and moves a power piston connected to the production plunger in the bottom-hole pump. Well fluid and the power oil become mixed and return to the surface storage through the smaller tubing. The ratio of fluid to power oil is usually one to one.

One advantage of the subsurface hydraulic pumping unit is that it may be used to pump several wells from a central source. The maximum capacity of such equipment is dependent upon size of the tubing and the ability of the well to produce. The system has been successfully used to lift fluid from depths greater than 15,000 ft.

The *casing-type hydraulic pumping unit* requires only one string of tubing set on a casing packer, and the power oil travels down the tubing string. The power oil and the production fluid mixes and returns in the casing tubing annulus.

Gas Lift

In fields where a large supply of natural gas is available and the amount of fluids expected to be recovered justifies the cost, gas lift may be used (Figure 8–8). Natural gas is introduced into the well in the annular space between the tubing and the casing. A series of gas-lift valves are placed along the tubing, and these close as the gas enters the next lowest valve. As the gas enters the production zone, it aerates the thick oil and lightens it so that is moves more quickly to the surface as the gas expands.

If the well is a good producer and is able to maintain a column of fluid above the point of injected gas, the well is under *continuous-flow gas lift.* In other wells where considerable time is needed for fluid to build up in the tubing, gas is injected into the well in batches that brings the fluid to the surface in slugs. This is an *intermittent-type gas lift system.*

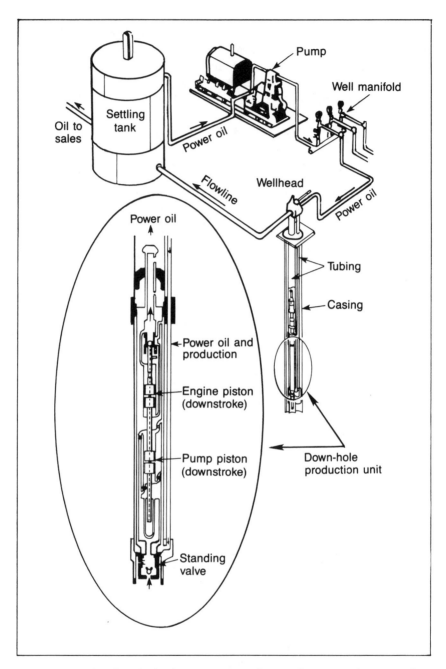

Figure 8-7 Subsurface hydraulic pumping installation. The power oil is pumped down the large string and a mixture of power oil and produced oil is pumped to the surface through the small string. The pump may be removed from the bottom of the well by reversing the direction of flow (courtesy PennWell Books, *Petroleum Production for the Nontechnical Person*, Gray).

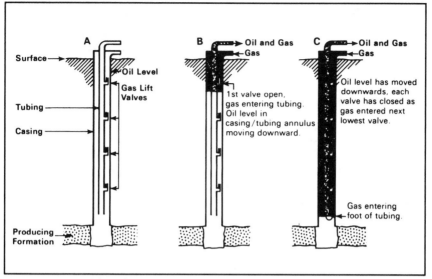

Figure 8–8 Gas-lift artificial lift.

Other Submersible Lift Systems

Other methods of lifting apparatus that have been tried include the *sonic pump*, the *ball pump*, and *plunger lift*. The sonic pump, as with the gas lift, is a series of valves installed in the tubing string which is suspended from a wave generator. Oscillations are produced that proceed down the tubing at the speed of sound (as it travels in metal). This results in resonation in areas of the tubing, causing it to expand and contract. This makes the check valves open and close, lifting the fluid to the surface through the tubing.

The ball pump utilizes two parallel strings of tubing through which a synthetic rubber ball is circulated as a means of lifting the oil to the surface through the production tubing.

The plunger lift uses a steel plunger, or swab, inside the tubing. Compressed gas is forced down the annulus between the casing and tubing and enters the tubing through a gas inlet valve at the bottom of the tubing and at intervals upward. The pressure of the gas forces the plunger to rise and a column of fluid is lifted to the surface where it is discharged into the flow line. Gravity then pulls the plunger to the bottom for another load. A valve mechanism controls gas input to the casing that makes the operation of the plunger automatic. This system has the advantage of keeping the tubing free of paraffin.

Which Method to Choose?

Each method of lift has its advantages and disadvantages (Table 8–1). The best approach is to review all the well testing data acquired and then use it to make a judgment. If no well testing was made, the operator of the well or field will have to weigh the information available, and decide what would work best for that situation.

Pumping Off

The term *pumping off* as used in Table 8–1 means that usually there will only be so much oil available at a given well. For example, a pump may be timed to run for 4 hours at a time. During this period, the oil at the bottom of the wellbore will have been pumped out. A rest period for the pump will then follow while a new supply of oil flows through the formation to the bottom of the wellbore. At a preset time, the pump will restart for another pumping cycle. If the pump were allowed to run continuously, the flow of oil through the formation could be stretched so thin that salt water could intrude and isolate pockets of otherwise recoverable oil. Thus, much of the total oil in place would be trapped in the formation, and the pump could also be damaged.

How long to operate each pumping cycle should be calculated by the reservoir engineer, who would normally be involved in making the potential test and other well tests. Such recovery methods were largely unknown, or ignored by, early operators who worked hard to produce as much oil as they could as rapidly as possible. Some regulating bodies set arbitrary production figures that were not based on sound scientific or engineering principles which, unfortunately, depleted wells prematurely. Many fields, therefore, were pumped dry and abandoned, when in reality they still contain a great deal of oil. On the other hand, where no excessive pumping methods were used, many wells are still maintaining a satisfactory daily level of production after years of operation.

When the daily level of production falls below 10 bbl, the well is usually called a *stripper well.*.

Primary Recovery

So far discussed have been the usual methods of primary recovery. This is the initial production phase of hydrocarbons from a well or field. During this phase of operation it can be expected that approximately 25% of the oil in place in the reservoir can be recovered. Primary recovery is the easiest phase of getting crude oil and natural gas from the formation to the surface because of the initial driving

Method	Advantages	Disadvantages
Free Flow	Produces by natural reservoir drive. No artificial lift necessary.	—
Standard Beam Rig	Proven and improved over many years of use. Motive power available from gasoline, diesel fuel, natural gas, or electricity. May be the cheapest lift, and its operation is familiar to field personnel. Pumps can usually handle sand or trash. It is possible to pump off.	Sucker rods subject to failure. Unit required at each wellhead on lease. Volume decreases as depth increases. Rods must be pulled to change pump. Unit is susceptible to free gas in pump.
Hydraulic Pumping	High volume can be produced from great depth. Power equipment on the surface can be centralized. Pumps can be changed without pulling tubing. It is almost possible to pump off.	Susceptible to free gas in pump and is also vulnerable to solids in pumps. Well testing is difficult because power oil is also in well stream. This also causes oil treating problems.
Submersible Electric Pumping	Very high volumes at shallow depth can be produced. It is almost possible to pump off.	Maximum volume drops rapidly as depth increases. Very susceptible to free gas in pump. Tubing must be pulled to change pump and cable. Unit required for each well.
Gas Lift	Takes advantage of the gas energy in the reservoir. It is a high-volume method. Can easily handle sand and trash. Valves may be retrieved by wireline or hydraulically.	A source of gas must be available. It cannot pump off. Minimum bottomhole pressure increases with depth and volume.

Table 8-1 Choosing a Method of Lift

forces present. This was evident when the first real gusher in the United States blew in a Texas swamp called Spindletop.

The recovery of petroleum fluids depends on the ability of those fluids to flow through the formation to the wellbore, and this of course is in turn dependent on the porosity and permeability of the formation and the driving force behind the fluid.

Enhanced Oil Recovery (EOR)

After the *primary recovery* period of a reservoir is over, the area may not be abandoned as was done often in the past. More efforts are concentrated on how to produce all the petroleum fluids possible through the techniques of secondary and even tertiary recovery. Not all *enhanced oil recovery (EOR)* efforts are delayed until primary recovery has peaked out. Some fields have EOR systems underway immediately after the reservoir is tapped and reserve estimates made. One large field in the Middle East began injecting natural gas into the reservoir soon after production began.

Secondary Recovery

It is estimated that more than 100 billion bbl of crude oil may still remain in depleted reservoirs in the United States alone. *Secondary* and *tertiary-recovery* techniques can increase or maintain the present levels of production for many years to come. *Horizontal drilling* in nearly depleted reservoirs has sparked new hope in aiding secondary recovery.

Secondary recovery actually consists of replacing the natural reservoir drive or enhancing it with an artificial, or induced, drive. Generally the use of injected water or natural gas into the production reservoir is the most common method. When water is used it is referred to as *waterflooding.*

The first known waterflooding was by accident; an abandoned oil well was being used as a disposal salt water well when it was noticed that production of nearby wells was increasing as more water was being dumped. Some of the first waterflooding was accomplished by drilling a well, or a series of wells, on the perimeter of the reservoir and injecting water under pressure (Figure 8–9). Another method was to drill a well in the center of a four-well producing location and inject water in the center well, known as a "five spot" waterflooding system. There are many other systems of locating injection wells that are engineered to fit a particular field.

Conventional secondary recovery methods do not completely

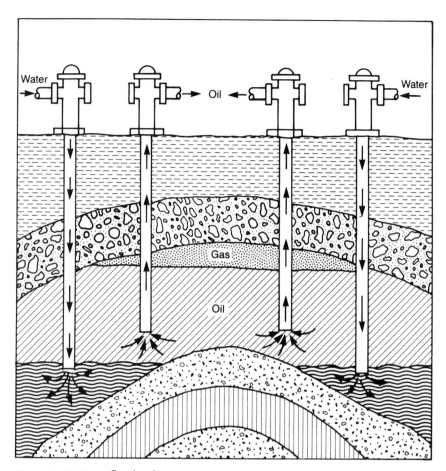

Figure 8-9 Water-flood technique.

deplete the reservoir. At or near the end of their effective life, tertiary recovery may be used to recover more oil from the reservoir.

Tertiary Recovery

Tertiary recovery methods go ahead where secondary recovery methods leave off. They are usually divided into three major categories: *thermal, chemical,* and *miscible displacement.*

Thermal processes include *steam stimulation, steam flooding,* and *in-situ combustion.* Steam stimulation involves injecting steam into a producing well for two or three weeks. The well is shut in so the heat can dissipate and transfer to the oil. As the oil warms up, it begins to flow more easily, increasing production. Steam flooding is similar to waterflooding. Input wells are located in a pattern for injection while the oil is produced from adjacent wells. The injected steam or hot water forms a saturated zone around the input well. As the steam moves away, its temperature warms the oil, moves it, and leaves water to displace the vacant pores.

In-situ combustion (also called fire flooding) involves igniting an air-injection well. It is usually applied to reservoirs that have low gravity crude oil. Heat is generated within the reservoir by injecting air and burning part of the crude oil. The combustion front moves away from the well, heating the oil so it moves more easily and pushing it toward a producing well. This method can be transferred back and forth between several wells until the reservoir is depleted (Figures 8–10, 8–11, and 8–12).

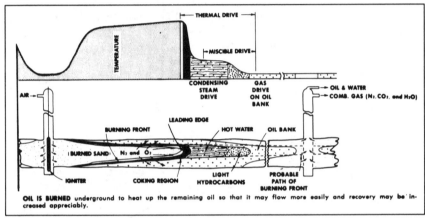

Figure 8–10 In-situ combustion (courtesy *Oil & Gas Journal*).

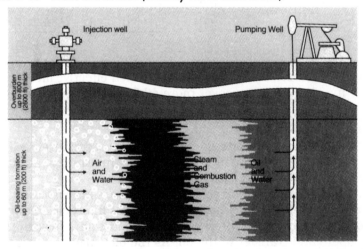

Figure 8–11 In-situ recovery process in the oil sands is based on principles similar to those used in the recovery of heavy oil. Heat thins the oil, enabling it to move toward producing well bores (courtesy Anne McNamara, Petroleum Resources Communication Foundation).

Modern Petroleum

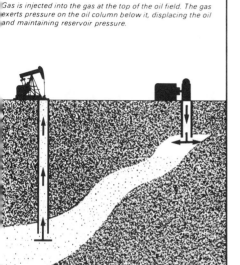

Gas Injection:

Gas is injected into the gas at the top of the oil field. The gas exerts pressure on the oil column below it, displacing the oil and maintaining reservoir pressure.

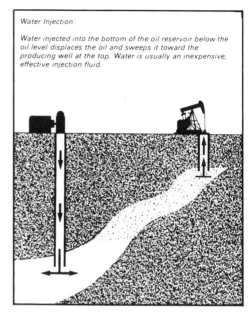

Water Injection:

Water injected into the bottom of the oil reservoir below the oil level displaces the oil and sweeps it toward the producing well at the top. Water is usually an inexpensive, effective injection fluid.

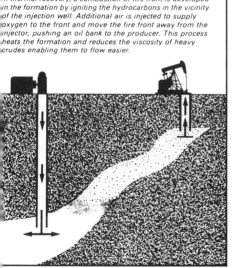

Thermal Recovery

In thermal recovery, a combustion or fire front is developed in the formation by igniting the hydrocarbons in the vicinity of the injection well. Additional air is injected to supply oxygen to the front and move the fire front away from the injector, pushing an oil bank to the producer. This process heats the formation and reduces the viscosity of heavy crudes enabling them to flow easier.

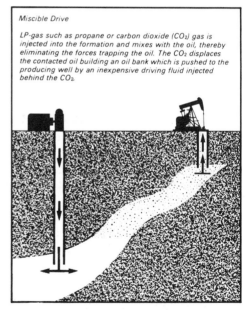

Miscible Drive

LP-gas such as propane or carbon dioxide (CO_2) gas is injected into the formation and mixes with the oil, thereby eliminating the forces trapping the oil. The CO_2 displaces the contacted oil building an oil bank which is pushed to the producing well by an inexpensive driving fluid injected behind the CO_2.

Figure 8-12 Secondary and tertiary recovery techniques (courtesy Canadian Petroleum Association).

Production

4-D Seismology

The new technology of 4-D seismic technology can aid in enhanced recovery. By comparing seismic measurements acquired at different moments in time (the fourth dimension) pore fill changes in reservoirs can be monitored during depletion. By taking into account the large changes in compressional velocity in oil as a function of temperature, thermal secondary recovery may be monitored.

Chemical processes include *surfactant-polymer injection, polymer flooding,* and *caustic flooding.* In the surfactant-polymer process, a surfactant slug is injected, then a polymer mobility slug is injected as a buffer. This process is repeated, providing a cleansing agent and then a moving agent. The polymer flooding process adds a thickening agent to the water in a waterflood system. The polymer reduces the volume of water required and increases sweep efficiency. Water always follows a path of least resistance, i.e., it doesn't spread out but tends to form rivers or channels. The polymer enhances water's sweep. Caustic or alkaline flooding changes the pH of water, making it more acidic. This helps open passages in the pores and increases production.

Miscible-displacement processes include *miscible hydrocarbon displacement, carbon dioxide injection,* and *inert gas injection.* Miscible hydrocarbon displacement uses a solvent that is injected and followed by an injection of a liquid or gas moving agent. The carbon-dioxide injection process mixes CO_2 with the crude oil. After a proper period, the CO_2 forms a miscible front that pushes the oil to the production well. Inert gas injection is usable for only a select few wells. An inert or almost totally unreactive gas is added to push the oil through the formation.

Microwave Technology

In 1990 it was reported that the underground application of electromagnetic microwaves in the Canadian heavy oil reserves near Leominister, Alberta increased production by 70% by heating the oil to 200°F. It was estimated that this technology could recover some 10 million bbl. The process promises to have the potential of being non-polluting when compared to the "huff and puff"method of steam injection, which produces considerable amounts of dirty water that must be treated before it can be disposed of.

Microbial Enhanced Oil Recovery (MEOR)

Research has been underway for more than a decade on various methods of using living microorganisms *(microbial enhanced oil recovery MEOR)* to enhance oil recovery. One method calls for the injection of bacteria and nutrients into the formation. The bacterial action results in the decomposition of part of the oil in place into detergents, CO_2, and new cells. These products then either mechanically or chemically release oil from the reservoir pores.

Researchers in Australia indicate that the use of MEOR can increase oil production from 10 to 20%. In tests conducted on low volume wells, production increased as much as 50%. Proponents of MEOR also point out that very little equipment is necessary and the cost is extremely low in comparison to other EOR methods, some of which add greatly to the cost of the recovered oil.

All of these processes are undergoing study and experimentation, and will be for years to come. Presently, they each apply to specific kinds of reservoirs, and they do not cleanse the formation of 100% of the hydrocarbons present.

Field Processing

Most of the petroleum produced from a reservoir requires processing. Crude oil and natural gas are complex mixtures of many different densities, vapor pressures, and other physical characteristics. Some well streams are a turbulent, high-velocity mixture of gases, oil, free salt water, salt-water vapor, solids, and impurities. As the stream flows, it undergoes continuous pressure and temperature reduction because it is leaving a high-pressure reservoir that is hotter than the Earth's surface and is coming in contact with lower temperatures and pressures as it moves upward.

Petroleum mixtures are very complex and are often difficult to separate easily and efficiently. As the well depletes and loses its driving force, more changes are apt to take place, and field processing equipment may have to be replaced to meet the change. One example is where a well initially has a high gas pressure, no salt water, and good oil production. As the well depletes, it loses practically all its gas pressure, and produces more salt water than crude oil. There is no longer any gas to fire the heater-treater or to fuel the internal combustion engine to pump the well, so it is obvious that replacement and changes are necessary.

When the oil or gas comes out of the wellhead, either by an artificial lift method or by free flow, it enters a flow line and is carried to the header. The header, which consists of different types of valves and fittings, is the junction connecting all the flowlines in a given area. It is at this point the oil and gas enter the separator.

Separators

Separators are closed vessels that remove natural gas from oil and water, and oil from water. The simplest form of separator is a closed tank in which the stream from the header enters the boot and the force of gravity separates the gas, oil, and water. The force of gravity has a greater pull on water because it is heavier, so it settles to the bottom of the tank. Gas is lightest and moves to the top of the tank and leaves through the gas outlet piping system for further handling. The oil, being lighter than the water will float on top of it. Thus separation is caused by the differences in *specific gravity* (Figure 8–13). Gravity separation does take time, and especially so when the ambient temperature is very cold. Often a chemical is added to the production to aid in separation. Slow separation may not be a significant factor for low-producing wells. However, some high-production wells may need special equipment for efficient separation.

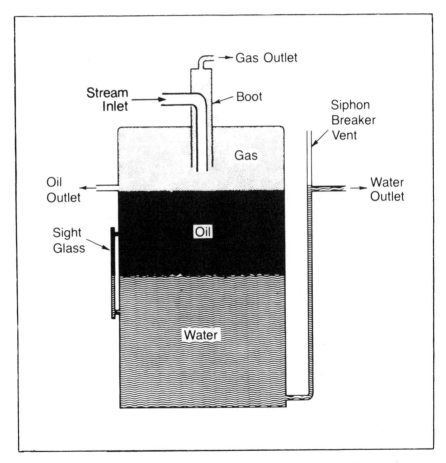

Figure 8–13 Oil and gas gravitational separator (courtesy PennWell Books, *Gas Handling and Field Processing*, Vol. 3, Berger and Anderson).

Separators are usually classified in two ways: the shape or position of the vessel, and the different number of fluids segregated. The three common vessel shapes are vertical, horizontal, and spherical. The number of fluids to be segregated is either two or three. The separator is referred to as a two-phase type when only gas and liquid are segregated. The three-phase type segregates gas, oil, and water.

Vertical separators may be either two-phase or three-phase, depending upon the field requirements (Figure 8–14). Vertical separators are often used in low to intermediate gas-oil ratio wellstreams. They are also used where large slugs of liquid are expected and liquid level control is not critical.

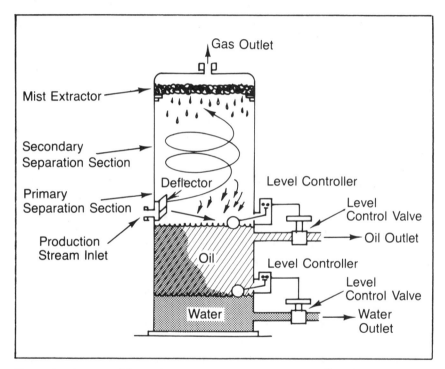

Figure 8–14 Vertical three-phase separator (courtesy PennWell Books, *Gas Handling and Field Processing,* Vol. 3, Berger and Anderson).

Vertical separators are normally equipped with an inlet diverter that causes the incoming fluids to swirl in a centrifugal motion, providing the proper momentum reduction to allow the gas to escape. The liquid falls to the liquid accumulator at the bottom and the gas rises to the top. Separation is not complete yet because some of the small liquid particles are swept upward with the gas stream. These small liquid particles are separated by either a baffle arrangement of knitted-wire mesh pad or a vane-type extractor positioned near the top, which permits the liquids in the accumulator to be separated by the differences in their specific gravity.

A *vertical separator* is much taller than the horizontal or spherical but occupies less foundation space. It handles large quantities of sand or other solid impurities and is easier to clean out. One disadvantage is that it is larger than the others of the same capacity and costs more.

Horizontal separators may be two-phase or three-phase and have a greater gas-liquid interface area that permits much larger gas velocities than other types (Figure 8–15). They are generally used for high gas-oil ratio wellstreams and are more efficient and economical for processing large volumes of gas. They are also used for handling foaming crude or for liquid-from-liquid separations.

The horizontal separator is easier to transport and connect with piping and fittings. It is more adaptable for skid mounting, and can be stacked for stage separation, such as on offshore platforms.

The interface area of the horizontal separator consists of a large, long, baffled, vane-type gas separation section. Gas flows horizontally and at the same time slants toward the liquid surface. The moist gas flows in the baffle surfaces and forms a liquid film which is drained away to the liquid section of the separator.

Horizontal separators of the two-phase, or double-barrel, design are used in some locations where a large amount of free liquid is in the wellstream. It has all the advantages of a conventional horizontal separator plus a higher liquid capacity.

Spherical separators are more compact than the vertical or horizontal separators, and make maximum use of all the known methods of oil and gas segregation. They usually cost less than the other two. The disadvantages are that the surge capacity is limited, and it is not as economical for larger gas capacities.

It is often desirable to use more than one stage of separation in order to obtain more complete segregation of liquids and gases. Each location or situation will indicate which shape separator will best fit the installation. When two-stage separation is used, the mixed stream of oil and gas is passed through a separator where the first separation of oil and gas is handled the same as it is with only one separator. This first separator will lower the pressure of the liquid that goes to the next separator because of the gas being discharged. The liquid from the first separator is then sent into a second separator where more gas is separated and discharged for

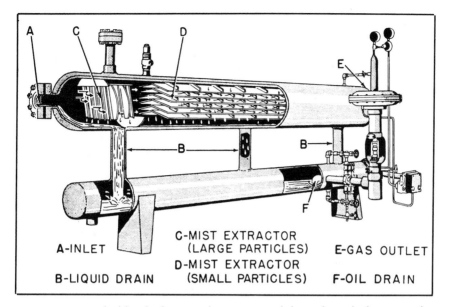

A-INLET	C-MIST EXTRACTOR (LARGE PARTICLES)	E-GAS OUTLET
B-LIQUID DRAIN	D-MIST EXTRACTOR (SMALL PARTICLES)	F-OIL DRAIN

Figure 8–15 A double-tube horizontal separator. Oil drops down the bottom and goes to the stock tanks. Gas leaves through the gas outlet (courtesy PennWell Books, *Gas Handling and Field Processing, Vol. 3*, Berger and Anderson).

further handling. Each stage removes more gas and results in a more complete separation.

Oil Treating

After leaving the field separator, some crude oil is emulsified and requires further segregation. *Emulsions* are mixtures of oil and salt water that appear like muddy-brown foam. They may not exist in the production formation, but are caused by agitation as the wellstream flows through all the piping and equipment. A great number of very small spheres of water are surrounded by a tough film caused by the difference in surface tension between the water and oil. The film must be destroyed in order for the water to drop out so that the oil is ready for sale and transportation. Saleable quality usually means oil containing a maximum of 1% *BS&W (basic sediment and water)*.

Certain actions tend to weaken the film so that the droplets can come together to form larger drops that will settle to the bottom. Heat weakens the film, as do certain *demulsifier* chemicals. After the emulsion is broken, settling time is required to complete the treating method. Some emulsions can be destroyed with chemicals and settling, while others can be destroyed by heat and settling time.

Chemical treatment is normally done by chemical injection equipment that applies the chemical downhole, in a flowline, or by batch treating.

Treating emulsions with heat requires special independent equipment usually called a *heater-treater*. These are usually either constructed as vertical vessels or horizontal vessels. Flames at the bottom of the equipment (usually from natural gas) heat the emulsion. The oil rises to the top and is carried to settling tanks. The water to the bottom and flows through pipelines to salt water tanks.

Another method of treating is done with electrostatic treaters. They are similar to horizontal heater-treaters except that high-voltage electric grids are added. Electricity is sometimes an effective means of breaking emulsions.

Most field separation and procession equipment will operate under pressure. The equipment is usually constructed of steel, and seams are welded. It is built according to strict pressure vessel specifications. Pressure relief valves or other suitable protection is provided for safety. Severe corrosion can be a problem because of the salt water present and its effect on steel. Heat is another factor that will accelerate corrosion. What makes this more difficult to cope with is that it is both internal and external corrosion. Corrosion-resistant materials or other means of mitigation are used whenever possible.

On–Stream

After the well has been drilled, completed, tested, and properly equipped, it is ready to go on-stream to join the thousands of oil and gas wells that are already producing. Each new well has an effect on the daily production of the world, no matter how insignificant it may appear. The world's daily production changes constantly as some wells are depleted and new ones brought on-stream. Any numbers quoted will only be approximate but will serve to give a broad idea of current worldwide crude oil and gas production.

Operation and Maintenance

When the well has been brought into production, the important job of operating, maintaining, and repairing it, along with replacing any needed equipment, will commence. Just how this job is performed depends a great deal upon who owns the well.

In many instances, when a well or group of wells is owned by individual investors, operation and maintenance are contracted out. For example, a contract *pumper* will operate the artificial lift and field processing equipment, gauge the production tanks, and make minor repairs. When other routine maintenance and repair is necessary, a contract *roustabout crew* is hired. If major work must be done, it is common practice to contract with companies that do well servicing and workover jobs.

When an established oil company owns the producing wells, they usually have a complete production department to operate their leases. Some of the field operating personnel are pumpers, roustabouts, mechanics, engineers, corrosion specialists, and supervisors. Each person has their own responsibility to keep the oil and gas wells producing around-the-clock, regardless of the weather or any other factors.

The pumper, or lease operator, checks the wells and field equipment, and gauges the tanks on a daily basis. Any change in normal production will be observed by the pumper first, because he is charged with close surveillance. The pumper will notify the production foreman, who in turn will take the appropriate action to correct any problem.

A head roustabout, *gang pusher*, or leader, directs each crew of roustabouts. It is the head roustabout's job to be sure the needed tools and equipment are obtained, and to do the work in a proper manner. Written procedures are sometimes provided where specialized tasks must be undertaken. The *head roustabout* takes responsibility for completion of the job and is also responsible for crew training and safety.

The roustabouts have a special *gang truck* that serves an assortment of needs, including transportation of personnel, hauling tools, and light duty hoisting. It is also equipped with first aid kits, fire extinguisher, safety apparatus, drinking water, and a two-way mobile radio.

General supervision of all operation and maintenance work in the field or district is under the control of a field superintendent. Plans for drilling and work-over operations, tank-battery plans and construction, and other specialized work are carried out by a field superintendent through personnel working in the department. Technicians, electricians, mechanics, corrosion specialists, wireline operators, and personnel of associated skills all coordinate their jobs with the field superintendent's office.

Equipment Failure

One of the most common production problems is equipment failure. Wear and tear has an effect on any piece of equipment that is under constant use. Packing wears out and leaks, pipe threads leak, bearings seize, valves break, belts slip, bolts break, paraffin builds up, and metal is subject to fatigue.

Special Problems

Most special problems associated with oil and gas production are directly associated with water being present in the fluids. This water is usually salty or acidic, or both, so that corrosion will occur when contact is made with the metal equipment in use.

Salt water produced from oil wells often contains many compounds that leave deposits in the form of scale inside tubing, or in surface equipment. This scale deposit restricts flow and causes other problems that result either in replacement or difficult descaling operations.

As many oil fields have grown older, more and more water has been produced, sometimes in large quantities. This water must be disposed of in a manner that will not damage land or pollute subsurface fresh water formations or other bodies of water.

Corrosion is another costly problem to be dealt with both in the wellbore and in surface equipment. Corrosion attacks both the outside and inside of casing and tubing as well as the outside of rods. Research has produced many chemicals that reduce the rate of downhole corrosion. However, no one type of chemical or group of chemicals can be completely effective. There are many different conditions that exist in various producing areas that each must be dealt with separately. A common impressed current system using rectifiers or a sacrificial anode system of anodes may be used to protect the external or outside of the casing, the internal or inside of it must be protected by chemicals. Cathodic protection and corrosion are covered more in detail in Chapter 15.

Sanding

Sand is usually produced with fluids from wells that produce from loosely consolidated sandstone formations. A small portion of the loose sand will be produced through the well head at the surface, but most of it will accumulate at the bottom of the wellbore. Accumulation of sand can reduce or even halt production. When sanding occurs it is necessary to use a pulling unit to pull the rods, tubing, and downhole pump, and then clean the hole out with a special type of bailer. If the sanding problem continues, corrective measures might be necessary, such as gravel-packing the well.

Subsurface Repairs

A pulling unit is required to pull rods when the rods part, or the down hole pump needs repair. The unit is also used to retrieve submersible pumps for servicing. Most pulling units are truck mounted, except for trailer-mounted special equipment used for servicing deep wells. Truck-mounted pulling units have a separate engine to power the winches and raise and lower the mast.

Workover

Workover operations are different downhole repairs than the normal subsurface repairs that require only a pulling unit. The equipment used is much larger and resembles a rotary drilling rig. It can either be truck-mounted or trailer-mounted.

Workover operations are usually necessary to repair mechanical failures of the primary cement job, casing, tubing, and packer leaks, and other bottom-hole problems.

Serious production impairment can be caused by primary cement failure, such as salt water invading a producing zone behind the casing and between the bad cement. Production fluids could flow through the faulty cement into a lower pressure zone, causing a drop in production. Indications of cement failures might include sudden increases in salt water production or an irregular drop in production rates.

More common downhole problems are casing, tubing, and packer leaks which can be caused by many sources including pressure, erosion, and corrosion.

Chapter References

Neal Adams, *Workover Well Control*, (Tulsa: PennWell Books, 1981).

Forest Gray, *Petroleum Production for the Non-Technical Person*, (Tulsa: PennWell Books, 1986).

Guntis Moritis, "CO_2 and HC Injection lead EOR Production Increase," *Oil & Gas Journal*, April 23, 1990.

9

Storage
PART 1

The Types of Tanks and Vessels Used and Their Construction.

Well fluids must be separated and treated before being sent on to a refinery or gas processing system. This first step in the handling of production usually takes place at a tank battery located near the wellhead, or at a location where the production from several wells is treated. This *lease tank battery* system is the first set of tanks and vessels that the crude oil, salt water, and natural gas will flow through (Figure 9–1).

Figure 9–1   A typical lease tank battery (photo by Ken Anderson).

The typical tank battery will be made up of one or more tanks, a separator, and often a heater unit. The more common and simplest form of an oil and gas separator is a vertical tank with a *boot* out its top, in which the force of gravity is used to separate the oil, gas, and salt water. This tank is called a *gravitational separator*, or, in some areas, a *receiver*. Salt water is heavier than crude oil or gas and settles to the bottom of the tank where it maintains a certain level in the separator, with the excess water flowing into a saltwater tank. Crude oil is lighter than water, so it floats on top of the saltwater, maintaining a certain level while its excess flows into an oil tank or tanks. The natural gas is lighter than either the water or oil, so it flows out of the boot into the gas-gathering system.

Gravity separation takes time, a factor which must be considered when treating fluids from wells with high production rates. Because of this, other types of separators are generally used to service wells with higher production.

After natural gas has been separated from the oil, and the crude oil has been separated and treated to remove the *basic sediment and water (BS&W)*, the oil goes to stock tanks. These are commonly referred to as the *tank battery* or *production tanks*. They will vary in size and number depending on the daily production of the lease and the frequency of pipeline runs. The use of *lease automatic custody transfer (LACT)* units and their acceptance by pipelines and producers has reduced storage requirements. The total storage capacity of a lease is usually three to seven days' production.

Tanks may be classified in various ways: first, by the way they were constructed, then whether they are to be used for production or storage, and finally by the type of liquid they are to contain: crude oil, fuel oil, gasoline, etc.

Production Tanks

Production tanks are used in the producing fields to receive crude oil that has been separated from the well fluids. They are also called *tank batteries, flow tanks*, or *lease tanks*. These tanks are used to receive the oil from individual wells and leases so that it can be measured and tested and the quantity and quality of the oil produced accounted for to the producer, royalty owner, purchaser, and the various governmental regulatory and taxing agencies. Production tanks represent the starting point for oil entering pipelines. This is also where the *pumper* turns his responsibility for the oil over to the *gauger*.

Production tanks may be constructed of bolted steel plates, welded steel plates, fiberglass, or wood.

Bolted Production Tanks

Bolted steel production tanks were developed to provide low-cost tanks that could be shipped knocked-down to areas where transportation is difficult, and then be erected in a short time by a field crew. Their design makes dismantling and subsequent re-erection at a new location readily possible. Since these tanks are manufactured to API standards, it is possible to interchange parts of tanks of the same diameter even though they may have been made by different manufacturers.

All of the joints in bolted steel tanks are secured against leakage by the use of rubber or composition gasket materials which are inserted in the joints. Available in capacities ranging from 100 to 10,000 bbl, bolted tanks are being replaced by all-welded tanks.

Welded Steel Production Tanks

Welded production tanks are the welded counterpart of bolted production tanks. The smaller sizes are usually shop-fabricated and then hauled to the lease. Larger size tanks are erected on site under field welding conditions. They are superior to bolted tanks in many ways and are less likely to leak.

Large field-erected steel production tanks do not need specially formed plates, but are constructed in the same manner as storage tanks. *Bottom plates* normally go to the site squared, as they are received from the mill, and are cut to shape on the job with an acetylene cutting torch. *Shell plates* are rectangular, with their larger dimension directly welded horizontally to the bottom plates. *Roof plates* may also be cut on the job and welded to the shell either with or without an angle ring at the joint. Roofs are usually cone-shaped with the proper hatches and pipe connections welded in place.

Wooden Production Tanks

Wooden production tanks are being replaced with steel or fiberglass tanks. Some are still in service in areas where they were erected to handle very corrosive sour crude oil. They are commonly made of redwood or western red cedar. Cast iron is used for the tank flanges, relief valve, and gauge hatch connections. These tanks could be made vapor-tight.

Fiberglass Production Tanks

Fiberglass production tanks are constructed in fabrication plants and then transported to the oil lease site. This puts a limit on their physical size. Tanks of a size that can be hauled by truck are in use where corrosion is a problem.

Storage Tanks

Storage tanks are designed for the storage and handling of large volumes of oil. These tanks are usually much larger than production tanks and are considered to a more permanent tank. Two general types are employed for oil storage: riveted and welded.

Riveted Tanks

Riveted steel tanks have been replaced by welded tanks in practically 100% of new construction. Since a few riveted tanks are still in service, and are likely to remain so for some time, they merit discussion here.

Tank size is one great difference between riveted tanks and welded tanks. Riveted tanks measuring 168 ft in diameter by 64.7 ft in height are considered large. By contrast, welded steel tanks measuring 400 ft in diameter are now in service.

Another principal difference between the two is in the manner of assembly. When building a riveted tank's bottom, it is necessary for the workers to reach the underside during riveting operations. To do this, the bottom was held above grade level by wooden supports which had to be spaced over the tank's bottom surface area so as to properly support the bottom plates. These supports remained in place until the first shell course was erected and tested. Then the wooden supports had to be carefully removed as jacks lowered the bottom to grade. This was a procedure that was both dangerous and time consuming.

Welded Storage Tanks

The art of fusion welding was developed during the 1930s. It improved rapidly, with the result that welding began to supercede riveting in tank construction. Welding offered definite advantages in tank construction, such as fewer joint leaks and permanent tightness. The industry requested specifications for *welded tank construction* from the API, which in 1936 issued the first edition of API Standard No. 12–C. This standard gained immediate acceptance and has enjoyed ever-widening popularity.

Tank Grade

A satisfactory tank foundation or *pad* is one which will provide a stable, well-drained surface with sufficient load-bearing capacity to support the tank and its contents. The foundation may be flat, or it may be *coned up* or *coned down*. Normally, the top layer of the pad under the tank is made of oiled sand, approximately 2 to 6 in. thick. Some pads have concrete rings which enclose and stabilize the entire area and provide a level surface for the tank shell. These concrete rings compensate for the extra load of the tank shell which rests on them, and they also assist in distributing the tank load onto the supporting soil.

It is important for the long-term life of the tank that its foundation be fully capable of supporting the tank and its full product load. In areas where the soil conditions are not known or are suspect, soil investigation and enhancement programs may be necessary. Soil samples are taken and analyzed to determine what improvements are required, and the appropriate methods to achieve these improvements. Often, it is necessary to compact the existing soil, or excavate non-conforming soil and replace it with a suitable backfill. In areas with extremely poor soil, a pile-driven foundation may be required.

The grade should ensure that the bottom exterior surface of the tank bottom will be kept dry to minimize tank bottom exterior corrosion.

Erection of Welded Storage Tanks

Once a suitable grade is prepared, the next step is to assemble the tank bottom. The rectangular steel bottom plates are laid out on the grade with their edges overlapping slightly. They are then joined temporarily with tack welds to hold them in place while the first course of shell plates are set in position and tack welded. The vertical joints of the first course are then welded followed by completion of the weld joint between the first course plates and the bottom plates. Then the final continuous welds between the tank bottom plates are made.

It is normal to test the tank bottom for tightness before the second course is erected. After the bottom has passed the necessary testing, erection of the remaining shell courses begins. The plates are placed with their longer dimensions horizontal, forming courses one above the other until the full tank height is reached. Butt-type welds are usually preferred for both vertical and horizontal shell joints. Shell plate thickness increases as diameter increases, and decreases as height increases. Normally, the plate used in the upper course, or top ring, would be much thinner than the plate used in the first course or bottom ring. However, field experience has shown that due to stress, the plate in the upper course should be just a little thicker than the plate in the course just below it.

The last item to be placed in the erection of the tank shell is the top angle. It is located at the top edge of the top course and its purpose is to connect the tank's roof plates to the tank shell.

Erection of Tank Roofs

Erection of the roof-supporting structure can begin while the final welding of the shell joints is being completed. This structure consists of a number of structural steel or steel pipe columns. Evenly spaced, they rest on the inside of the tank bottom and extend upward to support the roof plates. These plates are welded in place following the erection of the roof supporting structure. The finished roof is cone-shaped, sloping upward from the tank shell to the center.

Storage Tank Accessories

After and often during storage tank erection various accessories that are necessary for the proper operation of the tank are installed. Representative of these are:

1. One or more manholes, located near the bottom of the first course to provide access to the interior of the tank when cleaning or repairs are necessary.

2. One or two roof manholes for access from above.

3. Shell nozzles for making necessary pipe connections.

4. One or more drain connections.

5. Roof nozzles, usually flanged, set into the roof for attaching relief or vent valves, gauge hatches, and thief hatches.

6. Relief or vent valves.

7. Gauge hatches and thief hatches.

8. Steel stairways and access platforms.

9. Other accessories such as an automatic float gauge.

Painting Tanks

The exterior of storage tanks are painted to preserve the metal and prolong the service life of the tank. Paint also enhances the tank's appearance and can also provide advertising for the company. White paint reflects the sun's rays and helps to stabilize the internal temperature. This tends to reduce the *weathering away* of the lighter fractions of the oil being stored. Aluminum paint is also used, but its effect is less than that of white paint.

The importance of surface preparation cannot be overemphasized because the useful life of any paint job depends on the quality of the paint itself and on the adherence of the paint film to the tank surface.

All surfaces should be cleaned of dirt, rust, oil, grease, and mill scale before any paint is applied. These foreign substances are generally removed by sand blasting. The sand used for sandblasting should be dry, should be a sharp flint or silica abrasive type, and should be free of clay or other material which might become imbedded in the steel. After sandblasting, dust should be removed from all surfaces and the metal primed immediately. Some companies apply a phosphating solution after sandblasting to provide a sound foundation for all types of primers. This also tends to inhibit corrosion, and provides a suitable surface for the paint to adhere to. Other companies prefer to use a zinc chromate and alkyd coating. If the tanks are near the sea or in other severe environments, a primer of inorganic zinc silicate may be used. It is usually applied to a thickness of .075 in.

Following the primer are one or more coats of paint. The choice of paint, number of coats, and thickness of cover, is the tank owner's decision.

The type of paint used on the tank interior is basically determined by the product to be stored. For crude oil, a bitamastic or coal tar epoxy is used. There is a special coating used for the storage of jet fuel, and when motor oil or other products are to be stored, the interior may be left unpainted.

Special Roofs

In addition to these land-based cylindrical tanks with conical roofs, there are other special types of roofs for cylindrical tanks whose design are patented by individual manufacturers. These include the *breather roof*, the *balloon roof*, the *lifter roof*, and the *floating roof*.

The primary purpose of these roofs is to reduce the loss of the volatile parts of stored oil to the atmosphere, or,*"evaporation losses."Cone roof tanks* have a tendency to "breathe," especially when they are only partially filled and there is a wide temperature differential between day and night. As the tanks breathe, there are vapor losses large enough to be of considerable economic importance. Thus the extra cost of using special roof and tank designs as a means of reducing these losses may be justified. Each special roof design is unique and has a purpose for certain conditions.

Breather Roof

A breather roof is a modification of the cone roof, and is similar in construction, except that the roof supports are designed so that when the roof is resting on them it is an inverted cone. The shell and bottom of the tank are the same as any standard cylindrical steel welded tank. When the liquid level in the tank is low the roof rests on the supports. As pressure under the roof rises, because of temperature or from the tank being filled with more liquid, the roof rises from the supports as is suspended solely by the pressure in the vapor space. It can travel upward until it reaches normal liquid level. At this point the automatic control valve opens and excess vapors are allowed to escape.

The breather roof tank is adopted best to standing storage where the tank remains nearly full for long periods of time.

Balloon Roof

The *balloon roof* is similar in design and function to the breather roof. It is different in that the roof diameter is larger than the tank shell diameter. The roof overhangs the tank shell.

With both the balloon and breather types of roofs, the metal in the roof is actually bent when the roof moves to accommodate varying vapor volumes. This has led to cracking of the roof sheets and increased maintenance costs, so the lifter roof has supplanted these types of roofs.

Lifter Roof

The lifter roof is entirely separate from the tank shell and is free to move, with suitable guides and controls, through a vertical distance of several feet. It provides a very large variable vapor space by means of a seal between the moveable roof and the stationary shell. Two types of seals are commonly used. One is a liquid seal formed by having a metal skirt extending downward from the roof into a liquid-filled trough built on the upper part of the tank shell. The other type uses a flexible curtain of treated fabric that is impervious to oil vapors. This fabric forms a continuous ring around the tank top with one edge fastened to the moveable roof and the other to the tank shell.

Floating Roof

The floating roof actually floats on the surface of the oil or other liquid stored in the tank. It rides up and down inside the tank as the liquid level changes.

The first successful floating roof was designed and constructed in 1923 by Chicago Bridge and Iron Company (CBI). It was a pan type, with the underside of the deck plates in contact with the product on which the roof floated. This provided an answer to the urgent need for a safer and economical method of storing petroleum products. All tests that were run on this structure proved that floating roofs would effectively conserve vapors and minimize fire hazards. Since 1923 other floating roof designs and concepts have been developed and introduced by CBI. The floater virtually never wears out. Some floating-roof oil tanks have enjoyed trouble-free life spans of up to 50 years.

Along with the development of various types of floating roofs, many types of sealing rings were also developed. These sealing rings provide a system of flexible "shoes" to close the space between the edge of the roof and the inside wall of the tank shell (Figure 9–2). This type of primary sealing system is known as a *mechanical (metallic) shoe*. The other general type of shoe is known as the *resilient (non-metallic) filled seal*. It consists of a fabric band or envelope held against the tank shell by liquid or resilient pressure (Figure 9–3).

Open top tanks with floating roofs need additional support around the top ring on the outside to stiffen the tank shell and hold it "in round." These rings are known as *wind girders*. This is because that when the liquid level is down and the floating roof is riding low in the tank shell, there can be a lot of stress placed on the top ring, particularly during a strong wind storm.

The 1923 version of the floating roof was known as the *pan type* because it looked very much like a shallow pan. It was a single-deck that sloped to the center for drainage. The pan roof could tip and sink if it was exposed to excessive loads of water or snow. Also, heat from the Sun's rays could cause oil in contact with the roof to vaporize rapidly.

Pontoon-Type Floating Roof

In 1929 CBI offered a *pontoon type* floating roof that was virtually

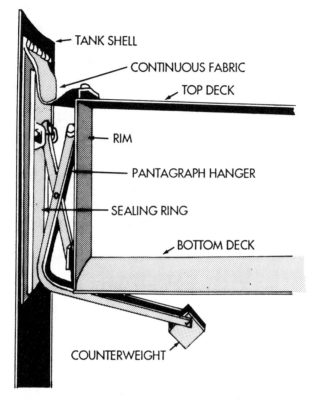

Figure 9–2 Seal of a double-deck floating roof (courtesy CBI Industries, Inc.).

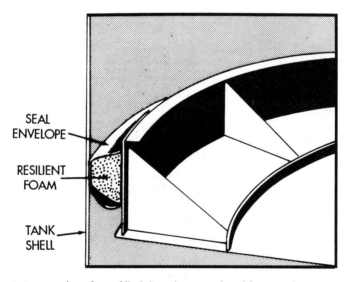

Figure 9–3 A resilient foam-filled, liquid-mounted seal for Weathermaster tanks (courtesy CBI Industries, Inc.).

unsinkable. It had a single deck with a annular pontoon around its outer perimeter. In 1954, the Horton Type-5 Pontoon Floating Roofs were introduced. The design of this roof provides adequate pontoon volume to keep the roof floating even with the single deck and any two pontoon compartments punctured. These roofs comply with the requirements of API Standard 650, Appendix C. The roof will retain a 10-in. rainfall and can be designed for even greater amounts of rain if required. The roof floats directly on the product. The underside of the pontoon slopes upward toward the center of the roof and the top roof slopes downward toward the center.

Double-Deck Floating Roofs

Small *double-deck floating roofs* were in use earlier, but it was not until 1946 that these became generally available for large diameter tanks. CBI developed the Horton Double-Deck Floating Roofs that are now available in diameters from 25 through 400 ft. The design of the roof provides sufficient pontoon volume to keep the roof afloat with any two compartments punctured. Emergency overflow drains are provided to accommodate a 10-in. rainfall occurring during a 24-hour period. The bottom of the roof floats directly on the product, and because of its design, temperature-generated condensible vapors are contained under the center of the roof. The double-deck floating roofs can be insulated to extend their use into ranges of higher or lower temperature (Figure 9–4).

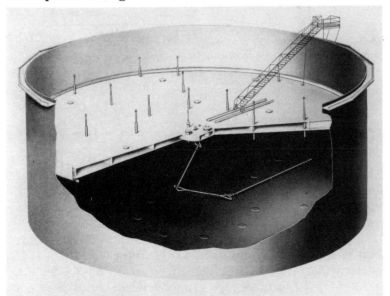

Figure 9–4 Typical double-deck floating roof storage tank (courtesy CBI Industries, Inc.).

Internal Floating Roof

The covered, or *internal floating roof,* tank, has been in operation almost as long as the floating roof. It is a fixed-roof tank with a floating roof inside. The covered floating roof tank can combine many of the advantages of both fixed roof and floating roof tanks. The covered roof provides shade from the Sun, protection from the wind, and also keeps rain and snow off of the floating roof. A single-deck, pan-type floating roof can be used, since it is protected from water and snow loads. Since the cover protects the internal floating roof from the weather, no drainage system is required. The CBI Weathermaster, a tank of this type, meets the requirements of API Standard 650, Appendix H, "Internal Floating Roofs," (Figure 9–5).

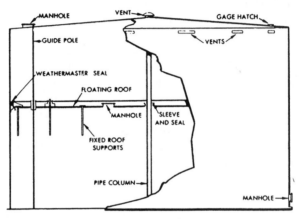

Figure 9–5 Cutaway through Weathermaster Floating Roof Tank showing standard components normally furnished unless otherwise specified (courtesy CBI Industries, Inc.).

Ultra Large Storage Tanks (ULST)

CBI and its affiliates have supplied more than 16,000 floating roof tanks to the world petroleum industry in capacities of up to 1.5 million bbl each. Each *Ultra Large Storage Tank (ULST)* requires approximately 3,500 T of steel to build and takes about 12 months to design, fabricate, test, and commission ready to receive oil. Figure 9–6 is a photo of a Middle East crude oil loading terminal comprised of some of the largest floating roof tanks in the world. These ULST's range in capacity from 1.25 million bbl (200,000 M^3) to 1.5 million bbl (240,000 M^3). The total storage capacity of this terminal is approximately 25 million bbl of crude oil.

Figure 9–6 Ultra large storage tanks (courtesy CBI Industries, Inc.).

Firewalls

It is customary to build an earthen dike or a concrete wall, called a firewall, berm, or fire bank around cylindrical tanks. In the event of a rupture, these walls are to retain any fluid leaked from the tank until it can be pumped into another tank. Another purpose is protect other tanks or buildings in case of a fire in the tank. There are codes and recommendations for the construction of firewalls and for the spacing of tanks.

Special Tanks

Thus far, only cylindrical tanks designed for low pressure liquids have been discussed. These storage tanks are able to withstand very little upward pressure under the roof. Other shapes are better for pressure storage of highly volatile liquids such as casinghead gasolines, butane, and butane blends, etc. Both *spheroid* and *sphere*-shaped tanks are used for storing highly volatile liquids.

Spheroid Tanks

The spheroid tank is suitable for working pressures of up to 25 psi. It has the appearance of a flattened sphere; its shape is patterned after the shape taken by a drop of water resting on an oiled surface. The smaller spheroid tanks follow the ideal drop shape. This shape is modified somewhat for structural reasons and for the effect of the vapor pressure above the liquid. For larger tanks, structural reasons have made it necessary to adopt the so-called noded form.

Sphere Tanks

The sphere tank resembles a large ball mounted on steel columns. It provides maximum storage capacity for the weight of the steel used in its construction. The sphere tank is a pressure vessel for the storage of volatile liquids and gases under pressure. It is built of steel plates that have been formed and cut to the proper radius in the shop, with their edges prepared for welding. The major attachments such as nozzles and supports, are usually welded to the steel plates during shop fabrication.

The sphere vessels are commonly used to store petroleum and chemical products, and can be designed for high or low pressure and high or low temperature applications, and they can be built in a wide range of diameters. Some have been built out of stainless steel plates six inches thick for 1,800 psi hydrogen service. The highest pressure spheres built thus far were designed for 5,500 psi. Sphere vessels designed for cryogenic storage are designed for as low as –425°F (–253°C), the temperature at which liquid hydrogen is stored.

Larger diameters and higher pressures are practical through the use of field postweld heat treatment and/or the use of steels with higher strengths (Figure 9–7).

Horizontal Tanks

Horizontal tanks consisting of a cylindrical shell with hemispheric heads are also used for the storage of petroleum products under pressure. These tanks are usually shop-fabricated and are limited in size.

Other Special Tanks

Other land-based special steel tanks include Liquefied Petroleum Gas (LPG), propane, butane, and Liquefied Natural Gas (LNG) storage tanks and any other tank or vessel used for petroleum products. Their sizes and designs can be altered to meet special needs.

Modern Petroleum

Figure 9–7 Sphere vessels at a refinery in Louisiana. These vessels are for various liquids and gases that must be stored and handled under high pressure (courtesy CBI Industries, Inc.).

In addition to land-based storage tanks there are special tanks built for offshore oil storage. Figure 9–8 shows a tank designed to store crude oil on the sea floor in 155 ft of water. It is shown being towed by two tugboats to its destination. On arrival at its exact assigned position, it will be submerged, anchored, and placed into service.

Figure 9–8 Offshore storage tank to be placed on the sea floor in 155 ft of water (courtesy CBI Industries, Inc.).

Underground Storage

Underground storage systems usually consist of caverns that have been constructed as a result of removing salt from a mine, either by solution mining or conventional mining. Other systems may be developed by conversion of coal or other solid-material mines to storage, or by repressuring depleted oil or gas fields.

Underground storage is especially beneficial for storing large volumes of liquids, such as crude oil. One of the largest underground crude oil storage areas lies in the salt domes of Texas and Louisiana where nearly 590 million bbl of oil is stored. The goal of the U.S. government is to eventually stockpile 750 million bbl.

Congress authorized the *Strategic Petroleum Reserve (SPR)* in 1975 following the Arab oil embargo of 1973. The *Energy Policy and Conservation Act* stipulates that use of the reserve can only be authorized by the president.

Underground storage of natural gas is an ideal way of meeting peak day load requirements for supplying gas to customers. Gas is stored during summer months when the demand is down and then withdrawn during the cold winter months. This practice was put into operation in 1929 when one major company repressured an old depleted natural gas field, and then withdrew gas as it was needed. The cycle was renewed again the following summer.

An ideal system is a depleted field that still has the capacity to be repressured and that is located near a concentrated area of customers. Usually, the old wells are plugged and two new ones drilled. One well is designated as an observation well, and the other is used for injection of gas to be stored, and the withdrawal of stored gas to be used.

Storage
PART 2

The Measurement, Testing, and Custody Transfer of Crude Oil and Petroleum Products.

Strapping Tanks

Before a tank is placed into service for the first time it must be strapped. *Tank strapping* is the term commonly applied to the method of taking and reporting the dimension of tanks. This is necessary for the computation of accurate *tank tables* which truly indicate the volume of liquid in the tank at any measured depth. Tank measurements are figured in barrels and fractions of barrels on the tank table. They give the capacity at any given height in the vertical section of the tank. The methods and measurements are taken in the field in compliance with API codes.

Measurements in feet, inches, and fractions of inches or in the metric system of measurement constitute one of the most important elements in the purchase or sale of crude oil. It is vital for the custody transfer and transportation of crude oil and refined products. Every barrel and fraction of a barrel handled has a definite money value.

Tank strapping encompasses the following measurements:

1. Depth
2. Circumference
3. The thickness of tank walls
4. Pipeline connection

Depth, or tank height, is the vertical distance from the top to the inside surface of the tank bottom. This measurement is usually taken with a steel tape with a plumb bob attached, and the measurements reported to the nearest ⅛ in. on steel tanks, and to the nearest ¼ in. on production tanks.

The oil height is the highest point to which a tank can be filled without overflow.

Circumference is the circular distance around the outside of the tank. Usually, a calibrated steel tape is used to physically measure each ring (course) of the tank, taking into account wall thickness and other vital factors. Of course, the inside circumference must be accurately calculated with whichever method is used for the outside.

The thickness of the shell sheets or plates is measured with an accurate depth gauge.

Deadwood is defined as any object within the shell of the tank that would displace fluid, such as columns and braces. The measurements and calculations must account for the volume of deadwood objects.

Pipeline connection measurements are taken on all production tanks and on some station working tanks. It is the distance from the bottom of the tank to the bottom of the outlet connection.

After all strapping measurements are made, it is sometimes necessary to fill the tank with water or product and then meter the liquid as it is drained from the tank as a check of the physical measurements and calculations.

Fluid Testing and Measurement

The lease pumper will usually take care of the gas metering equipment to make sure that it is operating properly and that gas deliveries are being recorded. This will include the changing of the recording charts and pens, and the forwarding of the charts to the leaseholder's office so that the amounts delivered are calculated and the recipients billed. Like so much of the petroleum industry today, this is another area where automation, remote control, and computerization of operations is taking hold.

In the operation of an oil lease, the pumper is required to make sure that the pumping unit is working properly and that the production tanks are being filled. When one tank is filled, the flow is then diverted to another. The pumper will measure the tank contents and fill out the daily *gauge report* (Figure 9–9). The final gauging before *running* or delivery will be made by the gauger. The gauger also makes the final gauge when the tank is empty, and when finished, seals the outlet valve so that the tank may be refilled. The pumper watches this process and verifies the measurements and tests. It is important that all tests, such as those for salt water content, temperature, and *API gravity*, along with the volume measurement are made because custody, or ownership, of the oil passes from one

Daily Gauge Report

ANDERSON PRODUCTION COMPANY

Date_____,19____

LEASE NAME	TANK NUMBER	TIME OF DAY	YESTER-DAY		TODAY		MADE TODAY	B. S. & W. DR'N	GA. AFTER B.S. & W. DR'N	
			FT.	IN.	FT.	IN.	INCHES		FT.	IN.

PIPE LINE RUNS

TANK NO.	BEFORE		AFTER		LEASE NAME
	FT.	IN.	FT.	IN.	

REMARKS

_____

_____

_____

Figure 9-9 Daily gauge report.

party to another at this point. The gauger records the test and measurement data on *run tickets*, and forwards copies to the proper persons involved.

Testing Procedures

Testing and sampling are usually done through the opening in the top of the tank called the thief hatch, although some samples may also be obtained through the sidewall of the tank if appropriate taps have been provided for that purpose. Several tools have been devised

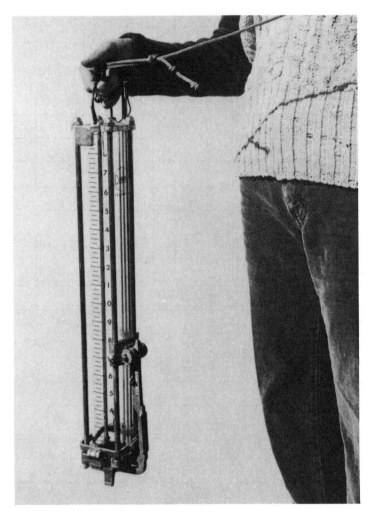

Figure 9–10 Tank thief.

to make the tests and measurements necessary before the sale or transfer of the oil. The primary tool lowered through the hatch to sample the oil is the *thief*, and its use is known as *thiefing the tank* (Figure 9–10). The tool is first lowered to the top third, then to the middle third, and finally to the bottom third. Then an average of the three measurements is used as a sample.

Other gauger's tools include a hydrometer and graduate for measuring the API gravity of the oil, a thermometer for measuring the temperature of the oil, a tank-gauging line for measuring the height of the oil, and a centrifuge for measuring *BS&W* content.

Specific Gravity

Not all crude oil will be the same, one lease may produce a very "light" crude, while the oil from another lease may be thick and viscous. The first step in scientifically measuring the difference between samples of crude oil is to determine the specific gravity of the samples.

Specific gravity is the ratio between the weight of a unit volume of one substance compared with the weight of an equal volume of another substance taken as a standard. For liquids, the standard is usually water, which is assigned a value of 1 (assuming it is demineralized). For gases, the standard is usually air, which is also assigned a value of 1 (assuming a temperature of 60°F). Since the temperature at the tank battery is obviously not always going to be 60°, temperature tables will have to be used to determine the correct specific gravity.

API Gravity

The API has adopted a standard method of expressing the gravity, or unit weight of petroleum products. This is an arbitrary method that had its beginnings in the chemical industry long before it was applied to liquid hydrocarbons or before much was really known about the relationships among the weights of substances. *API gravity* may be determined by placing a sample of the oil in a hydrometer and reading the value directly off the scale or by making a fairly simple computation.

The first step in computing the API gravity (API°) of a sample is to determine its specific gravity. The API formula is:

$$\text{API Gravity} = \frac{141.5 - 131.5}{\text{Specific Gravity at 60°F}}$$

Thus, if a certain sample had a specific gravity of 0.82, its API° would be:

$$\frac{141.5}{0.82} = 172.6 - 131.5 = 41.1°\text{API}$$

The temperature of the oil in the tank will usually be close to that of the outside air if it has been in the tank several hours. The usual method of testing is to lower a thermometer on a line through the thief hatch and take a reading. If it is other than 60°, then it should be correct to that figure by using a conversion table.

BS&W

Samples of oil are also taken to measure the basic sediment and water content of the oil. A sample is removed either by means of a thief or through a *sample cock (sidewall cock)* and placed in a centrifuge. The centrifuge causes the heavier components such as salt water to fall to the bottom of the centrifuge tube where the results can be read directly on the tube scale.

Standards

Sampling and testing are carried out in accordance with standards set up and adopted by the API and the American Society for Testing Materials (ASTM). Normally, samples are taken with a thief from production tanks and some small storage tanks. In the case of larger tanks, samples taken by thief, as well as those drawn in bottles from sidewall cocks are usually used. The thief is a round glass tube, usually 15 in. long, that is suspended from a chain attached to its top. At the bottom is a spring-operated sliding valve that can be actuated to seal it. It is lowered into the tank with the valve open so that the oil passes freely through the tube until the desired depth is reached. At that point the spring is released and the valve closes, capturing a sample of oil. The thief is then brought out of the tank and its contents tested.

A 1-qt metal bottle may also be used. The bottle (beaker) is weighted so that it will sink, and it has a cork stopper. A line is attached to the stopper so that it might be removed at the desired depth, and the bottle filled at that point (Figure 9–11).

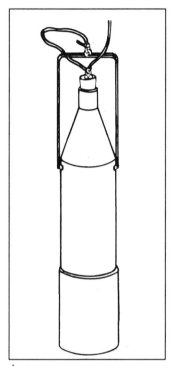

Figure 9–11 Beaker sampler

There are many different types of samples that may be taken, and an agreement should be reached by all concerned as to which type will be used, such as all-levels sample, upper, middle, bottom, composite spot sample, etc. (Figure 9–12).

A tap sampling may be done through sample taps, or from sample cocks properly placed in the shell of the tank. All sample cocks should be equipped with sealable valves and plugged inspection tees.

The importance of getting a good sample and an accurate API° reading cannot be overstressed because the API° of the crude determines the price the buyer will pay for it.

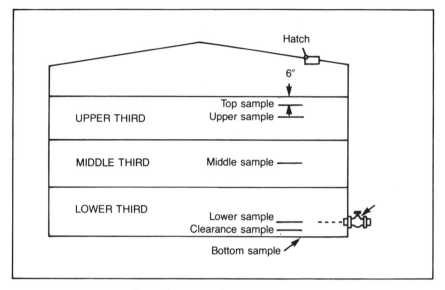

Figure 9–12 Location of samples in a tank.

Automatic Tank Batteries

Today most leases use tank batteries or production tanks standardized as to type, size, layout, and fittings. This simplifies both installation and operation. Tank battery operation is now becoming more automated. They are being fitted with overflow lines to prevent overfilling any one tank, automatic valves that divert the flow to an empty tank as the one in use is filled, and electrical, mechanical, or air-operated switches and valves that will shut down the well when all the tanks are full.

LACT Units

There are fully automatic *LACT (Lease Automatic Custody Transfer)* units in the field. They provide unattended transfer of oil or gas from the well to the pipeline. The unit takes samples, records temperatures, determines the quality and net volume, recirculates oil that needs treating, keeps the records necessary for production accounting, and even shuts itself down and sends an alarm in the event of trouble. LACTs also offer an additional advantage in that they require less tankage since custody transfer is more frequent.

Safety

A lease tank battery or storage tank farm can present many hazards to untrained persons. Therefore there are a number of definite safety rules which must be strictly followed in order to prevent serious injury or death as well as property damage and loss of production.

1. Do not smoke or carry smoking materials in the vicinity of petroleum tanks. There will likely be volatile materials with low flash points present.

2. Do not step onto or walk on tank roofs.

3. Keep your face and upper body averted when opening thief hatches. There will quite likely be a rush of accumulated gas and fumes when you do so.

4. Never, under any circumstances enter a tank unless you are wearing protective clothing and an approved breathing device and you have a "buddy" present on the outside.

In the field, some crudes are high in paraffin which tends to build up in accumulations on tank walls. As the walls thicken, their measurements are skewed and they must be periodically cleaned. Unfortunately, this task is often left to untrained and unskilled people who are unaware of the hazards. They do not know that on a hot summer day, the fumes present in an "empty" tank can overcome a person in a matter of seconds. One will enter to clean the tank, and when his buddy realizes something is wrong, he too will go into the tank to help his friend and also be overcome. Sadly, sometimes as many as three or four people have been killed this way in a single incident, and there are several such incidents each year.

In addition, some of the materials used to clean tank interiors can also produce hazardous fumes. Therefore protective clothing, breathing devices, and proper training are a necessity.

Chapter References

1988 Annual Report, CBI Industries, Inc. 800 Jorie Boulevard, Oakbrook, Illinois 60522-7001

CBI Bulletin 3200, Horton Floating Roof Tanks (Copyright 1981), Chicago Bridge & Iron Co., 800 San Jorie Boulevard, Oakbrook, Illinois 60522

Crude Oil Tanks: Construction, Strapping, Gauging and Maintenance, (Austin: The University of Texas, Petroleum Extension Service, February1972).

10

Transportation

Crude oil, condensate, and natural gas generally have a limited application in the producing field. With the exception of natural gas, where a fraction of it might be used as fuel for internal combustion engines, heater-treaters, or other equipment, the greatest demand for these petroleum fluids is in the consuming areas throughout the world.

Transporting petroleum fluids from the producing field to the customer requires a very complex network of systems. It may involve pipelines, transport trucks, railway tank cars, inland waterway barges, and ocean-going tankers or a combination of several or all of these.

In the United States, approximately 46% of the crude oil and products are transported by pipelines, while motor carriers move about 27%. Water carriers haul approximately 25% and railroads handle the remaining 2%. Of course these figures change constantly, and there is no method of accurately forecasting the changes. Other countries have their own methods of transportation that best fit their needs.

Generally, the first phase of transportation is moving the liquids to storage tanks. The next is moving them to the refinery. From the refinery, the final phase is transporting petroleum liquids and their byproducts, usually referred to as "products" to other refineries, petrochemical plants, or factories before the desired end result is sent to terminals for further distribution (Figure 10–1).

Railroads Grew with the Oil Industry

Colonel Drake shipped the crude oil from his well in wooden barrels that were hauled on horse drawn wagons to inland waterway barges or on railway flatcars. This method was slow and costly, and it was soon realized that improvements must be made. Attention was focused on the rapidly growing railway industry, because there was no better transportation

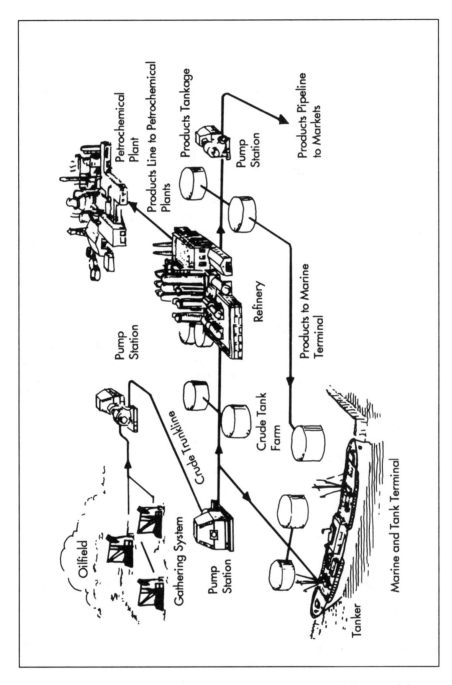

Figure 10–1 Pipelines serve to gather crude oil, transport it to refineries, and then distribute the products to process plants, retail markets, and to other forms of transportation.

system in the United States in 1859 when the Drake well was drilled.

Rail mileage in the USA increased during the 1850s from 9,000 to more than 30,000 mi by the eve of the Civil War. What had been a scattering of short lines from Maine to Georgia became an iron network serving all the states east of the Mississippi River. By 1857, the rail mileage of the United States was almost half of the world's total. The fastest and most important railway growth just before the Civil War was in the western states where mileage grew from 1,250 in 1850 to 11,000 by 1860.

One big change in rail freight shipping came with the end of the Civil War. Prior to the war, many railroad companies had different rail gauges, which meant the cars of one line would not operate on the tracks of another. Thus, goods had to be unloaded and reloaded several times if they were being sent any great distance. To facilitate shipment of goods in the North, and later when the destroyed southern lines were rebuilt, the present standardization of 4 ft 8.5 in. (1.4 m) was implemented. This new ability to transfer goods from one part of the country to another led to a rapid growth in rail freight.

In the half century between the Civil War and World War I, the railways in the United States enjoyed a Golden Age; they experienced supremacy and near monopoly between 1865 and the 1920s. During this time there was a major increase in construction, especially west of the Mississippi River, when public attention was focused on the construction of the first transcontinental road. The Union Pacific and Central Pacific railroads together laid nearly 1,800 mi (2,900 km) of line from the Missouri to the Pacific, completing the road on May 10, 1869 at Promontory Point in Utah Territory (Figure 10–2).

The years of near monopoly and rapid growth were also years of

Figure 10–2 Early 20th Century railway tank car with a capacity of 5,000 gallons owned by Canfield Refining of Yale, Oklahoma (© Gary Gibson, Yale, Oklahoma).

extensive corruption and rate discrimination within the railroad industry. Shippers complained of freight rates that changed almost weekly. Cheap rates were often followed by excessive rates, and special low rates were given one of the large oil companies, which brought allegations of discrimination that favored the big shipper.

Prior to World War II, the railways carried a large percentage of the nation's petroleum. But the use of rail shipment was expensive to the producers and marketers of oil. J.B. Saunders, Jr., one of the largest of the independent marketers was fond of saying that he didn't make any real money until he could switch from rail to barges and save one or two cents per gallon shipping costs. This marketer owned 900 tank cars at one time and still had to pay the railways for moving them from one point to another. Others who did not own their own cars had to pay shipping costs plus demurrage (the time a car sits idle at a terminal waiting to be unloaded).

New modes of transportation and more stringent federal regulations threatened the longtime rail dominance, thus the railroad industry soon entered a period of decline. Highways, motor trucks, more efficient water transportation, plus a growing network of pipelines began reducing the number of tank cars on the average freight train. Railways have lost their supremacy, but there is still a fractional amount of petroleum and petroleum-based products carried by rail.

The United States has approximately 200,000 mi of railroads, which is about 25% of the world's mileage. Russia is second with approximately 85,000 mi of government-owned railway system. Most of the railroads of the world are state owned and vary in type, mileage, and service. All freight railroads in the United States are privately owned. However, Conrail in the northeast corridor is government subsidized.

Barges

A barge is a flat-bottomed boat, with both ends square or slightly sloped. It is used to transport bulk freight, chiefly on rivers and canals or in harbors. Normally, it is towed by a tugboat or may be self-propelled by an engine or sails. Barges are constructed of steel or wood, with protective bulkheads, and are built in different models and sizes to accommodate their cargo and the conditions of navigation to which they may be subjected. Special barges are designed for the transportation of petroleum liquids and gaseous fuels that require closed containers.

Barges are in common use throughout the world and have been for many years. Many of the early towns and cities were built along rivers and harbors because waterways offered an easy means of direct point-to-point shipping. In the early days in the United States when roads were largely nonexistent, the rivers and canals were the only means of transporting bulky freight. At the time, canals had an advantage over river because they could offer two-way service. Their motive power was supplied by draft animals that walked along a tow path on the bank pulling the canal boats from one town to another and back. The canals were expensive to construct, and the ones in the north were subject to freezing over in the winter.

The rivers, on the other hand, were free for anyone to use, and many of them remained navigable for much of the year. Their big drawback was that since the vessels used were rafts, keelboats, or canoes that depended on the current for motive power, they were one-way avenues. The coming of the steamboat changed this, though, and for the first time goods could be carried both ways: from ocean ports such as New Orleans to river towns like Memphis or St. Louis and back.

Today, rivers still provide a great service in transporting petroleum products and nonperishable goods over great distances at low cost. The steamboat has given way to the diesel towboat that can push or tow a string of barges carrying the equivalent of literally hundreds of railway tank cars full of petroleum products. Inland towns such as Tulsa have become port cities due to the construction of waterways.

Many of our largest rivers have actually been turned into canals by the channelization work of the Army Corps of Engineers. Floods have been controlled by levees and dikes, locks built, and the most modern of navigational aids installed. This, however, has not been accomplished without controversy. The Corps has come under criticism from the railroad and trucking companies, who see the tax dollars they pay for land and highway use being spent to construct a competitive transportation system. Environmentalists claim that channelization has ruined rivers and turned them into sterile industrial canals that still flood periodically, and lack the aesthetic benefits of natural streams.

Hazardous conditions encountered in the open sea limit the practical use of barges for oceangoing service to near-shore operations, or coastal service. Many towboats and barges operate in the intercoastal waterway along the Atlantic and Gulf coasts.

Integrated tug-barges are a form of barge design that closely connects the stern of the tugboat to the front of a string of barges. This type of barge transportation has become important in coastal service, and is often referred to as a compoundable ship.

Environmental groups and others are concerned, and rightfully so, about the seriousness of an oil spill from barges to wildlife in and along rivers and canals, just as they are about oil spills from tankers. They feel that stricter regulations should be placed on the construction and operation of barges in the United States. In recent years there have been a growing number of accidents in congested areas, such as those adjacent to New York and New Jersey port facilities.

Tankers

A tanker is a tank ship designed to carry bulk cargo, usually petroleum or petroleum products. Small tankers are used to carry oil in coastal areas or river channels, while large ones cross oceans and seas. The increasing dependence on foreign sources of oil by the United States, and the energy demands of the world, have created a tremendous upsurge in the use of tanker ships for transporting oil by sea. Also, transporting petroleum by sea in the only logical means of supplying the United States, western Europe, and Japan from the producing areas of North Africa, Venezuela, and the Middle East (Figure 10–3).

Carrying oil by sea is certainly not new; the forerunners of our present tankers were the New England whalers. They would put to sea on voyages that would sometimes last for years in search of the great whales. Once a capture was made, the kill was brought alongside and the blubber stripped off and rendered into oil in pots

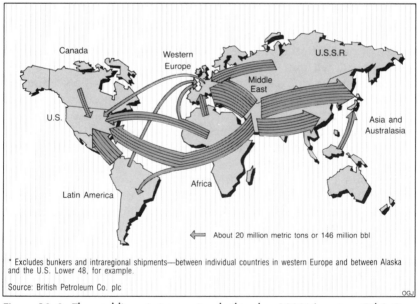

* Excludes bunkers and intraregional shipments—between individual countries in western Europe and between Alaska and the U.S. Lower 48, for example.

Source: British Petroleum Co. plc

Figure 10–3 The world's main interregional oil trade—1989* (courtesy *Oil & Gas Journal* and British Petroleum Company).

on deck. The oil was then stored below in wooden casks until the end of the voyage when it was sold for use in lamps. Later, of course, whale oil gave way to kerosene, and wooden ships were replaced by ships made of steel.

Petroleum was originally transported in wooden kegs stowed in the holds of sailing ships, just like the whale oil had been transported. In 1886 the first bulk tanker, *Gluckauf*, was built in England for a German company. It had large tanks in the forward part of the hull, and the engine room was placed at the stern of the ship. Our present tankers also have the engine room in the stern of the ship.

During the early part of the twentieth century, largely due to the urging of Winston Churchill, then Great Britain's First Lord of the Admiralty, warships began burning bunker oil as a replacement for coal. This prompted the need for tankers to transport fuel. During World War I there was a great demand for gasoline to power army tanks, trucks, and motorized equipment. New tanker ships were built to supply the increasing demands for bunker fuel and gasoline.

After World War I there was an oversupply of tankers and few new ones were needed, however, the sea movement of oil increased due to many factors, such as oil began replacing coal for heating and more automobiles were being manufactured. Tanker design had changed little during this period, and the U. S. Navy insisted on an improved tanker design. The outcome was a vessel having a carrying capacity of 18,300 *deadweight tons (DWT)*, with a speed of 18 knots. Later, World War II fleet tankers were 34,000 DWT. Soon after the war those ships were dwarfed by the *VLCCs (Very Large Crude Carriers)* of 150,000 to 300,000 DWT; and *ULCCs (Ultra Large Crude Carriers)* of greater than 300,000 DWT. Deadweight ton is a term used to rate tanker capacity. It indicates the amount of cargo that can be carried. A deadweight ton (2,240 lbs) is equal to about 7 to 7½ bbl of crude oil (42 U.S. gal per barrel), depending on the specific gravity of the crude (Figure 10–4).

Either class of modern tanker can carry many times the cargo of a 34,000 DWT tanker, thus fewer ships need be at sea at a given time, which reduces the chances of collision or other accident. Also these new ships carry the very latest in navigational and communication equipment as well as all the latest safety devices. They can be built for a lower cost per ton than smaller vessels and do not require a much larger crew to operate them. They have more efficient engines that use thousands of barrels per day less fuel than would be needed to operate a fleet of smaller vessels required to haul an equal amount of oil.

During the late 1960s many operators rushed to construct super tankers as the industrialized nations came to depend imported oil.

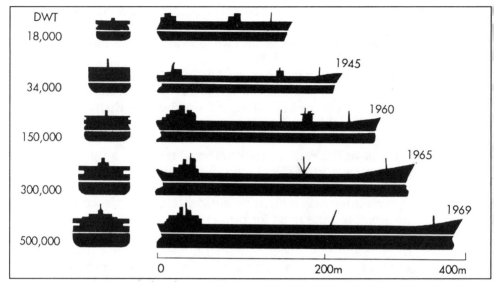

Figure 10-4 How tankers have grown.

Both of these became issues of concern to many groups and started a controversy over big ships. Environmentalists and commercial fishermen raised concerns over the impact caused by a rupture and the subsequent spilling of hundreds of thousands of barrels of oil, which could do serious harm to marine life and bird populations along the beaches.

What could happen did happen in March, 1989, when the *Exxon Valdez* hit a reef near Prince William Sound, Alaska, and spilled almost 11 million gallons of crude oil. This was the worst spill in the history of the United States. Then in June, 1989, the *World Prodigy* hit rocks near a reef in Narragansett Bay, at Newport Rhode Island, and spilled about 400,000 gal of heating oil. In 1988–89 there were 43 spills throughout the world from tankers of 10,000 DWT or more.

These accidents were dwarfed into insignificance in January, 1991, when Saddam Hussein of Iraq, in an act of environmental terrorism, dumped millions of gallons of Kuwaiti crude into the Persian Gulf.

The tanker accidents reopened the debate on double-bottom tankers, and some federal and state legislators introduced bills to require double-bottom tankers; believing that ship owners should build double hull tankers, others were less stringent in their demands, and asked instead for ships with segregated ballast tanks.

There is a difference between a *double-bottom tanker, double-hull tanker,* and *segregated ballast tanks.* A double-bottom tanker has an outer bottom hull and an inner skin that is tied to the bottom by structural members. There is a void space between the two bottoms, similar to the dry-cargo vessel construction used several years ago. The inner skin provides a smooth deck for cargo, and the space between the two bottoms can be used for carrying fuel or other liquids. The double bottom covers only the bottom of the ship and leaves the sides unprotected.

A double-hull covers the bottom and both sides, and is, in effect, a ship built inside the hull of an outer ship. Double-hull ships are more expensive to build, but in case of accidents, are less likely to rupture than double-bottom ships.

Segregated ballast tanks (SBT) are ships with cargo tanks dedicated to carrying water ballast. A certain percentage of ships's deadweight is available for use as a segregated ballast system. The ballast tank system is independent and cannot be connected to the vessel's cargo-handling systems, and it must have its own pumping and piping facilities.

Double-bottom tankers have more appeal to legislators, who appear to agree with many maritime professionals that these ships are safer and that spillage during an accident would be mitigated. Also, improved stripping and tank cleaning provisions offered by double-bottom tankers could limit ocean pollution and reduce the operational costs of tank cleaning. They would no longer need to mix ballast water in a dirty cargo tank, except when it might be necessary to ride out a severe storm at sea. Normally, double bottom construction has all the structural members such as web frames and stiffeners placed under the cargo tank. This leaves the tank bottoms clean and smooth. The lack of internal structural members decreases tank discharge time and simplifies the cleaning process.

Double-bottom ships must be built larger than vessels with single bottoms in order to carry the same DWT. This increase in size could require the dredging of some ports and increase the number of tugs or other assisting vessels needed for docking.

Some tanker authorities believe that double-hulled vessels are safer and less polluting than double-bottom ships. They believe double hulls would have prevented or minimized 28 of the 43 recent spills but would have made no difference on 12 others. Double-hull construction costs would be approximately 25% greater than those of single hulls.

Opponents of legislation calling for double-bottom tankers claim that it would make a large portion of the world's merchant fleet obsolete, and that the cost of replacing these ships would be prohibitive.

Tanker Emissions

The *Exxon Valdez* oil spill and the subsequent legislative appeal for double-bottom or double-hull tankers overshadowed another legislative measure to reduce pollution from tankers. The U.S. Coast Guard put together proposals designed to mitigate the problem of inert and hydrocarbon gases being vented to the atmosphere when a tanker loads a cargo of crude oil.

During the past few years tankers have been obliged to fill empty crude oil tanks with inert gases to prevent a buildup of dangerous accumulations of hydrocarbon gases from the residue of previous cargoes. The most cost-effective method of providing enough volumes of inert gas for this purpose is by using scrubbed CO_2 from the tanker's boilers.

At the loading terminal, inert gases are vented to the atmosphere as the tanks fill with crude oil. Hydrocarbon gases are given off from the crude oil being loaded at the same time. This too, is vented to the atmosphere throughout the loading operation. Some of these gases vented are considered to be contributors to the "greenhouse effect" that today is causing great environmental concern.

The proposed regulations would require construction of onshore vapor emission control systems which would move inert gases by pipeline to a processing plant. Each onshore plant would cost millions of dollars, and tanker conversion could cost approximately $1 million per ship.

Petroleum Industry Response Organization

The *Petroleum Industry Response Organization (PIRO)* is an organization of American oil companies formed to enhance their ability to prevent pollution of the nation's coastal waters. When it is fully operational, it will be the world's largest oil spill response organization. Its goal is to have the right people and the right equipment at the site of a spill immediately, and its services will be available around the clock.

PIRO will have regional response centers located along the nation's coastlines. Company members will fund the entire operation without the help of public funds. They will also undertake an oil spill cleanup research program to find better operating procedures and equipment.

Because of tanker spills, pressure is mounting in Congress from Oregon and the state of Washington to Florida and other areas to ban leasing and drilling in U.S. offshore frontiers. Most notable was a

House Appropriations Committee vote on June 29, 1989, to halt almost all Outer Continental Shelf leasing except in the western Gulf of Mexico for one year or longer. Opponents argue that the answer to tanker spills does not lie with moratoriums on offshore drilling, because that would increase rather than decrease tanker traffic by forcing the United States to import more oil.

Superports

The trend toward very large and ultra large crude carriers has prompted a need for *superports* to accommodate them. The prime consideration of the design of superports is the draft of the tankers that it can safely handle. Some attention also must be given to the length and width of the super carriers. Length is important when designing the diameter of turn basins, radius of bends in the channels, and the distance required between berths. Another factor is the stopping distance required for the ships. For example, when a 200,000 DWT ship is entering the channel at a speed of 5 knots, the approximate stopping distance, with the engines slow astern, is more than 1 mi (1.6 km). The same ship, at the same speed, with the engines completely stopped, but not put astern ("in reverse") would have a stopping distance of approximately 5 mi (8 km).

Channels in the open sea leading to the superports are also of major importance. A fully loaded 200,000 DWT tanker has a draft of about 68 ft (20.7 m) and a 500,000-550,000 DWT tanker has a draft of 94 ft (28.6 m). There are many other very important factors to be taken into account such as navigational aids, and the availability of tug boats and safety equipment. A relatively small percentage of ports throughout the world will accommodate the super large crude carriers that are now in service. However, no harbor in the United States can handle a fully loaded ULCC!

Lightering

One alternative to the shortage of port facilities is to anchor the large tankers offshore, and transfer the oil to barges or smaller tankers that would later offload in ports. This method, called *lightering*, is time-consuming, and can tie a ship up for days. It is prone to spills, and the operation is at the mercy of the weather.

Single Point Mooring Systems

Another solution is to use *single-point mooring systems (SPM)*. The SPM is a buoy attached to the bottom of the sea and connected to shore tank farms or pipeline terminals by an undersea pipeline. They are located in deep water way from shoals and other ship traffic and may even be out of sight of land. The tanker attaches a flexible hose to the SPM and is free to swing with the current as her cargo is pumped into shore tanks. In the event a sudden storm should blow up, the ship can quickly disconnect and move away (Figure 10–5).

While SPMs have been used in many parts of the world, there has been environmental resistance to them in the United States. One project that has been used to off-load at sea is the *Louisiana Offshore Oil Port* or *LOOP*. Here, tankers can unload into what is essentially an offshore pipeline terminal. The undersea pipeline connects with onshore lines, which then carries the oil directly up into the midsection of the country to serve areas of demand.

Gas Cargo Carriers

Liquefied natural gas (LNG) is transported in seagoing vessels from nations that have a surplus of gas to those that do not. Demand for LNG has been stimulated by increasing concern about the environment, which has focused attention on clean-burning natural gas.

Since natural gas is in a gaseous state, it must be changed to a liquid in order to be shipped by tanker. To be changed to a liquid, its temperature must be reduced below –259°F. This is done in cryogenic refrigeration plants. Cryogenic comes from the Greek words *kryos* meaning "icy cold," and *genes* meaning "to produce" (Table 10–1).

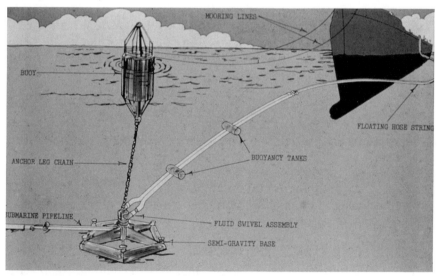

Figure 10–5 Single-point mooring system (courtesy API).

Cargo tanks in LNG carriers are normally double-wall independent tanks that are well insulated. They may be free-standing or of the membrane type. There are several designs of tanks utilizing special materials, including stainless steel, nickel steels, and aluminum. Also, many insulating methods are available using fiberglass, polyurethane foam, polyvinyl chloride, perlite, balsawood, and other suitable insulating materials.

The expansion rate from LNG to a gaseous state is 1 to 625, so that it is very important to maintain a temperature below the boiling point of the cargo. In low-temperature designs, the boil-off of cargo may be reliquefied by refrigeration. However, reliquefaction equipment is large and costly. The boil-off from LNG is usually burned as fuel for vessel propulsion.

Liquefied Petroleum Gas (LPG) is butane, propane, and other light ends separated from natural gas or crude oil by fractionation or other processes. Most LPG in the United States is propane, and about 10% is butane or butane-propane mixes. As a fuel, propane has some unique characteristics that have found favor in many applications. It can be liquefied at reasonable temperatures and pressures for ease of transport in pipelines, tankerships, and tanker trucks. The cost of equipment used in storing and transporting LPG is considerably less than that used for LNG.

Propane will readily vaporize at ambient temperatures, making it easy to burn in stoves or as fuel for internal combustion engines. Propane is also widely used as a chemical feedstock.

Motor Trucks

Motor tank trucks play an important role in transporting petroleum fluids in general, and specifically, in cases where small lease tank batteries are not connected to pipelines. Trucks have the advantage of being able to go almost anywhere and thus isolated tank batteries can be served. Today's large truck tractors and semitrailers can carry as much or more than early-day railway tank cars did.

When a new oil field in a remote area is first brought into production, semi-trailers are often used to store the crude oil on location before permanent tank batteries are installed. If the well does not prove to be economically feasible to produce, the semitrailers can be moved at a fraction of the cost of moving a tank battery. In many instances where the well or wells are profitable, but laying a pipeline would be too costly, tanker trucks haul the oil to refineries or to large storage tanks.

Trucks also serve the other end of the system as well. Finished

	Metric ton liquid	Cu ft liquid	Cu m liquid	Bbl liquid	Gal liquid	Cu ft gas	Cu m gas	Million Btu	Million kJ
1 metric ton liquid	1	83.26	2.358	14.83	622.8	52030	1397	52.6	55.5
1 cu ft liquid	0.01201	1	0.02832	0.1781	7.481	624.9	16.78	0.632	0.667
1 cu m liquid	0.4241	35.31	1	6.290	264.2	22070	592.5	22.3	23.5
1 bbl liquid	0.06743	5.615	0.1590	1	42.0	3509	94.22	3.55	3.74
1 gal liquid	0.001605	0.1337	0.003785	0.02381	1	83.54	2.243	0.0845	0.0891
10^6 cu ft gas	19.22	1600	45.31	285.0	11 970	10^6	26 850	1010	1070
10^6 cu m gas	715.8	59590	1687	10 610	445 800	$37.24(10^6)$	10^6	37700	39700
1 million Btu	0.01900	1.581	0.04478	0.2817	11.83	988.3	26.54	1	1.054
1 million kJ	0.01802	1.500	0.04247	0.2672	11.22	937.3	25.17	0.948	1

LNG Properties (Based on pure methane)

Boiling point @ 1 atmosphere (storage temperature)	−259°F	−161°C
Liquid density @ normal boiling point	26.47 lb/ft³	424.1 kg/m³
Gross heating value:	1010 Btu/ft³	39 700 kJ/m³
Gas to liquid ratio:	625 to 1	593 to 1

Note: Gas volumes in cubic feet are for gas at 14.73 psia and 60°F.
Gas volumes in cubic meters are for gas at 1 atmosphere absolute and 0°C.

Table 10–1 LNG Conversion Factors (courtesy of CBI Industries, Inc.).

products; gasoline, kerosine, jet fuel, and many petrochemicals leave the refineries in trucks en route to the customer. And even though a product may be shipped from the refiner to the jobber via pipeline, rail, or barge, chances are that it will leave the jobber's terminal by truck on its way to the neighborhood filling station. Such large vehicles, particularly those used in interstate trade carrying flammable materials, have to operate within the pertinent federal (Interstate Commerce Commission) and various state safety regulations. Some states will permit a truck tractor with a semi-trailer and one additional trailer, while other states only permit a truck tractor with a semitrailer. Load capacities also vary from state to state; the average load is 8,800 gal of gasoline.

Modern motor tank trailers are made of steel, aluminum, or stainless steel with an attractive paint job. They are most always decorated with the owner's logo and are successfully used as a means of advertisement. Petroleum trucks are visible throughout all major highways and most off-roads in the United States, Canada, and the other countries of the world.

Pipelines

Petroleum pipelines serve a dual role. Initially, they gather crude oil from the producing field and transport it to the refinery; then they transport the refined products to various markets. Although natural gas is a hydrocarbon that is produced from an underground reservoir, often with crude oil, pipeline transportation of natural gas is a separate industry from the pipeline transportation of crude oil. Pipelines are vast networks of gathering, transporting, and distribution systems comprised of hundreds of thousands of miles of pipe.

The exact details of the world's total pipeline mileage, diameters, and throughput capacities are not available, and with the constant revamping and new construction, they are difficult to estimate. With the records on all interstate pipelines regulated by the federal government in the United States, however, it is possible to obtain an approximation of the current mileage. There are about 38,000 mi of crude flow and gathering lines, almost 58,000 mi of trunk lines, and more than 76,000 mi of petroleum products lines. There are about 73,000 mi of natural gas field gathering lines, and almost 4,600 mi of miscellaneous piping, and approximately 197,000 mi of gas transmission pipelines. The overall total mileage is in the range of 447,000 mi, which includes a growth of about 27,000 mi in the last decade.

The "grandaddy" of all pipelines is, of course, the *Alyeska* or *Trans-*

Alaska Pipeline System, which runs more than 800 mi from the Prudhoe Bay producing field on Alaska's North Slope, to the port of Valdez on Alaska's southern coast. Here, it terminates in large storage tanks overlooking the harbor where the oil awaits transfer to tankers which will carry it to the lower 48. Much of the oil is destined for the eastern and central United States; however, the supertankers carrying it will not fit in the Panama Canal. Thus, it must be offloaded on Panama's Pacific coast and pipelined across the isthmus to port facilities on the Atlantic Coast where it is reloaded into tankers for shipment on to the United States (Figures 10–6, 10–7, and 10–8).

Laying this network of pipelines is a grand accomplishment considering that the first cross-country oil pipeline in the United States was 109 mi of 6-in. diameter line laid from Bradford to Allentown, Pennsylvania in 1879. The early pipeliners were bona fide pioneers who worked long hours without the aid of mechanized equipment. Construction was largely performed by hand labor. This included clearing the right-of-way, digging the ditch, screwing the pipe ends together, lowering the pipe in the trench, and then backfilling. Horse-drawn wagons were used for stringing the pipe, and hauling in supplies over rugged trails where there were no roads.

In 1895, steel line pipe became available and replaced the wrought iron pipe that had been previously used. The first steel pipe joints were screwed together via a threaded collar. By 1917, lines were being welded together by the oxyacetylene welding process, and by

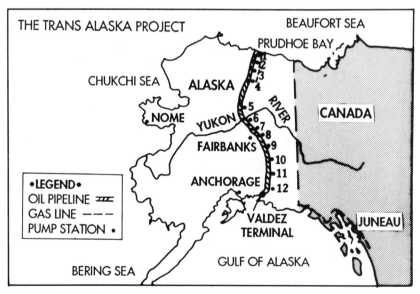

Figure 10–6 Route of the Alaskan pipeline, showing pump stations (courtesy API).

Modern Petroleum

Figure 10–7 The end of the Alaskan pipeline at the Valdez Harbor (courtesy CBI Industries, Inc.).

Figure 10–8 Oil from Alaska is transported to Panama by tanker and then pipelined across the Isthmus (courtesy CBI Industries, Inc.).

Transportation

1920, welded joints became commonplace on large crude oil lines. In 1928, both the electric arc-welding process and 40-ft lengths of seamless line pipe were developed. Welding technology has progressed to keep pace with the use of new pipe steels, increased pipe diameters and pipeline lengths. All of these advances have made possible the laying of today's offshore and arctic pipelines.

Pipelines are generally identified by their application, size, location, and the specific liquid, gas, or solid being transported. Crude oil pipelines include *flow lines, gathering lines, trunk lines,* and *distribution systems* (also called *product lines*). Natural gas pipelines include flow lines, gathering lines, and *transmission lines.* There are also *two-phase pipelines* that carry both a liquid and gas, LNG pipelines, CO_2 pipelines, and coal slurry pipelines.

Crude Flow Lines

A flow line is the pipe that petroleum fluids flow through from the wellhead to the tank batteries. It is the first surface pipeline that the fluids actually travel through. They are usually very short in length, from only several feet to a mile in length, depending upon the location of the tank batteries. In an oil field with relatively low pressures and production, the pipe is 2 ⅜ in. OD (commonly called 2 in.). In fields with higher pressures and production, larger flow lines with increased yield strength will be installed. Another consideration for use of high strength pipe is that the consequences of uncontrolled well flow can be so severe, especially offshore, that all piping must be designed to withstand any surge of high pressure. Proper safety systems can sense conditions on the lease or offshore platform and shut in the well or wells when conditions very from preset limits. In recent years there has been a growing use of flexible pipe.

Inland flow lines should be sized to allow for moving low gravity oil during cold winter months. Lines too small may become plugged and shut down production during cold winter months. In some areas it is necessary to inject chemicals into above-ground flow lines, which face the danger of cold-weather plugging.

Other additives are used to reduce friction in lines, particularly long ones. Their use not only allows more oil to be moved in a given period of time, but they also reduce the amount of energy needed to move the oil through the line.

Still other additives contain antimicrobial agents. These are very effective in protecting oil lines from bacteria that form slime and from biofilm that forms on pipe walls. They kill biological growth that diminishes flow rates, and they also inhibit anaerobic sulfide-producing bacteria that occur between the biofilm and the pipe wall and cause corrosion.

Specialized additives are also available for natural gas pipelines. These combat sulfate-reducing bacteria that cause highly corrosive sulfides and eliminate other bacteria that secrete organic acids.

Crude Gathering Lines

Gathering lines are used to transfer crude oil from tank batteries of individual leases and other production units to a central location. They often consist of pipeline branches flowing into a pump station or other facilities where oil is transferred to a trunk line system. Gathering lines are the second phase of the transportation network for moving oil, except in cases where motor tank trucks are used to pick up oil from tank batteries.

As a general rule, gathering lines are longer that flow lines, but shorter than trunk lines. They may range from 4 in. through 12 in. diameter, depending upon the quantity of oil transported. A gathering system could be quite extensive and include several pump stations, or it could collect oil produced in only one field. Testing must be done on the quality and quantity of the oil coming in from each lease for proper custody transfer before entering the trunk lines.

Crude Trunk Lines

The next phase of transporting crude oil is by the trunk lines that form an artery which moves the oil from the gathering system to processing points or terminals for further handling. Generally these pipelines are much longer and larger in diameter than gathering lines. Usually trunk lines are buried in the ground, except where conditions exist that warrant putting them above ground. An example are portions of the Trans-Alaska Pipeline that zig-zag across the barren snow-covered North Slope on above-ground supports. Their design converts pipe thermal expansion and movement from other forces into a controlled sideways movement on the supports (Figures 10–9 and 10–10).

Almost all crude trunk lines are welded steel pipe that is also coated and wrapped for corrosion mitigation. It may be double-jointed, coated and wrapped at a yard, then hauled and strung along the right-of-way; or it can be coated and wrapped over the ditch by the crew.

Although the term pipeline seems to refer only to the pipe and fittings, a modern trunk line system is a highly complex work of engineering involving countless components. Facilities used to

Figure 10–9 Portions of the Alaskan pipeline built above the tundra must zig-zag to compensate for thermal expansion (courtesy API).

Figure 10–10 Supports of the above-ground portions of the Alaskan pipeline are really heat exchange devices that protect the thermal integrity of the soil supporting the line (courtesy API).

transport crude oil from the gathering depots to refineries and other terminals include pumps and pumping stations, methods of controlling oil flow, means of measuring and accounting for the volume, storage tanks, and various other devices used in quality control.

Crude trunk lines must cross roads and streams, traverse wildlife areas, and other regions that require environmental permits to cross. Obtaining these permits can be a complex and lengthy process that often delays construction for months. The Trans-Alaskan oil pipeline required an act of Congress to speed environmental review in the courts so that permits could be granted.

Products Pipelines

The products pipeline system's role, especially in the United States, is primarily one of distribution of the petroleum products from refineries to the storage and distribution terminals in the consuming areas. The refineries on the coast of the Gulf of Mexico constitute the focal point of the largest gathering operations in the United States as well as its largest refining center. Fifty percent of the country's refining and sixty percent of the petrochemical industry is concentrated along the Gulf Coast from Brownsville, Texas, to New Orleans, Louisiana. Although both inland water transport and deep-sea tanker movements are available from the coasts of Texas, Louisiana, and Mississippi, all major refineries in the area depend heavily on the services of products pipelines service, in addition to water-shipping.

Products shipped include several grades of gasoline, aviation fuel, diesel, and home heating oils. Products pipelines can transport several different products in the same pipeline in *batches.* Segregation of the shipments is sometimes achieved by use of physical batch separators, however, in most cases no separator is required because the difference in density of the two materials maintains the separation during turbulent flow, with only a short length interval in which mixing occurs. The extent of mixing and the position of each batch can be monitored at points along the line by measuring the densities of the products (Figure 10–11).

Normally line *scrapers (pigs)* and spheres are used for pipeline cleaning. Scheduled scraper runs will remove water and sediment from the pipeline and assist in preserving product quality. Booster stations along the pipeline are designed to automatically pass the scrapers and spheres. Input stations, terminals, junctions, and points where pipe diameter changes, are usually equipped with launchers for inserting and receivers for receiving spheres and scrapers.

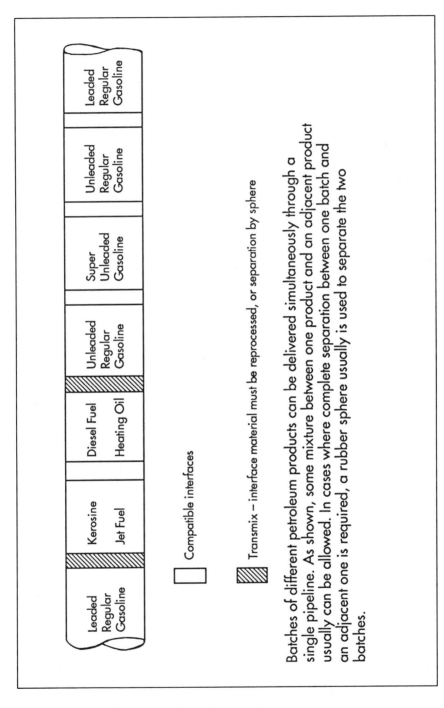

Leaded Regular Gasoline | Kerosine Jet Fuel | Diesel Fuel Heating Oil | Unleaded Regular Gasoline | Super Unleaded Gasoline | Unleaded Regular Gasoline | Leaded Regular Gasoline

☐ Compatible interfaces

▨ Transmix – interface material must be reprocessed, or separation by sphere

Batches of different petroleum products can be delivered simultaneously through a single pipeline. As shown, some mixture between one product and an adjacent product usually can be allowed. In cases where complete separation between one batch and an adjacent one is required, a rubber sphere usually is used to separate the two batches.

Figure 10–11 Batch transport of petroleum products.

Gas Pipelines

Natural gas may be produced from a gas reservoir, or it may be produced from a formation that produces both crude oil and gas. When it is produced with crude oil, both share the initial surface flow line from the wellhead to the gas separator. From the separator onward, the natural gas is transported in its own pipeline system, except in some special installations where two-phase pipelines are used.

Technically, a gas pipeline system is similar to both crude oil and products pipelines with respect to the actual pipe and fittings used, and the methods of constructing the system. However, there are certain pertinent differences between them. First, in the United States, a gas line built to transport natural gas in interstate commerce is not considered a common carrier, but a public utility. As public utilities, gas pipeline owners enjoy the right of eminent domain when it comes to acquiring right-of-way, but at a going market price. If the company and the property owner cannot agree on a figure, then it is set by a court.

Gas pipelines in the United States are regulated by the Federal Energy Regulatory Commission (FERC) which approves tariffs and other service rates pipelines can charge in interstate commerce. This has been an area of controversy in recent years, with some pipelines wanting regulation concerning rates maintained, and producers wanting them removed. Again, this represents a problem not easily solved. The users, of course, want to purchase gas as cheaply as possible. The producers say that low rates do not allow them to carry out exploration, development, and production of new sources of natural gas. FERC also issues permits governing the economic and environmental aspects of interstate gas pipelines.

Secondly, gas is moved through the pipeline system by compressors and compressor stations rather than by pumps and pumping stations. Gas pipeline systems usually operate at higher pressures than crude or products lines.

Natural gas pipelines systems must maintain a certain pressure in order to supply a volume of gas to the consumer. This is made possible by many factors including an adequate supply, proper size pipe, regulating equipment, and compressor stations. Flowing high-pressure wells must have their pressure reduced at the well by a choke or regulator before entering a low to moderate pressure gathering system, so the lower pressure wells can flow into the system. On the other end of the range, a compressor can be installed at or near the well to boost the pressure enough to permit it to flow into the gathering system. Flow lines connect individual gas wells to a gathering system that eventually run to a processing plant where the gas is treated. Unwanted water is removed so that the specifications for dry gas can be met. Also, the plant typically removes varying amounts of propane, butanes, ethane, and heavier hydrocarbons that have a marketable value.

From processing plants, dry natural gas enters the gas transmission system for delivery to industrial and domestic customers. Most often delivery to the individual consumer is handled by utility companies with their own distribution piping system. Utility companies supply low pressure gas through small, metered pipes to their customers. Although steel pipe is used for practically all oil and gas gathering systems and main lines, plastic pipe is often used in gas utility systems.

Gas transmission pipelines, like crude oil trunk lines, are made of steel pipe with welded joints. They are cathodically protected and buried below ground. The newer systems are constructed of larger pipe that cover vast geographical areas and can be hundreds of miles in length.

The relationship between pipeline size, capacity, and economy is obvious. A 36-in. diameter pipe has more than eight times the cross section area of a 12-in. diameter pipe and can transport many times its volume, but right-of-way, construction, and operating costs do not increase at the same ratio. Pipe sizes very according to estimated throughput of gas and can be as large as 60 in. in diameter, such as a pipeline installed in Russia.

The growing demand for natural gas, advances in pipeline technology, and the campaign to move gas to markets have touched off a wave of pipeline construction around the world. Some of the longer gas transmission pipelines planned or in operation include a line from Alaska's North Slope for a distance of about 4,800 mi; a 36-in. diameter line, called the Trans-Canada, that extends 2,300 mi from the Alberta-Saskatchewan border to Montreal, and others, including a 2,800-mi pipeline from Russia to Europe.

Offshore Pipelines

Offshore flow lines and gathering lines serve the same purpose as those in an ordinary oil field onshore. That is, to transfer the oil and gas from the wellhead to a storage or handling facility. There are, however, fewer flow lines because most offshore wells are directionally drilled from central platforms where the wellheads are assembled on the platform. Other remote wells are connected by undersea pipelines to above-water platforms. Some platforms with a number of wells may be connected by undersea lines to separate platforms where the oil and gas is further processed or stored.

Modern Petroleum

Generally, the oil and gas gathered at a platform is transported to onshore facilities via underwater pipelines. The exception being that very few offshore platforms have a tanker loading capabilities where crude oil is loaded into a tanker rather than being piped ashore.

Where undersea pipelines are used to transport petroleum fluids from offshore platforms to onshore handling or storage, it could be economically feasible to use the two-phase pipeline instead of laying one line for oil and one for gas. Where a line is designed to carry a liquid, the presence of gas can cause a problem in flow and pumping rates, one reason being that liquids are almost non-compressible while gases are compressible. Special design of the system can make a two-phase line workable.

Some of the petroleum industry's greatest developments have been in offshore pipelining during the past few years. New-generation lay barges and ships are now able to lay large-diameter pipelines in deep water at surprisingly fast rates. The larger barges are constructed with several welding stations, an X-ray control station to check the welds for imperfections, and pipe coating and cementing stations.

Coating is applied for corrosion protection and a coating of concrete is added to provide "negative buoyancy" or the weight needed to keep the pipe on the sea floor or in the trench. A small number of barges now use dynamic positioning that accurately and continuously feeds the barge's position into a computer that controls the operation of the positioning thrusters, thus eliminating the use of anchors. Welding, X-ray inspection, and coating stations remain in the same position on the barge but the pipe moves through these stations and on to a stringer as it is lowered to the seabed.

Offshore reel vessels are now being used more frequently to lay small-diameter pipe that is specially constructed and placed on a reel onshore for unspooling at sea. Special straightening rollers are used as the pipe is unreeled into the water to return the pipe to its original straightness.

Most offshore pipelines must be buried below the ocean bottom to protect the pipe from damage by ship anchors and natural hazards. In certain areas offshore, pipelines are buried according to government specifications, all depending on the country in whose waters the pipeline is located. In Tokyo Bay, Japan, a pipe was originally required to be buried to a depth of 16 ft, and North Sea lines were required to be buried with 10 ft of cover. However, more flexible and realistic requirements are now forthcoming.

Plowing and jetting are the two most common methods of burying underwater pipelines. Plowing can be done before the pipe is laid, or it may be done after the pipe is in place by opening a trench alongside the pipe then moving the pipe into the trench afterwards. Jetting is done after the pipe is in place by use of a jet sled with powerful pumps and jets that are directed under the pipe that force soil from beneath the pipe and permit the pipeline to settle into the ditch.

New technology in underwater line welding procedures have been developed that permit pipe sections to be replaced or to hot tap an existing line with ease. Welding methods utilizing inert gases are in use that are both safe and also permit superior weld quality.

Various Type of Pipelines

There are several other types of pipelines besides oil, gas, and product lines. Some carry non-petroleum products, and others transport petroleum-related products. The non-petroleum lines include coal slurry pipelines and carbon dioxide pipelines. Petroleum-related products pipelines include liquefied natural gas (LNG) and liquefied petroleum gas (LPG) lines.

Coal slurry pipelines have been in operation in the United States for many years. They are designed to carry finely ground coal solids in water. However, most of the coal in demand today is located in western states where water is scarce. The water problem has largely been worked out, and it is possible that more lines will be in operation in the near future.

The use of carbon dioxide (CO_2) for enhanced oil recovery has prompted the construction of pipelines from areas where there is a supply of naturally occurring CO_2 to oil producing areas. It is estimated that the use of carbon dioxide injection could recover additional millions of barrels of oil from existing fields.

Liquefied natural gas pipelines are in operation where LNG tankers are loaded and unloaded. These pipelines are made of special steels and require super insulation to keep the product in a liquid state. Prospects of long-distance LNG pipelines are becoming more realistic as extended studies of their feasibility are being made.

Liquefied petroleum gas (LPG) is much easier to transport than LNG because the minimum temperature is in the –50°F range and more conventional materials and construction designs can be used. In some of the Middle East countries where liquid-hydrocarbon-rich associated gas is produced with oil, there are pipelines that carry propane and butane to refineries and petrochemical plants for feedstock.

Line Pipe Standards

The current API standards applicable to line pipe are covered in the latest editions of API Standard 5L which covers seamless and longitudinally welded steel pipe in Grades A and B. Grade A line pipe has a minimum yield strength of 30,000 psi; Grade B a minimum yield of 35,000 psi. API Specification 5LX applies to high-test line pipe, both seamless and longitudinally welded, in Grades X42 through X70. X42 indicates pipe made of steel with 42,000 psi minimum yield strength, etc. API Specification 5LU covers ultra-high-test, heat-treated seamless and welded pipe in grades U80 and U100. API Specification 5LS applies to spiral-weld pipe in Grades A and B. API Specification 5LP is for Thermoplastic Line Pipe (PVC and CPVC) which is suitable for use in conveying gas, oil, or water in underground petroleum service. API Standard 1104 is the standard for welding pipelines and related facilities, and API Standard 6D is applicable for pipeline valves.

Although much of the line pipe is manufactured under one of the several API specifications, that does not inhibit purchasers and producers from using products made to other specifications.

Inland Pipeline Construction

While all pipelining techniques have some common characteristics, the environment will dictate the final cost, the complexity and technology, and even the choice of contractors to construct the line. Planning a pipeline begins with both a supply of, and a demand for, the oil or gas that will economically justify the cost of building a line. Tentative routes are selected on the basis of aerial mapping and careful survey on the ground. After a route has been adopted, the pipeline company obtains a permit, if necessary, from the government agency involved.

Right-of-Way

The company must obtain a *right-of-way* easement covering a strip of land 50 to about 100 ft in width and the length of the property involved, before the actual clearing and grading begins. The term *clearing* includes removal of fences, trees, brush, crops, boulders, and other obstructions. *Grading* means leveling the ground surfaces as needed to permit transit of trucks, trenching machinery, and other associated equipment.

Stringing the Pipe

There are no set rules for the sequence of hauling and stringing pipe. Sometimes the line pipe is strung before the ditch is dug and sometimes after. A contract might specify that the pipe be made available at pipe mills, storage yards, or pipe coating and wrapping plants. In these cases, the pipe is hauled to the right-of-way and strung according to scheduled construction progress. In other contracts, trucks and handling equipment unload pipe from railroad cars, barges, or other common carriers and haul and string it. Sections of pipe may be *double jointed* into lengths of 40 ft, or even welded into longer sections, all depending upon the type of roads and bridges that must be used.

Ditching

Ditching includes all excavation work that is required to provide a ditch of the specified dimensions and depth, usually 3 ft or deeper. The most common method of ditching is done by use of a ditching machine of all the soil conditions are favorable. Ditching may also be accomplished by a backhoe, plows, drag lines, or other methods when necessary (Figure 10–12).

Each pipeline project often involves crossing roadways and other obstacles. A variety of techniques may be used, depending on the length of the crossing and regulations. Crossing roadways usually can be done by either ditching or boring. When ditching, the roadway must be repaired to its original condition upon completion. This method is seldom permitted. More usually, a horizontal boring machine is used to drill a hole under the roadway without disturbing traffic of the road surface. The hole is sized to allow for a conductor pipe to be installed and the pipeline is then placed inside the conductor (Figure 10–13).

Bending and Laying Pipe

Every time the ditch changes direction or elevation, the pipe must be formed to fit it. Thus in hilly country, many joints of pipe must be bent or curved. Small pipe can be formed around a bending shoe, but large diameter pipe must use special mandrels and bending machines to keep the pipe round and distortion free. After the pipe is properly bent, it should be prepared for laying. Laying will include swabbing the pipe, buffing the beveled ends, and lining it up for welding. Lining up consists of placing lengths of pipe in position with welding clamps and leaving a space between the faces of the pipe ends so that proper welding can be achieved.

Welding and Inspection

After the beveled pipe ends are lined up the first pass is made. This is followed up by the welding gang, which makes the final welds. The number of welds made at each joint will depend on the method of welding and the thickness of the pipe walls. A complete weld may take from two to five passes to fill up the beveled space between the two joints and thus provide a continuously smooth surface the

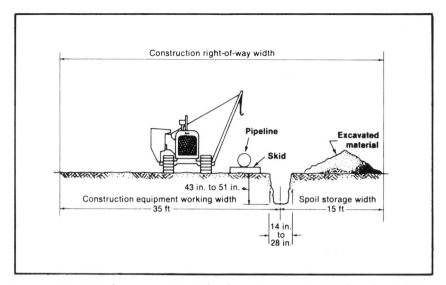

Figure 10–12 Pipeline construction right-of-way (courtesy PennWell Books, *Oil & Gas Pipeline Fundamentals*, Kennedy).

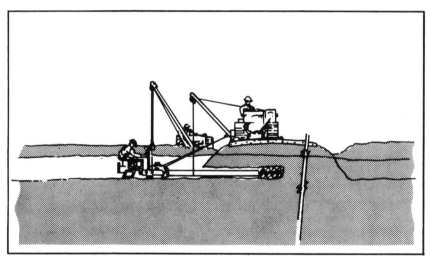

Figure 10–13 Boring under a highway crossing and pushing the casing in place.

length of the line. In between each pass, the weld just made is wirebrushed clean to provide a good surface for the succeeding weld (Figure 10–14).

Either stick or wire welding may be used. Sticks are the common electrodes or welding rods used in arc welding. Wire welding uses a wire-type electrode supplied to the welding gun on a spool. Wire welding may be enhanced by the use of CO_2 during the process.

Visual inspection of welds and observation of welding operations are relied upon to some extent. However, some contracts call for all or part of the welds to be X-rayed. X-ray or radiographic examination is a nondestructive method of inspecting the inner structures of welds and determining or inferring the presence of defects. Sources of the radiation may be X-ray machines or radioisotopes.

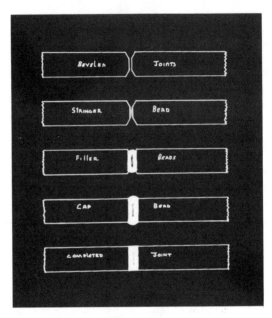

Figure 10–14 How pipeline joints are welded together to form a continuous line.

Coating and Wrapping

After welding, the pipe is ready for cleaning, priming, coating and wrapping. If the pipe has been previously prepared in a *pipe yard*, only the welds need to be attended to. If not, then the entire line must be field-coated and wrapped. Cleaning and priming machines scrub the pipe to remove all foreign matter and apply a coat of primer. When the primer has dried, a second machine moves along the pipe and applies a coating and then wraps the pipe with either plastic or kraft paper to combat external corrosion of the pipeline.

An electronic holiday detector, sometimes called a *jeep* is then passed over the pipe to locate any hidden flaws. This is a device which places an electrical potential between the pipe and an electrode, which is typically a coiled spring that fits around the pipe, in contact with the outside of the coating. The electrical potential is high enough to produce an arc when a *holiday* or gap appears in the coating, thus producing a sound which alerts the operator that there is a defect in the coating.

Lowering and Tie-In

Lowering pipe into the ditch requires using powerful forces with precise skill. The pipe may be lowered into the trench as part of the coating and wrapping process, thus reducing possible damage to the coating, and eliminating an extra step in handling. Or, the coating may be allowed to harden before the *side boom tractors* pick the pipe up from the wooden skids it has been resting on and lower it into the trench.

There may be numerous spaces left in the line where it was interrupted for various reasons. Short sections of pipe must be fully prepared and put in place across roads, rivers, creeks, and under other pipelines and underground objects. Installation of valves and other pipe fittings requires breaks in the continuity of the line. The welds connecting these various pipe sections into continuous pipelines are referred to as *tie-in welds*.

Backfilling

A variety of *backfilling* procedures may be performed to comply with specifications regarding protection of the pipe and coating. In areas of solid rock or rocky spots, it is necessary to pad the pipe with loose soil before backfilling. Frequently backfill soils need to be compacted to avoid later settling that would leave surface depressions. The right-of-way should be restored as closely as possible to its original state, with topsoil replaced so that vegetation will soon grow again. Fences and gates must be restored and trash removed from the right-of-way.

Swampland and Rivers

Construction activities in wet coastal areas or swamp lands is much the same as those on land except that most of the operations have to be carried out aboard barges. Dredges or clamshells aboard barges are used to open canals and trenches, and if access roads are not available, canals deep enough for supply barges may have to be opened. Most barges are designed with floating crew quarters and are large enough to handle pipe and other necessities.

Flotation devices may have to be provided for the pipeline until it is ready to be lowered from the lay barge, and then weighting is attached to hold it in place on the bottom. Construction in such an environment combines elements of both land and offshore construction and is handled by specialized contractors.

Crossings of rivers, canals, and other bodies of water are more expensive than normal construction. Special equipment has to be used and trenching is generally much deeper than inland. If the waterway is navigable, scheduling of construction must sometimes take interference with navigation into account. Directional drilling, similar to the techniques used to directionally drill oil and gas wells, has been used with success in special areas. A pilot hole is first drilled, then reamed to the desired size, and the pipe is pulled through upon completion. Control of the drilling is achieved by use of an instrument that measures the direction and inclination of the bit and transmits the information back to the operator so that necessary adjustments can be made.

Testing

Internal hydrostatic testing of the pipeline may be regulated by government agencies. These tests are most often performed by construction contractors or by companies specializing in testing. The pipeline must be tested after the trench is backfilled and construction completed. However, river crossings and other critical sections should be tested while they are still accessible, so that any needed repairs or corrections can be made before completion.

The purpose of hydrostatic testing is to locate any part of the system that might not withstand the designed working pressures plus pressures that are added for the test as a safety factor. Records of each test must be kept, and must include the date, proper identification of the pipeline or portion of the line tested, and the signatures of those responsible for the testing.

Chapter References

1988 Annual Report, CBI Industries, Inc., 800 Jorie Boulevard, Oakbrook, Illinois, 60522-7001.

John L. Kennedy, *Oil and Gas Pipeline Fundamentals*, (Tulsa: PennWell Books, 1984).

Oil Pipeline Construction and Maintenance, (Austin: The University of Texas, Petroleum Extension Service, 1975).

Roger Vielvoye, "Rising Demand Sparks Burst of Tanker Orders," *Oil & Gas Journal*, July 30, 1990.

11
Refining

Crude Oil

Crude oil is of little practical use in its natural state as it comes from the wellhead.

Crude oil is a very complex chemical compound consisting of many elements. It contains impurities such as sulfur, oxygen, nitrogen, and certain metals that must be removed. Its two principal elements are hydrogen and carbon, so the name hydrocarbon is often used when referring to crude oil, which we also know as petroleum. Hydrocarbons are compounds, which must be separated and processed to be of value.

The most important behavioral characteristics of crude oil is what happens to it when it is heated. When it is put in a still or other proper container and heated to a certain temperature, and this temperature is held constant, only part of it will evaporate. Each of the compounds of crude oil has its own boiling temperature, and this is what makes refining possible.

It has taken crude oil millions of years to develop into the state in which we find it in various underground formations. It is in a liquid state in reservoirs, which is about the only accurate statement that can be made about it in general, although it may also occur as almost solid *heavy oil* or as *tar sands*. Seldom do two reservoirs contain the same crude oil. Each stream produced has its own composition. Each well in the reservoir may have different compositions, and can change composition as the reservoir is depleted. Each type of crude oil has a unique distillation curve that helps characterize what kinds of chemical compounds it contains.

History of Refining

The practice of refining is far older than the petroleum industry. It began centuries ago with the simple distillation of many different raw materials. In Europe during the eighteenth century, scientists experimented with coal, shale, tar from seepages, and whale oil to see if new products, principally for lighting purposes, could be obtained. This early work led to the achievements of James Young of Scotland, who in 1847 was granted British patents for his process of distilling coal oil from bituminous shale and cannel coal. Young built an improved lamp which could burn the distillates made from these raw materials.

Samuel M. Kier, who had been selling his Pennsylvania crude as "Kier Rock Oil," a medicinal cure-all, became interested in refining in the early 1850s. He had far more crude than he had customers for his medicine, so he decided to see if the oil could be processed into kerosine or coal oil. He sought the aid of J.C. Booth, a Philadelphia chemist who redesigned an old iron kettle into a crude still. Fired with coal, its vapor line, surrounded by a water-cooled coil as a condenser, yielded a few gallons per day.

Kier began to market this product as medicine, too, but the design of a new lamp created a ready market for it as a fuel. At first, he used crude skimmed from brine wells, but later began buying crude from the Drake well. Others followed suit, and by 1859 there were some 80 coal-oil plants in the United States; but many of these quickly shut down after the advent of the Drake well and the resulting rapid developments in the infant industry.

In 1854, the developers of the Pennsylvania Rock Oil Company sent a barrel of their salt-skimmed crude to Prof. Benjamin Silliman, Jr. of Yale University and asked him to analyze it. Silliman, who at the time was the nation's leading chemist, believed that he could separate the oil into various components by distilling it at different temperatures. His first run produced a thin, clear liquid with a strong odor (gasoline). He heated the remainder to a higher temperature and this time condensed a straw-colored liquid (kerosene). After experimenting with 13 temperatures up to 518°F in his laboratory, he placed the remaining material in a copper still and heated it to 750°F, at which point a thick, dark oil condensed and formed white crystals as it cooled. Silliman recognized these crystals as paraffin, which he reported could be used for candles; the thicker oil could be used for lubricant, and the kerosene for an illuminant.

The first new refinery after Drake's discovery was built near Oil Creek in 1860 by William Barnsdall and William H. Abbott. Soon, they were followed by many more. By the end of the Civil War, more than 100 plants were using 6,000 bbl of crude oil per day. By 1870, a basic operating pattern had developed: produce as much kerosene as possible to light the lamps of the world and make as much lubricant as possible from the remainder of the crude. This was truly

Modern Petroleum

the age of mechanization, and the new machines created too much friction for the lard and tallow greases of the day to handle.

Gasoline, which because of its lower boiling point, "cooked off" first when the crude was heated in a still, was a dangerous waste by-product. Refiners considered themselves lucky if they could sell it for a penny or two per gallon. But mostly it was either burned off, or more likely, dumped in the nearest creek.

However, as the new century dawned, a series of events began that would change not only the oil industry and the art of refining, but the world itself.

In Europe, Gottlieb Daimler and Karl Frederich Benz, working independently, were perfecting internal combustion engines fueled by gasoline, and using them to power wheeled vehicles. In the United States, Thomas A. Edison had perfected the incandescent electric lamp and society was converting to electricity for lighting at a rapid pace. And in Detroit, Henry Ford, a mechanic at an Edison plant, was working on his own to realize his dream of building an inexpensive, reliable automobile that everyone could afford.

Not only did the coming of the automobile revolutionize American society, but the new device was fueled by gasoline, which previously had been the waste product of the oil industry, at almost the same time electricity was replacing lamp oil, the refiner's mainstay.

Thus, the refiners of the first half of the twentieth century were faced with a lessening need for kerosine and a growing demand for gasoline. It was during this same period that Herman Frasch of Canada developed a process that enabled the use of sour crude oil to make kerosene. Previously, sour crude oil had been considered objectional because of its odor and poor burning characteristics. The *Frasch process* opened the market for supplies of sour crude oil from fields in Canada and Ohio. This was important for it was feared that production from the Pennsylvania and West Virginia fields had peaked.

In the early typical batch run, the crude oil would begin to vaporize at 180°F. The temperature of the still would gradually be brought up to 1,000°F. The lightest product or fraction (the first portion of the crude oil to vaporize) was gasoline of 72–74° API gravity. The next was a 62–65°API naphtha or benzene. There was almost no market for these products, so they were either burned or dumped as waste.

The next cut or fraction was 40–50° API gravity kerosene, which was the principal product of the day. The residue left after gasoline, naphtha, and kerosene cuts were made was treated with acid and naphtha and run again with steam-refined stock. This was further treated with a lubrication material known as bright stock. Other stocks were obtained from 32–36° API gravity distillates. Petroleum greases were made with fatty oil and wax.

The early kettles soon gave way to *shell stills* (Figure 11–1). These consisted of a battery of vessels through which the crude oil or feedstock passed. Each vessel had a higher temperature than the previous one. The distilling process broke the crude oil down into different cuts. The lighter ends or aromatics were those that vaporized first at lower temperatures. The shell stills in some areas were replaced by large cylindrical stills called cheese boxes. These could hold up to 1,000 bbl of crude oil, had greater heat efficiency, and saved on labor costs.

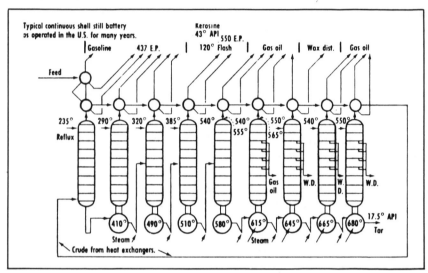

Figure 11–1 Diagram of Shell still refinery (courtesy *Oil & Gas Journal*).

Not long afterward, the first attempts at *fractionation* were made. Empty towers called fractionating columns were placed on the vapor lines from the stills to the condensers. These caught the heavier liquids carried by the vapors and returned them to the still. However, the fractionating pipe still as it is known today did not make its first appearance until 1917.

As we have seen, refiners of the first half of the twentieth century were faced with a lessening need for kerosene and growing demand for gasoline. As the demand for product increased, the oil industry was faced with a choice: either find new fields and use all the crude oil they could to get the desired end result or develop new processes that would increase the yield of product from existing supplies. They did both because new oil fields were discovered, and refiners did discover new and more productive techniques.

They reasoned that the place to start was with the residuum that was left after the traditional processing had taken place. The residuum (resid) was a thick, tough deposit left in the bottom of the still. It could be sold for use as fuel in place of coal, but after a few runs of crude through the vessel, the layer of thick and tarry resid that had built up had to be chopped out by hand.

Dr. William Burton, a Standard Oil of Indiana chemist, developed a successful cracking process that marked a milestone for refiners because now they were able to get a yield of 70% distillates, of which half was gasoline. The *Burton process* was in vogue from 1913 to 1920, when newer processes made it obsolete (Figure 11–2).

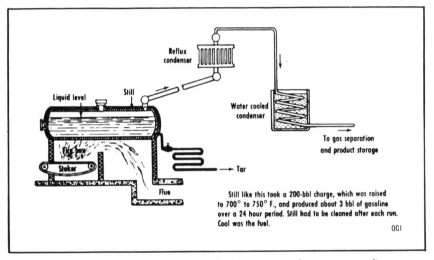

Still like this took a 200-bbl charge, which was raised to 700° to 750° F., and produced about 3 bbl of gasoline over a 24 hour period. Still had to be cleaned after each run. Coal was the fuel.

OGJ

Figure 11–2 How the Burton process worked (courtesy *Oil & Gas Journal*).

An entrepreneur aptly named Carbon Petroleum Dubbs developed a cracking process basically designed to cut down on the coke deposits that had plagued the Burton process (Figure 11–3). With the *Dubbs process*, fractions heavier than gas oil could be cracked, coke deposits were greatly reduced, and the unit could run for days without having to be shut down for cleaning.

Eugene J. Houdry, owner of a French structural steel firm, sought a catalyst that would crack hydrocarbons, then a way to remove the carbon that subsequently formed during operation. After three years of testing, he found that one of the reactors in his laboratory that was charged with heavy fuel oil was making a clear distillate of good gasoline qualities. The catalyst in the reactor was aluminum silicate. He solved the problem of the carbon deposits by burning them off, which became the key to the success of the *Houdry process* (Figure 11–4).

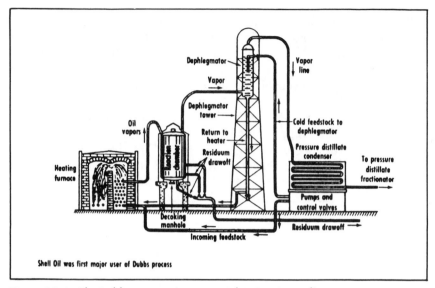

Shell Oil was first major user of Dubbs process

Figure 11-3 The Dubbs process (courtesy *Oil & Gas Journal*).

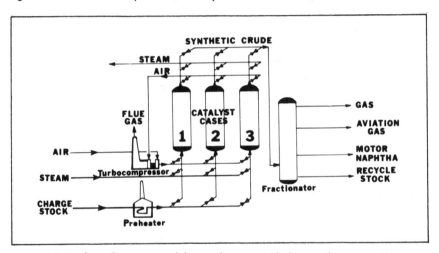

Figure 11-4 The industry entered the catalytic era with the Houdry process (courtesy PennWell Books, *Refinery Operations*, Vol. 2, Berger and Anderson).

In 1936, the first commercial plant went into operation. This was a fixed-bed process using activated bentonite clay. After 10 minutes it would be taken off stream, the vaporized oil sent to another reactor, and the catalyst regenerated with oxygen and gas.

In 1938, researchers discovered that gasoline of a much higher octane could be produced by treating the hydrocarbons with sulfuric acid. This process, called *sulfuric-acid alkylation*, provided most of the high-octane aviation fuel used during World War II and became one of the refiner's most important tools.

The catalytic process offered the opportunity for providing more and better gasoline components. By the time World War II finally broke out, there were 12 plants in the United States providing 132,000 B/D. By the end of the war, some 34 plants were in operation with a capacity of 500,000 b/d. *Cat cracking* was the major refining method until 1960.

The next innovation was *hydrocracking*, which utilized both a

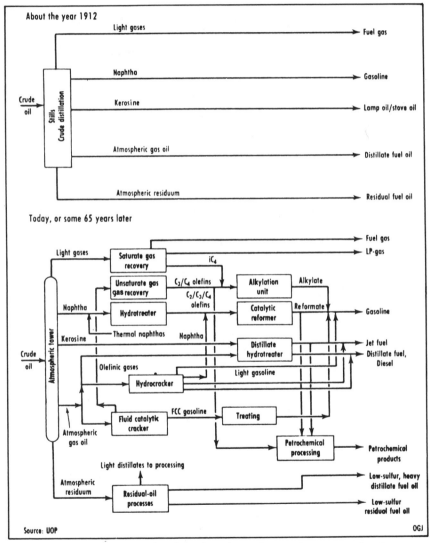

Figure 11–5a Comparison of early and modern refineries (courtesy *Oil & Gas Journal*).

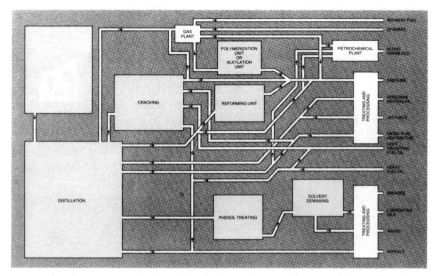

Figure 11–5b A variety of physical and chemical processes are used at refineries to turn a single substance (crude oil) into many substances (courtesy Canadian Petroleum Association).

catalyst and hydrogen to process residuals or product in the middle-boiling range to high-octane gasoline, jet fuel, and high-grade fuel oil. This came at almost the same time as kerosene reemerged as a demand product. The jet engine had been introduced during World War II, and in the two decades following most piston-engined military and commercial aircraft had been replaced by jets, which were fueled by a close cousin of the old lamp oil, kerosene.

Other modern refining technology includes reforming, which is the use of catalysts and heat to rearrange hydrocarbon molecules without altering their composition. The use of unleaded gasoline in all new automobiles is calling for increased refinery capacity and capability for this product at the same time the use of leaded gasoline is ending in the United States. Also in demand today are fuel oils of low-sulfur content (Figures 11–5a and 11–5b).

The average product yield from a barrel (42 U.S. gal) of crude oil has changed among gasoline, jet fuel, distillates, residual fuel oil, lubes and others almost every year. Gasoline has changed from 26% yield in 1930 to about 50% in 1975 and approximately 46% now. Jet fuel was non-existent in 1930, but now equals about 10% of product (Figure 11–6).

Basis for Refining

Refining is the breaking down of crude oil into the products desired. This is possible because crude oil is not a single chemical

Modern Petroleum

	GALLONS PER BARREL	% YIELD
TODAY		
Gasoline	19.5	46.4
Jet Fuel	4.2	10.0
Kerosine	.3	.7
Gas oil and distillates	8.6	20.5
Residual fuel oil	2.8	6.7
Lubricating oils	3.0	7.0
Other products	3.6	8.7
Total	42.0	100.0
1930		
Gasoline	11.0	26.1
Kerosine	5.3	12.7
Gas oil and distillates, residual fuel oil	20.4	48.6
Lubricating oils	2.4	5.7
Other products	2.9	6.9
Total	42.0	100.0

Figure 11–6 What is a barrel of crude oil?

compound but a mixture of hundreds of hydrocarbon compounds, each having its own boiling point. Since there is a whole range of boiling points, when a sample of crude oil is heated to successively higher temperatures, a boiling point or crude-distillation curve results (Figure 11–7).

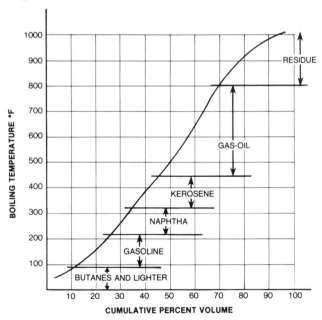

Figure 11–7 Crude oil distillation curve and its fractions (after Leffler).

As the temperature is raised, a point is reached where boiling starts. This is known as the *initial boiling point (IBP)*. Then boiling continues as temperature continues to be increased. Figure 11–7 shows that as boiling temperature increases, we first move through the butanes-and-lighter fraction of the crude oil. This starts an IBP and ends at just below 100°F. The fractions boiling through this range are known as the butanes and lighter cut.

Then the next highest fraction or cut starts at just below 100°F and ends at about 220°F. This is called straight-run gasoline. And starting at 220°F and continuing to about 320°F is the naphtha cut. The kerosene cut ranges from about 320°F to about 450°F, and the gas-oil cut is from about 450°F to 800°F. The so-called residue cut includes everything boiling above 800°F. The temperatures at which the various distilling products are separated are called the cut points.

It is not possible to continue the boiling process until all liquid has boiled away. If we heat the residue to temperatures much above 800°F., cracking begins to take place. This means that heat breaks down the heavy hydrocarbon molecules in the residue to smaller molecules, hydrogen, and carbon. If we wish to boil the residue to any extent, we must reduce the pressure on the residue by creating a vacuum. Then the residue can be boiled at a lower temperature with no danger of cracking. This process is called flashing and will be discussed later.

Crude Distillation

If we changed product containers at each cut point, we would be able to recover each cut during the crude-boiling process. If we went through the same steps in a large refinery, the process would be called *batch distillation*, and the vessel holding the boiling crude would have to be recharged with fresh crude oil each time the boiling vessel was emptied.

This process was actually used in the early days of the refining industry, but it became too time-consuming and was very inefficient. So refiners developed what is now called the *continuous crude-distillation process*. Crude is continuously pumped into the *distillation tower* and, as Figure 11–8 shows, products are removed at various positions in the tower.

This is made possible by the characteristics of the distillation column. This tall, large-diameter column is a cylindrical hollow steel tower containing flat steel trays welded perpendicularly to the column sides every few feet. Slots and holes in the trays permit vapor to flow up the tower and liquid to flow down. Vapor flows upward

Modern Petroleum

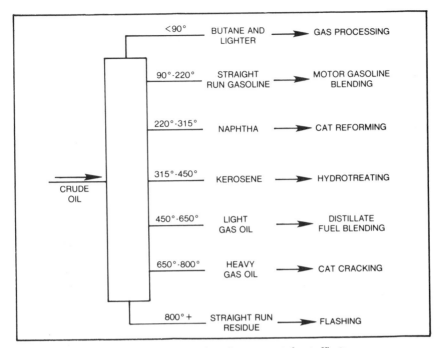

<90°	BUTANE AND LIGHTER	→	GAS PROCESSING
90°-220°	STRAIGHT RUN GASOLINE	→	MOTOR GASOLINE BLENDING
220°-315°	NAPHTHA	→	CAT REFORMING
315°-450°	KEROSENE	→	HYDROTREATING
450°-650°	LIGHT GAS OIL	→	DISTILLATE FUEL BLENDING
650°-800°	HEAVY GAS OIL	→	CAT CRACKING
800°+	STRAIGHT RUN RESIDUE	→	FLASHING

CRUDE OIL

Figure 11–8 Distilling crude and product disposition (after Leffler).

through the slots and liquid flows downward through *downcomers*, pipes that direct the liquids from higher trays to the trays below where they are either reheated and revaporized, or drawn off as liquids which have achieved their boiling points.

There are several ways to provide orderly flow of vapor through the slots. One of the most common is called a bubble cap. A small pipe is welded over a group of slots on the upper surface of the tray. The height of the pipe is equal to the desired level of liquid above the tray when the tower is in operation. The bubble cap, placed on the pipe, has slots in its side to permit vapor to come through the slots, through the pipe, down through the liquid, through the slots, and up through the liquid on the tray (Figure 11–9).

A downcomer to the tray below is placed opposite the downcomer from the tray above, so that liquid must pass in one direction and then is reversed. Liquid thus flows back and forth across column diameters as it moves down the tower.

When liquid reaches the bottom of the tower, it is removed and sent to a heater. There all but net bottoms product (straight-run residue) is vaporized and reentered into the column to provide a continuing source of vapor throughout the column.

When vapor reaches the top of the column, it is removed and all but

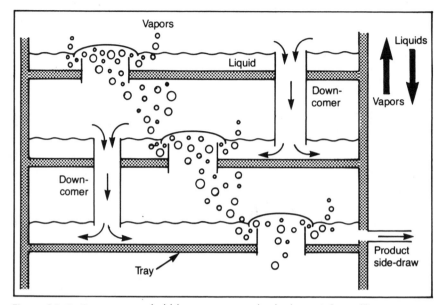

Figure 11-9 Downcomers, bubble-cap trays, and side draws (after Leffler).

the net overhead product (butane and lighter) is condensed and put back into the tower as a continuing source of liquid throughout the column.

The continuous admixture of liquid and vapor through the tower establishes a temperature pattern such that the tower-tip temperature corresponds to the upper cut point for butanes and lighter. The tower-bottom temperature is the upper cut point for gas oil. The cut points for the intermediate fractions may then be found with the lighter cuts in the top portion of the tower and the heavier cuts in the bottom portion.

Depending on temperatures, the intermediate cuts are withdrawn from the tower at the appropriate tower locations through what are known as side draws (Figure 11–9). Each of the cuts shown in Figure 11–8 is further processed in other parts of the refinery.

Flashing

The liquid recovered from the very bottom of the crude tower is subjected to vacuum flashing. As stated earlier, further boiling would lead to cracking or thermal breakdown of the very heavy residue. So a physical chemistry phenomenon is called into play; as pressure decreases, the boiling temperature for any given liquid also decreases (Figure 11–10).

If the effective tower operating pressure is lowered, then the residue could be heated further at a lower temperature without

Modern Petroleum

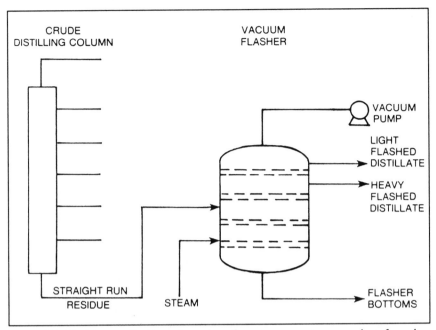

CRUDE
DISTILLING COLUMN

VACUUM
FLASHER

VACUUM
PUMP

LIGHT
FLASHED
DISTILLATE

HEAVY
FLASHED
DISTILLATE

STRAIGHT RUN
RESIDUE

STEAM

FLASHER
BOTTOMS

Figure 11–10 Vacuum flashing allows the refiner to recover more products from the straight-run residuum (after Leffler).

cracking. This is accomplished by lowering the pressure, or operating at a vacuum.

Atmospheric pressure is about 14.7 psi (pounds per square inch), and this is the approximate pressure on the crude-distillation tower. The pressure in the flasher is about 5 psi, which is created by a vacuum pump at the top of the vessel.

It takes heat to boil or flash liquids into vapors, so superheated steam is introduced into the flasher. This also adjusts the partial pressure of hydrocarbons in the vessel and allows close pressure control.

Several streams can be taken off the flasher: light flasher distillate, heavy flasher distillate, and flasher bottoms. The distillate streams go to other refinery locations for further processing or fuel-oil blending. The flasher bottoms may be used in residual fuel, may be blended to asphalt, or may go to a thermal cracker where the big hydrocarbon molecules are broken down to make thermally cracked stock for still further processing.

At this point we have now separated several cuts from the crude oil in the distillation column, and more in the flasher. These cuts are now ready for further processing and blending into refined products.

Catalytic Cracking

The two cuts just lighter than straight-run residue from the crude-distillation column are light and heavy gas oils (Figure 11–8). Prime use of the heavy gas oil is especially as a feedstock for the catalytic cracking unit. Light gas oil may also be fed to this unit, either as a separate stream or as a mixture with heavy gas oil, or it may be blended with other stocks into a heavy-distillate fuel oil. Light flashed distillate from the vacuum flasher may also go to the cat cracker.

In the United States, most gas oils are fed to the cat cracker for gasoline production. In other countries, the bulk of the gas oils goes to distillate fuel oil. The difference is that the United States with its large car population has a high demand for gasoline production, and the cat cracker is a gasoline maker.

In the catalytic-cracking process, a gas oil is subjected to heat and pressure in the presence of a catalyst. A catalyst is a substance that causes or enhances a reaction, but which itself is not changed in the reaction. In the cat-cracking process, the catalyst is developed and selected to convert the gas oil largely to gasoline, though the stream from the reactor contains a full range of hydrocarbons, methane through residue.

The heart of the cat cracker is the reaction chamber or reactor. Field gas oil is heated to about 900°F, mixed with a catalyst (usually a very fine powder), and introduced into the reactor (Figure 11–11). The time required for reaction is only a few seconds. Then the spent or used catalyst is separated from the hydrocarbon and sent to the regenerator. Here it is mixed with air and the carbon deposited during the reaction process is burned off at about 1,100°F under carefully controlled conditions.

Fresh (regenerated) catalyst exits from the bottom of the regenerator and is again joined with more incoming feed before entering the reactor. As the result of the combustion of carbon in the regenerator, flue gas from the regenerator is a mixture of carbon dioxide and carbon monoxide. Due to the heat released in the burning, this stream is very hot, and some heat is generally recovered in some other parts of the cracking process.

The cracked product from the reactor is pumped into a fractionator where five streams are recovered: C_4 and lighter, cat gasoline, cat light gas oil (LGO), cat heavy gas oil (HGO), and cycle oil. The overhead light stream goes to the cracked gas plant. Cut gasoline goes to gasoline blending. The cat-cracked LGO goes to light-distillate fuel oil blending, while HGO goes to heavy-distillate fuel oil. Cycle oil usually is sent back to the reactor feed; thus, it is recycled to extinction.

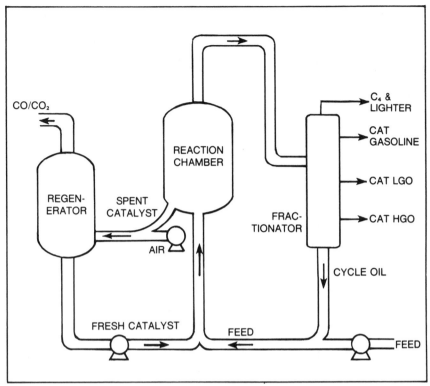

Figure 11–11 Cat-cracking unit adds greatly to the refinery's gasoline yield (after Leffler).

During the summer, some of the lightest LGO is moved over to gasoline (gasoline mode). In winter, some of the heaviest gasoline is moved into the LGO to maximize fuel-oil production (heating-oil mode).

Some typical cat-cracker yields are shown in the table below:

		% Volume
Feed:	Heavy gas oil	40.0
	Flasher tops	60.0
	Cycle oil	(10.0)*
Yield :	Coke	8.0
	C_4 and lighter	35.0
	Cat-cracked gasoline	55.0
	Cat-cracked light gas oil	12.0
	Cat-cracked heavy gas oil	8.0
	Cycle oil	(10.0)*

*Recycle stream not included in feed or yield total

Refinery Gas Plants

Almost all refinery processing units generate some gas (butanes and lighter). This gas is handled in a gas plant. The bulk of refining gas is saturated (no olefins), the principal stream of this type being the crude distillation overhead. The saturated gas is collected in the *sats gas* plant for further processing. Those streams containing olefins or unsaturates go to the cracked-gas plant for further processing.

Sats Gas Plant

Saturated gases from various parts of the refinery are collected at the sats gas plant. They are under very low pressure and must be compressed to higher pressure for processing (Figure 11–12). Following compression and cooling, some of the gas is liquefied and the two-phase mixture is sent to a gas-liquid separation drum. Overhead gas goes to an absorption system, while the liquid bottoms go to a fractionation train for separation into several products.

In the processing scheme, liquid from phase separation is joined by rich absorption oil from the absorber processing the gas stream. This brings in propane and heavier liquids, absorbed from the gas stream. The combined liquids feed a debutanizer fractionation

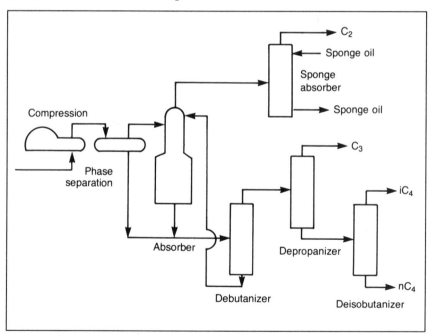

Figure 11–12 Sats gas plant recovers light hydrocarbon products (after Leffler).

column in which all butanes and lighter are taken overhead. The bottoms stream is recirculated to the absorber as lean oil.

The overhead stream from the debutanizer feeds a depropanizer, from which a propane product is taken overhead. The bottom stream from the depropanizer feeds a deisobutanizer, which makes an isobutane overhead product and a butane bottom product.

The vapor from phase separation goes through the aforementioned absorber for removal of propane and heavier liquids in the absorption oil, which then delivers these materials to the liquids fractionation train. Gas from the absorber goes through the sponge absorber for removal of lean oil lost from the absorber column.

Product gas from the sponge absorber contains methane and ethane. These usually go to refinery fuel, though the ethane can be recovered for petrochemical processing.

Propane recovered from the depropanizer goes largely to LPG use. Isobutane goes to alkylation, and normal butane is used almost entirely as a gasoline-blending component for octane improvement and vapor-pressure control.

Cracked Gas Plant

The *cracked gas plant* is similar to the sats gas plant except that the gases contain olefins or unsaturates: ethylene, propylene, and butylenes. Ethylene usually goes with methane and ethane to the refinery fuel system, though it can be recovered for petrochemical processing if desired. Propylene and butylenes usually are sent to the alkylation plant.

Olefins are one of a class of unsaturated hydrocarbons such as ethylene, which have many chemical potentials. In the chemical compound ethylene, C_2H_4, there is a double bond between the two carbon atoms to make up for the deficiency of hydrogen atoms. The double bond holding the two carbons together is actually weaker than a single bond. This makes the compound unstable and with relative ease it can be chemically reacted with some other compound or element to form a new compound, eliminating the double bond. That is why ethylene is so popular for making more complicated compounds.

Alkylation

The *alkylation* reaction joins an unsaturate with isobutane to form a high-octane component in the gasoline range. Propylene and butylenes are the unsaturates used. The reaction requires a catalyst, and either hydrofluoric acid or sulfuric acid may be used for this purpose. Figure 11–13 shows the flow diagram for a sulfuric-acid alkylation process.

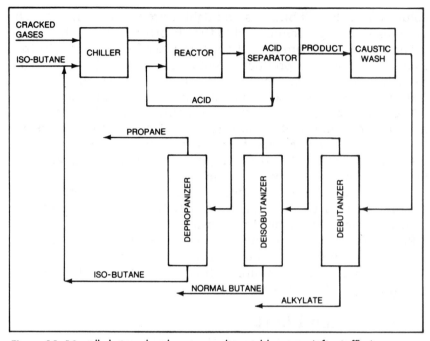

Figure 11-13 Alkylation plant boosts gasoline yield, octane (after Leffler).

Cracked gases, fresh isobutane, and recycle isobutane (unreacted) are joined in a chiller, which cools the mixture to the 40°F reaction temperature. Then the feed liquids are joined with the acid in the reactor system. A residence time of about 20 minutes is needed, so a battery of large reactors is used. Mixers ensure intimate contact of feeds and acid.

After sufficient residence time, the mixture moves to the acid separator. Here, there is no agitation, and the heavier acid quickly settles out and is withdrawn for return to the reactor. Hydrocarbon liquid from the acid separator is washed with caustic to remove any remaining traces of acid.

Then the product stream is passed through a fractionation train to recover alkylate, normal butane, isobutane, and propane streams. Butane and propane are handled as before (sats gas plant), and isobutane goes back to the chiller for another pass.

Catalytic Reforming

The *catalytic reforming process* operates on naphtha from the crude tower. The feed may also contain minor amounts of naphthas from coking, thermal cracking, and hydrocracking operations. Unlike the cracking processes where large molecules are whittled down to smaller ones, the cat-reforming process merely rearranges the naphtha-sized molecules without actually breaking them down.

Typically, there are high concentrations of paraffins and naphthenes in the feed naphtha. The cat reformer causes many of these materials to be transformed to aromatics, which have much higher octane numbers and some isomers. Much hydrogen is also produced which is used as needed in other sections of the refinery.

A rather unusual catalyst is needed for this process. It contains alumina, silica, and platinum, and some of the more sophisticated catalysts contain the metal rhenium as well. Great care is taken to keep track of this catalyst, because each unit has several million dollars' worth in its inventory.

Figure 11-14 shows flow through a typical fixed-bed unit. The naphtha feed is pressured to 200–500 psi pressure and heated to 900–975°F. It is then charged to the first reactor, where it trickles through the catalyst and out the bottom. This process is repeated twice in the next two reactors.

Product is then run through a cooler where it is liquefied. Then the accompanying hydrogen-rich gas stream is separated out and part of it recycled. The rest is sent to the gas plant for hydrogen recovery.

Liquid product feeds a stabilizer that takes off butanes and lighter (to the gas plant), leaving a stabilized gasoline-blending component.

Periodically, coke deposits on the catalyst cause a decline in activity and the catalyst must be regenerated. A fourth reactor (not shown on the plan sheet) is used, and one regenerator is being regenerated at all times with three onstream.

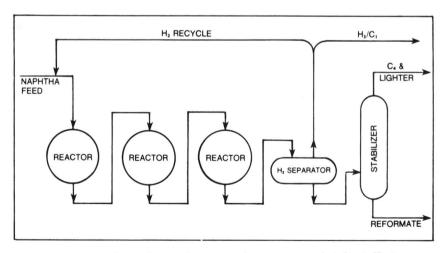

Figure 11-14 Catalytic reforming boosts naptha octane greatly (after Leffler).

After two or three years, the catalyst has collapsed enough to cause sufficient activity drop. The catalyst must then be removed and new catalyst substituted.

Residue Reduction

The processing units outlined thus far tend to concentrate on mid-boiling-range materials that have comparatively large volumes of heavy materials. Basically, two processes are used to process the heavy materials and convert appreciable amounts of residuals to lighter materials: thermal cracking and coking.

Thermal Cracking

Feed to a *thermal-cracking* unit is usually flasher bottoms, although cat-cracked heavy gas oil and cat-cracked oil may also be used.

If a broad range of feeds is to be processed, the lighter materials are kept separate from the heavier stocks. Each is fed to a separate furnace, since temperature requirements to cause cracking are higher for the light products. The furnaces heat the feed to the 950–1,020°F range (Figure 11–15). Residence time in the furnaces is kept short to prevent appreciable coking in the furnace tubes. The

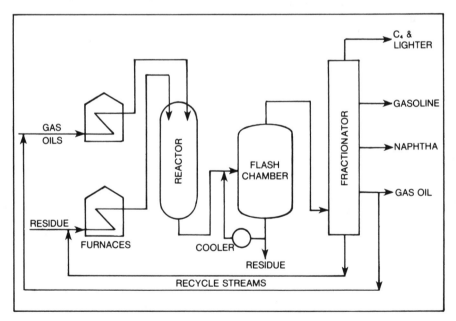

Figure 11-15 Thermal cracker helps reduce residual yield (after Leffler).

heated feed is charged to a reaction chamber, which is kept at high enough pressure (140 psi) to permit cracking but not coking.

Reactor effluent is mixed with a cooler recycle stream, and cracking is then stopped. The combined stream goes to a flash chamber, and the lighter materials flash overhead to a fractionator recovering light products: butanes and lighter (goes to the gas plant), gasoline (goes to gasoline blending), naphtha (to catalytic reforming), and gas oil (to cat cracker). The net heavy residue usually is blended into residual fuel.

Coking

Coking is severe thermal cracking. The feed is heated at high velocities to about 1,000°F and charged to a coke drum, where the actual cracking takes place (Figure 11–16). The lighter cracked product rises to the top of the drum and is drawn off. Heavier product remains and cracks to coke, a solid coal-like substance.

Vapors from the top of the drum are sent to the fractionator, which produces butanes and lighter, gasoline, naphtha, and gas-oil streams.

When the drum is full of coke, the vessel is taken offstream, cooled, and opened up. Decoking usually is accomplished by a high-

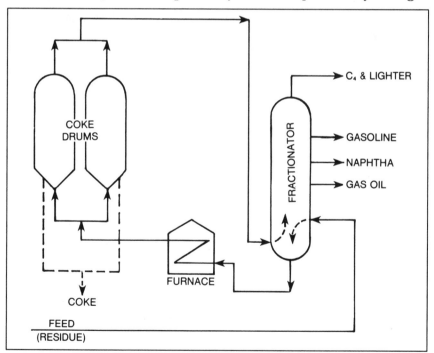

Figure 11–16 Coking reduces residual yield, enhances recovery of lighter materials (after Leffler).

pressure water jet that delivers lumps of coke to trucks or railway cars for shipment to factories which use it as fuel.

Hydrotreating and Hydrocracking Hydrogen

Many refinery streams have sulfur, nitrogen, and heavy metals in them. Treating with hydrogen, or hydrotreating, can effectively remove these materials and as well bring about some other benefits. In the hydrotreating process, the stream to be treated is mixed with hydrogen and heated to 500-800°F (Figure 11–17). The hydrogen-oil mixture is then charged to a reactor filled with pelleted catalyst. Several reactions take place:

1. The hydrogen combines with sulfur to form hydrogen sulfide (H_2S).
2. Nitrogen in some of the nitrogen compounds is converted to ammonia.
3. Any metals entrained in the oil are deposited on the catalyst.
4. Some of the olefins, aromatics, and naphthenes get hydrogen saturated and some cracking takes place, causing formation of butanes, propane, and lighter.

The stream from the reactor goes to the hydrogen separator, from which hydrogen is recycled to the reactor. The remaining materials

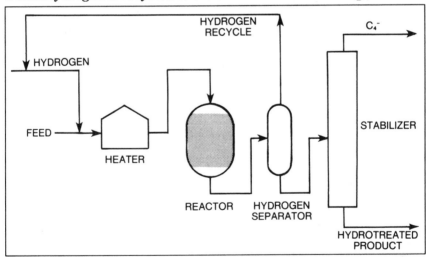

Figure 11–17 Hydrotreater removes sulfur, nitrogen, and heavy metals from refinery streams (after Leffler).

go to a stabilizer where light ends, including propane and lighter, hydrogen sulfide, and a small amount of ammonia are taken overhead. *Hydrotreated* product will exit at the bottom of the tower. Many of the refinery intermediate streams that require further processing are hydrotreated. Some of the product streams are also treated as well. Residual fuel, jet fuel, kerosene, and light and heavy distillate fuel are often hydrotreated.

Hydrocracking increases the overall refinery yield of quality gasoline-blending components. The process really is catalytic cracking in the presence of hydrogen. It can take as feed low-quality gas oils that otherwise would be blended into distillate fuel. The hydrocracker thus permits a fairly wide swing in a refinery's operation, from ultrahigh yield of gasoline in summer to high yield of fuel oil in the winter.

The hydocracking process features two or more fixed-bed reactors (Figure 11–18). Feed is mixed with hydrogen vapor, heated to 550–750°F, pressurized to 1,200–2,000 psi, and charged to the first-stage reactor. Here, about 40–50% of the feed is cracked to gasoline-range material.

The stream from the first-stage reactor is cooled, liquefied, and run through a separator. Hydrogen is recycled, while the liquid is charged to a fractionator. Here, such streams as butanes and lighter, light hydrocrackate, heavy hydrocrackate, and kerosene are taken off. The bottoms stream then goes to the second-stage reactor.

Conditions in the second stage include higher temperature and pressure. Outlet stream is sent to the hydrogen separator and thence to the fractionator. The product gasoline-blending components are the light and heavy hydrocrackates.

The hydrocracker, in addition to providing high-quality products, also has a 20–25% volume gain over the feed, as shown in the table below:

		Volume balance
Feed		
	Coker gas oil	0.60
	Cat-cracked light gas oil	0.40
		1.00
Product		
	Propane	
	Isobutane	0.02
	Normal Butane	0.08
	Light hydrocrackate	0.21
	Heavy hydrocrackate	0.73
	Kerosene range	0.17
		1.21

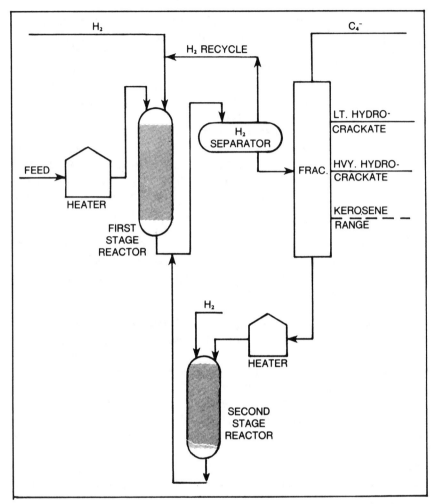

Figure 11-18 Two-stage hydrocracker process increases total amount of high-quality blending components for gasoline (after Leffler).

Hydrogen

The normal source of hydrogen is the *cat reformer*. The light-ends stream from the reformer column (stabilizer) is deethanized to produce a high-concentration hydrogen stream. However, the refinery may require more hydrogen than the cat reformer produces. Then a stream-methane reformer will be installed. As indicated, methane (the chief component of natural gas) and water are the two feedstocks required.

First, methane and water react at about 1,500°F to form carbon monoxide and hydrogen. Then more water reacts with the carbon monoxide to give carbon dioxide and hydrogen. Next, a solvent extraction process removes all but traces of the carbon dioxide. Finally, in a reaction called methanation, the remaining traces of carbon monoxide and carbon dioxide are removed by conversion back to methane and water.

Gasoline Blending

Vapor Pressure

These discussions on refinery processing have indicated a number of components available for gasoline blending: normal butane, *reformate* (94 RON), reformate (100 RON), light hydrocrackate, heavy hydrocrackate, alkylate, straight-run gasoline, straight-run naphtha, cat-cracked gasoline, and coker gasoline (*RON* stands for research octane number). All of these must be blended together to give the required vapor pressure, give the required octane number, and use up all of the available stocks.

The vapor pressure of gasoline is that pressure it exhibits to vaporize. Each component has a different vapor pressure, and for all practical purposes gasoline-blending components blend linearly by vapor pressure.

Required vapor pressure, related to the ability of gasoline to vaporize in an automobile carburetor, varies from one place to another and from summer to winter under government air quality regulations. And blending to vapor pressure is a function of ambient temperature in any one location. The practice is to blend various grades of gasoline according to octane-number requirements. Then normal butane or other octave additives are added as necessary.

Octane Number

Blending for octane number in the refinery again calls for a knowledge of the amounts of various available stocks and their octane numbers. Octane number is a measure of the knocking characteristics of the gasoline. The higher the octane number, the less tendency an automobile engine burning the gasoline will have to knock.

The term octane number comes from the test used to determine gasoline knocking characteristics. A test engine uses a mixture of two standard fuels, and tests the mixture against a given product gasoline. One of these two fuels is iso-octane (100 octane); the other is normal heptane (0 octane). Two octane numbers are determined on a gasoline blend: the *research octane number (RON)* and the *motor octane number (MON)*. The research octane number simulates driving under mild conditions, while the motor octane number simulates driving under load or at high speed. The number posted at the service station pump is (RON + MON)/2.

One other factor that needs to be covered in gasoline blending for octane is the use of *tetraethyl lead (TEL)*. Often, the refiner is faced with an octane deficiency when considering the various grades of gasoline that must be made. Until 1974, the refiner was free to use TEL as needed (up to 3 cc/gal) to suppress knock and thus enhance the octane number.

In 1974, EPA mandated a gradual phasedown of lead content in gasoline that started in 1975. Some smaller refiners as well as some larger plants had trouble meeting octane-number specifications during the change period. As of today, TEL has, for all practical purposes, been almost phased out of U.S. production.

With several blending stocks exhibiting different octane numbers and vapor pressures, making gasoline blends as required to meet specifications for several gasoline products is very difficult to work out. Probably the most successful technique for coping with all these variables (plus using up all available stock) is linear programming on a large computer to simulate refinery operations. The linear-programming technique solves all these requirements at maximum profit.

Reformulated Gasoline

Several major refiners have introduced *reformulated gasoline*, which is gasoline modified to contain more oxygen and to reduce hydrocarbon emissions. With reformulated gasoline, there are less olefins and aromatics, which are known *carcinogens* that can emerge unburned from the tailpipe. Reformulated gasoline will also reduce the amount of vapor emissions from automobile gasoline tanks.

Auto exhaust emission standards for the past two decades have focused primarily on vehicle design as a means of reducing pollutants. Little, if any, attention was paid to fuel composition. Now emphasis is also being placed on what goes into the gas tank. Fuel composition plays a variable role in emission reduction, depending on the emittant and the vehicle it is used in.

Joint research programs by both the automobile and the refining industry will provide a basis to identify clean fuels eligible for lower emissions. The main goals to be achieved by lower emissions are the reduction of ozone and carbon monoxide levels and the reduction of airborne toxic compounds.

Motor vehicles are believed to account for almost all carbon monoxide emissions, about 40% of all ozone emissions, and a significant amount of the volatile organic compound emissions in the urban areas of the United States. Thus, certain vehicle emissions are targeted for reduction including hydrocarbons, benzene, carbon monoxide, and oxides of nitrogen. Carbon dioxide emissions have also been studied, but they are not targeted at this time.

Carbon monoxide emissions from older vehicles that lack modern emissions controls can be reduced by blending gasoline with an oxygenated compound such as methyl-tertiary-butyl-ether (MTBE), ethyl tertiary-butyl-ether (ETBE), tertiary-amyl-methyl-ether (TAME), methanol, ethanol, or other alcohols. Lower emissions could also be achieved by a narrower gasoline boiling range.

Some refiners chose to take a positive position in the efforts to improve air quality in the United States. They began marketing reformulated gasoline that has been shown to reduce emissions or a few pollutants in several urban areas.

As a result of the present and proposed future U.S. EPA rules, and the Clean Air Act Amendments of 1990, specifications for the reformulation could be: vapor pressure limited to 8.5 psi RVP (Reid vapor pressure, the standard unit of measure for gasoline volatility), aromatics content limited to 25 V% (volume percent), benzene content limited to 1.0 V%, and that the reformulation have a minimum 2.7 wt% (weight-percent) oxygen content.

Distillate Fuels

Distillate fuels are those blended from light gas-oil range streams. Diesel oil and furnace oil are the two most often referred to. However, in many refineries, these are one and the same, and are sold out of the same tank.

There are several grades of diesel fuel. Regular diesel runs about 40–45 cetane; premium runs 45–50. The cetane number is determined much like octane number for gasoline. The test blends for comparison are mixtures of cetane (100) and alphamethylaphthalene (0). All light gas oils are candidates for diesel-fuel blending. Straight-run gas oil is prime, with a cetane number of 50–55.

The gas-oil hydrocarbons can also be blended into furnace oil, the most popular petroleum heating oil. It has a higher heating value than the lighter hydrocarbons, and its ignition characteristics are safer. Furnace oil also is easier to handle and more pollution-free than residual fuels. Furnace oil is called several other names as well (Number 2 fuel, distillate fuel, two oil).

Flash-point and pour-point specifications are most important considerations for furnace oil. The flash point is the lowest temperature at which enough vapor is given off to form a combustible mixture with air. This must be high enough to preclude danger of fire in the heating system.

In areas of the country where furnace oil is used for heating, distributors can avoid carrying over excess winter inventories by selling it for diesel use during the warm months. If the cetane number needs adjustment, this can be accomplished by blending in additives formulated for that purpose.

Residual Fuels and Asphalt

The refiner has two alternatives to handle the very heaviest high-boiling materials (residue) in the crude: residual fuel or asphalt. When asphaltenes occur in the residue, a strong, stable asphalt can be made. Asphalts are characterized in four ways: straight run, blown, cutbacks, and emulsions (Figure 11-19).

Straight-run asphalts come from the deep flashing of crude oil. Two important tests for these asphalts are softening point and penetration. Softening point is that temperature at which a standardized object will start to sink in the asphalt. Penetration refers to applied asphalt and measures the depth of penetration of a needle at standard test conditions.

Consistencies of softer grades of asphalt can be changed by blowing in hot air, which causes a chemical reaction. The resulting asphalt product is harder and more rubbery.

To reduce the severity of application conditions, a thinner (cutback) may be added. Naphtha or kerosene is usually used. After application, the diluent will evaporate, leaving a hard, durable asphalt.

Emulsion asphalts are also used to facilitate handling during application. The asphalt is emulsified with water and applied. Then the water evaporates, once again leaving a hard, durable asphalt.

Crude residue may also be used as residual fuel. It must be heated at all handling points to prevent solidification. Specifications include viscosity (how thick and sticky it is), sulfur content, and flash point.

Generally, flasher bottoms (crude residue) must have some kind of diluent to meet the maximum viscosity specifications. Cat-cracked heavy gas oil, relatively low in viscosity, and of low value as fuel for conversions, and other processes, is most often used as a diluent.

Residual fuels usually have at least some sulfur, and many may contain a great deal. The high-sulfur resids often are blended off to meet sulfur specifications, usually a 1% sulfur maximum.

Flash point is important because the residual fuel must be heated to flow. And, some relatively high flash-point material may have ended up in the resid. Usually, flash point is the limiting factor as to what can be dumped into the resid.

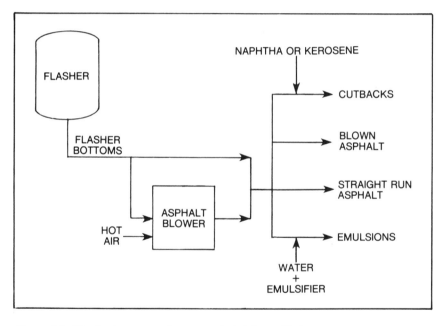

Figure 11-19 Options in production asphalt (after Leffler).

Chapter References

Guidance Manual for Operators of Small Gas Systems, (Washington, D.C.: U.S. Department of Transportation, Research and Special Programs Administration, 1985).

Safety Requirements for Gas Pipeline Systems Conducted for Materials Transportation, Bureau and State Agencies, 2nd Ed. (Washington D.C.: U.S. Department of Transportation, Transportation Safety Institute, Third Printing, January, 1987).

12
Marketing

Marketing is a term that is often misunderstood. To many, it means merely selling an item or a commodity. To an experienced businessman, however, marketing is a true science, one that includes researching a potential market to determine if there is a demand or need for a product, and then devising ways to fill that need. In some cases, marketing may even include creating a "demand" by consumers.

The marketing of petroleum products is an area far too broad to be covered in its entirety in a book designed to present a basic overview of the industry. It would take several volumes to describe in detail the marketing of all of the many products coming out of a modern refinery. Indeed, one would have to write separate volumes on the sale of asphalts, fuel oils, motor oils, jet fuel, and petrochemical feedstocks to name but a few. And, even then, another book or two would have to be written to cover the antics of the futures market in various crude oils and gasoline.

Instead, the authors have elected to present the story of the retail marketing of gasoline, since that is what the major portion of every barrel of crude is used for. Also, we wished to tell this story in a manner that would help you, the motorist—the retail consumer of gasoline—understand the many events which have to take place before you can wheel your car into the nearest gas station and "fill 'er up".

Thus marketing, for our purposes here, means the point where everything we have discussed earlier finally comes together. Where all the geology, exploration, leasing, drilling, and refining at last results in retail products. In this case, gasoline and motor oils, being offered to a customer for use in her automobile. "Her" automobile, because today women are the majority purchasers of gasoline at the pump, a phenomenon that has not escaped the attention of the marketers and has brought about new trends in the industry.

While perhaps less "hardware oriented" than other phases of the industry since no massive drilling rigs or ultra large crude carriers are involved, marketing is one of its most essential elements. During the past decade there have been sweeping changes in marketing just as there have been in all the other endeavors which make up the oil and gas business. Up until now, the major profit in oil and gas has traditionally been between the wellhead and the refinery. Today, the area of profit is shifting downstream, and the money is being made between the refinery gate and the nozzle placed in the customer's gas tank.

These changes have affected both branches of marketing: the actual retail marketers of petroleum products and the manufacturers of retail marketing equipment. Each change to one has brought about yet more changes to the other.

The Marketers

Integrated Companies

Over the years, two methods of selling petroleum to the public have evolved. The first is by an integrated company: a company that is large enough, either by itself or together with its affiliates, to engage in all aspects of the petroleum industry. Such a company prospects, drills, produces, refines, transports, distributes, advertises, and sells consumer products at retail at a station bearing the company's emblem, thus insuring a demand and outlet for the company's products (Figure 12–1).

A variant of this method are the independent dealers who own their stations but who sell the name-brand products of an integrated company which they purchase either directly from the company or through a jobber. In the early days of this century, the integrated companies set up their own distribution and sales networks. As roads improved and the number of automobiles increased, thousands of stations sprang up. The greatest period of growth in the number of gas stations was in the late 1950s and 1960s. During this time, vast sums were spent on advertising and on the development of brand names and trademarked merchandise. Generally, the local jobber was an independent businessman who contracted with a particular company as their agent to sell their brand of product at wholesale in a given trade area.

Thus the filling station operators selling the name brand product of a major, integrated oil company fell into two categories: they either operated a company-owned station, or they owned their premises and sold the company's product through their own pumps. Either way, owning or operating a filling station was part of the American dream, because it presented an opportunity for an individual to go into business for himself and to become his own boss for a very

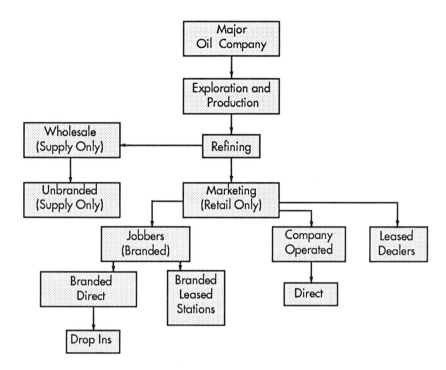

Figure 12-1 Various ways in which gasoline may reach the motoring public. Note that independent marketers buy at wholesale direct from the refinery and sell their gasoline as unbranded, or under private "labels". A "drop in" is where a gasoline marketer owns the fueling equipment on someone else's property, and the business operating the fueling system collects the money for the product. Then the business and the marketer split the profit on some predetermined ratio such as 60/40, etc.

modest investment, one that was usually within the reach of most people. A person could "own" a filling station (in actuality leasing a jobber or company owned location) by purchasing his or her initial inventory, paying a rent based on the number of gallons pumped, and agreeing to operate the station in accordance with company rules.

In recent years, there have been growing numbers of disputes between the independent station owners who sell major brands and the companies who produce those brands. Many of these operators claim they are being squeezed out of business by unfair tactics; that the majors furnish product to the operators of company-owned stations at prices far lower than those they charge the independent operators, thus placing the independents in a position of not being able to compete.

The majors spent vast sums to ensure that their stations were built to look alike so that the motoring public could easily recognize their favorite brand of gasoline no matter where they were. Also high on the major's list of marketing strategy was maintaining the cleanliness of their stations, and training their employees in courtesy and salesmanship. By mid-century drivers knew that they could pull into a company-owned station anywhere, use a clean restroom, have their tank filled with a quality product they trusted, and at the same time have their oil, water, battery, and tires checked, their windows washed and the car swept out. If they needed a new battery, tire, wiper blades, or lubricant, they could purchase these items bearing the same brand as the gasoline at the pump and have them installed while they waited.

Many of the operators who owned their own stations generally offered these services, but there were those who offered petroleum products for sale as a sideline, such as a country grocer who would have a pump or two out front and did little or no service work. Despite the number of both types of stations, there was still room for yet another type of marketer and station: the independent.

Independent Marketers

The need for independent marketers grew from a few simple economic facts. First, petroleum is and is not a seasonal product. It is distinctly different from a farm crop that matures at a given time, is harvested, stored, and then sold at a more-or-less given rate of demand all year long.

Petroleum can be produced all year, but the demand for certain products fluctuates seasonally. For example, in the North and the East, home heating oil is needed only in the winter; more gasoline is sold during summer vacation months and on holiday weekends. But traditionally, if all the refineries had operated at capacity year round, there would have perhaps been an excess of gasoline that company stations could not possibly sell. Long-term storage of gasoline is impractical for two reasons: first, gasoline is a highly-refined, complex product that tends to break down over time and suffer a loss in quality. Secondly, as a bulky product, it is expensive to store. Understandably, the major oil companies were interested mainly in building stations in areas of high traffic volume. This left many potentially profitable sites in areas that the majors did not wish to serve, but where, nevertheless, service stations were needed by the public. It was out of these factors that the need for independent marketers and unbranded stations grew.

If the majors had built too few refineries, they could have found themselves running out of product at critical times. If they built too many, they would have had an excess of products on their hands. Plus, there were tens of thousands of customers who either lived outside the majors' trading areas or who were more concerned with price than with brand name. However, as noted below, a system of using independent marketers evolved that became a safety net for the majors for it allowed them to run their refineries at a maximum profit.

The independent marketers stepped into the picture and helped to alleviate the refiner's predicament by buying up all of the excess or overstock that was produced on the spot market or by buying fuel on the open market or unbranded under contract. Some of them eventually grew so large that they were contracting for the entire output of certain refineries. They either resold to small, independent jobbers who supplied a few filling stations in their community, or they set up their own system of retail and distribution outlets. It was not unusual for an independent to sell to the public under 20 or 30 "unbranded brand" names such as "Taxi," "Raceway," "Checkered Flag" and many others.

While the independents brought stability to the marketplace, the picture has changed over the years. The number of refineries in the United States has shrunk from 311 to 204 in the past decade. However, the ones that have dropped by the wayside were mostly old, inefficient *teakettles* that were not capable of producing the types of product in demand today and could not be brought up to current environmental standards. Even though the total number of refineries in the United States has declined, refining capacity is up, and the survivors are technologically current. And, as we see shall see in a few pages, the "unbranded brand" names have given way to either the names of supermarket or discount or convenience store chains or to even no name at all save "Gas."

In general, the old, unbranded independent stations were small, no-frills affairs that offered the customer gasoline a few cents per gallon cheaper than the majors and sold motor oil in bulk rather than in 1-qt cans. One independent, who began business in the Great Depression with one $500 tank car load of product, became in 20 years the largest independent marketer in the world with his own pipeline and fleet of railway tanks cars and barges. This was reflective of the growth period of the independents in the 1950s and early 1960s; a marketing movement that was followed by the growth of self-service stations in the late 1960s and 1970s.

Market Research

Market research is one of the functions that keeps a company growing. It is a study of the potential users of a company's products and of the prospects for future growth.

Marketing research and marketing trends have not been ignored by the major companies, most of whom are fully integrated, and most of whom are fully or partially international in scope. Sound research endeavors tell management when it is time to add new products to meet public demands and when to remove old products from the marketplace.

Early Methods of Gaining Market Share

Today, advertising to many people means visual ads run on television or in magazines and newspapers. Over the years, oil companies engaged in many activities to capture customers, such as advertising always clean restrooms, or giving away free road maps, trading stamps, or premiums. Some went into the travel club business, offering trip routing, bail bond insurance, and towing services. Some companies were quick to spot the mass appeal of early-day aviators and sponsored endurance flights and air races to promote the quality and popularity of their gasoline. Not only did this bring the company's name before the public, it also gave them credence with the young aviation industry, since the connection allowed for research and development of flight-oriented products. Other companies made similar connections with boating and auto racing enthusiasts.

Credit

Oil companies were one of the true innovators in the consumer credit movement. Back in the days of the Great Depression the majors learned they could get repeat business and attract customer loyalty by offering certain customers the privilege of charging their gasoline purchases on a monthly basis by using a cardboard "courtesy card." After World War II, as consumer credit grew, so did the use of gasoline credit cards. Customers could charge TBA (tires, batteries, and accessories), merchandise, and pay for them on extended terms.

To meet competition, companies extended credit privileges into other areas. Some companies became affiliated with hotel and restaurant chains, and suddenly a motorist could charge an entire vacation on a gasoline credit card. To attract more and more customers, mass mailings of cards were made to graduating college students and other groups. And the companies also learned they could increase their income not only from service charges and interest on the outstanding accounts but also by selling their mailing lists of customer names or by selling merchandise of all kinds via direct mail.

In the 1970s, the practice of removing existing service stations in favor of self-service outlets led to the widespread use of bank cards and travel and entertainment cards. Through the turbulent years since the Arab oil embargo, the majors have vacillated between accepting outside cards or limiting credit purchases to their own system. For a time, some stations even began charging a surcharge to customers wishing to make credit purchases on a card issued by their own company. Today, almost any station will accept five or six different cards, although more and more a third-party computer network such as the one operated by Sears Payment Systems or J.C. Penney will perform all the credit functions from issuing cards to approving purchases to billing for merchandise, leaving the oil company to take care of sales. How today's generation of automated marketing equipment handles credit sales is discussed later in this chapter.

Repairs

Many stations, and companies, built customer bases by offering repairs and needed mechanical services. Depending on their equipment and the skill of their mechanic, many could offer the driver the same service as an auto dealership, and thousands of stations hung out signs reading "Mechanic on Duty." However, the number of these has shrunk in recent years. Modern cars have become increasingly more complicated, and their electronic and computer-operated systems can only be serviced by using expensive electronic diagnostic equipment which must be operated by trained technicians. Many cars today require parts that are available only from the manufacturer, who will supply them only to his dealers, and the repair of many new cars requires specialized tools available only at a franchised dealer for that marque.

Petroleum Marketing Equipment and the Advent of Self Service

The first gasoline pumps were hand-operated devices that dispensed a more or less accurately measured amount of the liquid into the customer's tank or container. As underground storage tanks were provided to the growing number of filling stations, pumps began to become standardized. A glass container marked off in 1-gal increments and holding a total of 10 gal was placed atop a tall metal column. A long lever at the side was used by the attendant to pump up 10 gal out of the underground tank into the glass globe. Once it was filled, then the desired amount was drained by gravity through a hose and nozzle to the customer's gas tank. If this sounds crude, it should be noted that many early cars, such as the famous Model A Ford which replaced the ubiquitous Model T in 1928, were not equipped with fuel pumps. Instead, the gas tank was actually behind the dashboard, and the filler cap was located just in front of the windshield. This arrangement placed the tank above the height of the engine and allowed the gasoline to flow by gravity down to the carburetor without the necessity or expense of a fuel pump.

The first real breakthrough in marketing equipment came at the heyday of the Model A when the Veeder-Root Corporation developed an electro-mechanical computer that revolutionized the retail marketing of gasoline. The old glass tank pumps only measured, none too accurately, the volume of gasoline delivered at each sale. The amount of the sale had to be computed with paper and pencil, not too difficult a task considering that the globe held a maximum of 10 gal and gasoline cost 18 cents or less per gallon. The new generation of pumps, available from a number of manufacturers who built the flow meters and housings, were designed around the Veeder-Root computer. They measured not only the volume of the sale, but also the dollar amount since the computer could be mechanically set to the current retail price per gallon and then calculate both the volume in tenths of a gallon and the dollar amount as the gas was being pumped from the *underground storage tank (UST)* to the customer's car. These devices also kept a running total of the gallons delivered to aid in inventory and cash control. While perhaps crude when compared to current pumps, these early models nevertheless laid the groundwork for today's self service marketing of motor fuel (Figure 12–2).

The next major breakthrough in retail delivery systems came in the late 1960s and early 1970s when the forerunner of today's Krown Systems of Oklahoma City introduced the first true electronic computer-controlled pump. These computers were controlled by an electronic circuit board measuring approximately 10 by 15 in. and consisted of dozens of transistors and other electronic devices.

Today's Equipment

Today's equipment, Krown's current generation true electronic computerized pump, has brought about the second great revolution

Figure 12–2 Today, traditional full-service gasoline stations are fast giving way to outlets such as this busy neighborhood convenience store (courtesy Krown Systems, Inc., © Don Krone).

in marketing by providing a true, point-of-sale terminal. Instead of merely measuring gasoline volume, today's pumps can:

1. Accept bank debit cards just like an automatic teller machine.
2. Accept credit cards. Not only will the internal computer run an instant check to make sure the card is valid, it will check the region of issue. The bank card companies charge the merchant a discount on each transaction, usually 3% of the amount of the sale. However, there are five regional bank card centers across the United States and each adds its own discount. Therefore, a card issued in Massachusetts being used in California could conceivably cost the merchant a 15% discount. These pumps can be programmed to recognize an out-of-region card and inhibit the transaction if the merchant does not wish to allow a higher discount.
3. Allow the customer to also pay cash inside the store or station prior to pumping the gas. The sale is rung up on the same computer system driving the pumps, and the customer will receive a printed receipt bearing a PIN or personal identification number. He or she then enters that number on the keypad at the pump, and the exact amount is delivered to the car. If the gas tank won't take the amount purchased, the computer makes note of the balance due, and at any time within a period

set by the seller, the customer may return and use the same PIN to obtain the balance of his or her fuel (Figures 12–3 and 12–4).

4. For the merchant, display the amounts of sales, the inventory, the money collected, make up the bank deposit, and even be programmed to order additional inventory automatically. They can automatically raise and lower prices as for a station in a rural area that must pay additional help in order to remain open over a weekend. Their salaries can be offset by having the pump increase the price per gallon at, for example, 5 P.M. on Friday, and then lower it again at 6 A.M. on Monday.

5. Offer the customer five different octane blends from just two underground storage tanks.

6. "Boot up" each time they are activated, which means the system checks all its components to make sure they are operating properly, even to checking for leaks at each sale. The device emits a series of electronic beeps while this is going on. These sounds are not really necessary, rather they serve to let the customer know the system is operating.

The inventory feature, or concurrent inventory of tank volumes day-by-day, which keeps track of the fuel present in the storage tanks, is an extremely important feature since running out of fuel, or losing track of inventory, could result in severe regulatory penalties since such an event could indicate a leak in the storage

Figure 12–3 This printer, attached to the store's cash register, prints out the customer's receipt and PIN number (courtesy Krown Systems, Inc., © Don Krone).

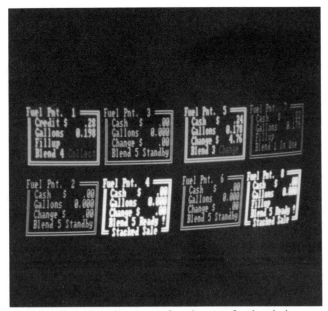

Figure 12–4 The VDT displays the status of each pump for the clerk or manager (courtesy Krown Systems, Inc., © Don Krone).

system. The EPA requires that all fuel placed in the underground tanks must be accounted for. Also, in high-volume stations, the disappearance of just .5 in.[3] of gasoline per day due to faulty equipment could result in a monetary loss of $15,000 per year.

The heart of these "smart" pumps is a chip about the size of a postage stamp. Since, as in many other fields, trained technicians are hard to come by, the pumps have been designed so that in the event of a malfunction, the entire circuit board is removed and replaced, a simple process that almost anyone can perform.

The Rise of Automated Marketing

In short, in a very few years automation has impacted the retail marketing of gasoline just as it has exploration, drilling, and nearly every other facet of the petroleum industry. Today, environmental regulatory requirements and oppressive labor costs have led to automation, which, according to the leading manufacturer of automated equipment, has opened up markets for retailers that never before existed.

The concept of self-service came about as a direct result of high labor costs. At today's station, a $4 per hour clerk can handle gasoline sales of $5,000 per shift. A unique aspect of self-service is that while a self-service station and a full-service station both require the same amount of investment, the self-service station has more through-put of products because it forces the customer to become an employee. And some studies show that customers would rather service their own vehicles. One can easily see what an impact self-service has had on the industry by noting the statistics; in 1975 less than 20% of all filling stations were self-service. By 1989, nearly 80% of all stations were self-service.

Who Controls America's Gasoline?

One of the most significant trends in recent years has been the move to foreign ownership of or investment in U.S. retail outlets. Some of these firms have long been familiar to North American motorists, such as Shell, the largest seller in the United States, part of the Royal Dutch Shell family. BP—British Petroleum—controller of half of the North Slope production in Alaska, became visible in American markets in the late 1960s, and since has become increasingly active.

Over the past 10 or 12 years, foreign presence in American retailing has grown. Today, 50% of Texaco's east coast marketing assets including refineries, terminals, and stations are owned by the Saudis. And CITGO—formerly Cities Service—and Champlin are both owned by the Venezuelan state oil company. German and Canadian investors have a market share of American retailing, and even the convenience stores (C-stores) are changing hands. In the spring of 1990, ownership of 7-11 passed to the Japanese. Foreign concerns now own 20% of U.S. refineries, own or sell through one-third of the branded stations, and control one-quarter of the gasoline sold in the United States.

Keeping Up with the Trends

Keeping pace with the public's wishes and devising new techniques to keep old customers and lure new ones is a highly developed science among petroleum marketers. Research at one major, for example, claims that most drivers buy their gasoline at one of two stations, and that these stations may be competitors. Some marketers pay close attention to their restrooms, claiming that while restrooms themselves may not draw customers, dirty ones will certainly cause a customer not to return. Therefore, some operators like to put theirs inside the station where their condition can be more closely monitored. Managers of other chains debate such "small" points as customer preference for hot-air hand dryers over paper towels.

Many station operators now offer give car washes with a fill-up, and those that do report a corresponding increase in sales.

The advent of C-stores in the 1960s and their resounding success in the late 1960s and early 1970s when their decision to sell gasoline doubled and then tripled their cash flow brought a move in the same direction by the major marketers, almost all of whom now have "marts" in place of the traditional filling station. These are clean, upscale stores that sell groceries, fast foods, snacks, sundries, beverages, and even in some cases, offer haircuts, long-distance phone banks, and game areas. It should be noted that all of this was made possible by the emergence of self-service.

In place of a grease-covered teenager on his way up or a middle-aged manager in a company uniform, today's gasoline mart may well be run by a woman, since researchers claim that women may be more likely to run a clean, well-maintained business, pay more attention to details such as timekeeping, scheduling, and bookkeeping, be more honest, and have a more pleasant attitude toward their customers.

Where Do We Buy Our Gasoline?

In 1980 there were 230,000 "true" service stations in the U.S.—a figure that had shrunk considerably during the tumultuous years of the 1970s and 1980s. As of 1990 there were approximately 115,000, a decrease that has given rise to fears that the majors were squeezing out the independents and that soon only a few major brands would be available to consumers.

But these "lost" stations have been replaced by thousands of gasoline pumps at C-stores and more recently at supermarkets and large discount retailers. In fact, the number of locations where a motorist can purchase gasoline has actually increased.

Since the supermarkets and discounters already have vast paved areas available, it is relatively easy for them to install underground tanks and Krown computerized pump systems. Thus when a grocery customer goes through the checkout line at Biggs Supermarkets in Cincinnati, one of the chains that has installed a Krown system, she merely has to give the cashier some money for gasoline at the same time she pays for her groceries, receive her PIN number, load her groceries in her car, drive to the pump in the parking lot, enter her PIN, fill her tank, and be on her way (Figures 12–5 and 12–6).

Such one-stop shopping offers benefits both to the store and the customer. By being able to change the prices and the electronic display signs instantaneously through the computer, the merchant can offer gasoline as a "loss leader"—several cents below his competitor, thus bringing in customers who will purchase other items while in his store. Or, if he keeps his prices at the local market level, he has another profitable item to sell to his established customers who appreciate the convenience of one-stop shopping. And, since the pumps are wholly unattended, the merchant does not have to hire any additional help.

Figure 12–5 Today consumers may buy their groceries and gasoline at the same time in modern supermarkets like this one in Ohio (courtesy Krown Systems, Inc., © Don Krone).

Figure 12–6 After paying for her gas and getting her PIN number inside at the checkout, the customer fills her gas tank at the pumps located on the supermarket parking lot (courtesy Krown Systems, Inc., © Don Krone).

Environmental Concerns

Concern for the environment has also had heavy impacts on marketing, the most visible of which has been the switch to unleaded gasoline and the phasing out of leaded fuels. In some areas, where there is still an abundance of pre-1975 automobiles, leaded fuel is still in demand. However, the amount of lead in gasoline today is less than 0.01%, and is on the verge of disappearing altogether. The next most visible effect of new environmental regulations are vapor recovery nozzles that are supposed to prohibit hydrocarbon emissions as the gasoline is being delivered. So far, few if any, are effective.

Not visible to the consumer except in increased fuel costs, but of great importance to the marketer, are the underground storage tank program regulations of the EPA. The Resource Conservation and Recovery Act (RCRA) requires that the EPA, or designated state or local agencies, be notified of all underground tanks that have been used to store regulated substances since January 1, 1974. These regulated substances are those defined as hazardous in Section 101 (14) of the Comprehensive Environmental Response, Compensation and Liability Act of 1980 (CERCLA), with the exception of those substances regulated as hazardous waste. It includes petroleum, e.g., crude oil or any fraction thereof which is liquid at standard conditions of temperature and pressure (60°F, 14.7 PSIA). Failure to notify and register underground storage tanks can bring penalties of up to $10,000 for each tank.

Some states and localities have enacted their own regulations that are much stiffer than those of the EPA. The major concern is older tanks that leak hydrocarbons into the ground and thus raise the possibility of polluting water supplies. Penalties for tank owners can be extremely severe, and some states such as Oklahoma have created their own "superfunds" to protect both the public and the marketers.

All existing USTs must be upgraded to the new standards for tanks by 1998. However, this regulation is governed by a timetable that calls for the upgrading of older tanks first and newer tanks by the final date. Therefore, some older tanks, depending upon the date they were placed in operation, may already be in violation. It is generally cheaper to start over and build a whole new station than it is to bring a new one up to current EPA specifications. These environmental regulations have resulted in the elimination of many sales locations and small gasoline marketers.

The Clean Air Act Amendments of 1990

The Clean Air Act Amendments of 1990, signed into law by President Bush in November 1990, have the following provisions:
Smog-producing emissions from automobiles must be cut by 30 to 60% by 1990. And gasoline sold in areas with smog problems must be 15% cleaner by the end of the decade.
More than 100 cities that are now failing to meet Federal air-quality standards must impose pollution-control plans that will bring them into compliance within the next 5 to 17 years.

Other portions of the law are aimed at industrial air pollution. It will require more efficient emission controls on automobiles and cleaner-burning motor fuels. It particularly requires fleets in non-attainment areas to turn to less-polluting fuels.

Other Regulations

Aside from alcohol and tobacco, few other consumer products are as carefully watched over as motor fuel. Each gallon is taxed by both federal and state governments, and the federal budget debacle of 1990 brought yet more fuel taxes. It used to be that such taxes were not applied to fuel used for offroad purposes, such as farming, construction work, and commercial fishing. In recent years, the IRS has taxed all fuel at the time of sale, leaving exempt purchasers to try to get a rebate at the end of the year by filling out complicated forms.

State fuel taxes are an important source of revenue that can be used for highway construction and repair, school support, and other public projects. The states guard this rich source of income so jealously that interstate truckers have to post a fuel bond just to drive through a state even though they might not have to purchase any gasoline or diesel fuel while passing through on that particular trip. State fuel taxation became a paradox during the fuel "crisis" of the 1970s when the federal government urged the public to conserve motor fuel by driving slowly, switching to smaller cars, carpooling, and making other efforts at conservation. The American motorist responded so effectively that almost immediately the states began complaining about lost tax revenues because gasoline sales were down.

The Future

Undoubtedly, the future of motor fuels, as well as the future of petroleum marketing and the oil companies themselves will be increasingly impacted by the growing use of alternative fuels such as gasohol, LPG, and compressed natural gas.

13

Petrochemicals

All living things depend on the physical resources of the Earth for their existence. And people are no exception, because we more than any other living creature have learned to exploit the resources of the Earth. Our ability to find and use mineral and energy sources have allowed us to control and modify our environment. Yet, at the same time, we are ultimately dependent on a continuing and, so far, usually expanding, supply of energy and mineral materials.

More than anything else, our ability to find, produce, and process crude oil has accounted for the standard of living that we enjoy today.

There is far more in a barrel of crude oil than gasoline for your car (Figure 13–1). More, in fact, than all the other traditional petroleum products such as oil, grease, wax, and asphalt. About 4% of the natural gas and crude oil in the United States is used as feedstock, or raw materials, by chemical companies to make products valued at about $50 billion annually. Thus, a single barrel (42 U. S. gal) of crude oil may be turned into consumer products worth hundreds of dollars. In fact, a cubic meter of crude oil (6.3 U. S. bbl) may give us all of the things shown in Figures 13–2 and 13–3.

Petrochemical Summary

Petrochemicals are an important part of our lives, and in many cases they have become necessities of life. There are so many consumer products made from hydrocarbons that it would take a large book just to list them. Just a few are aspirin, clothing, fertilizers, telephones, insecticides, plastics, paints, packaging and insulating materials, and inks (Table 13–1).

These materials have had a dramatic impact on our food, clothing, shelter, and leisure. Synthetics tailored for particular uses actually perform better than products made by nature.

To understand how synthetic materials are formed, it helps

Figure 13–1 Things that come from oil (courtesy Imperial Oil Ltd.).

to know that everything in the world is made of molecules. A molecule is the smallest part into which any substance can be divided and still keep its original characteristics. Molecules in turn are composed of atoms.

Carbon is the backbone of petrochemicals. Carbon can be obtained from other sources, such as coal or wood, but petroleum is the main source because it is the least expensive and can be used most easily.

There are two basic units used in making petrochemicals: olefins and aromatics. Olefins are molecules of carbon and hydrogen that appear as a chain, linked together in a long line. Aromatics are carbon and hydrogen molecules that are shaped like rings. Both olefins and aromatics are quite versatile because chains, rings, or even simple atoms can be added to either of them in any combination.

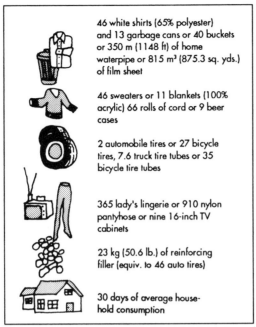

Figure 13–2 Products derived from 250L of crude oil, approximately one-and-a-half barrels of crude oil (courtesy Imperial Oil Ltd.).

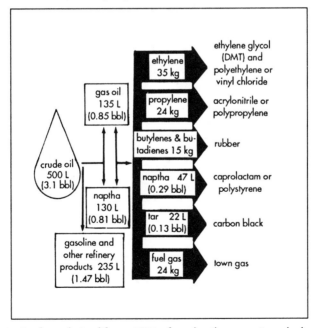

Figure 13–3 Products derived from 500L of crude oil, approximately three barrels of crude oil (courtesy Imperial Oil Ltd.).

Petrochemicals   *357*

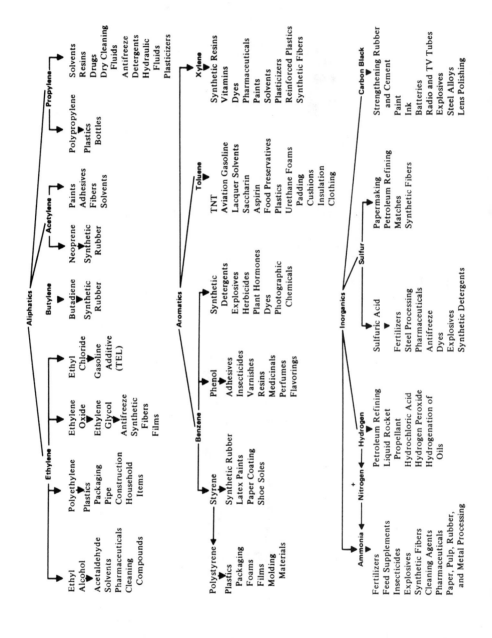

Table 13–1 The Petrochemical Family Tree

Modern Petroleum

By separating molecules, rearranging them, or adding different combinations, new chemical compounds are formed. These are the basis of our petrochemical products. Different processes are used to change the components until they are useful. Certain fractions of petroleum are selected and heated until their molecules change to form olefins and aromatics of a kind that can be used to form petrochemical products. They are separated by distillation and extraction processes.

Finished petrochemicals are used by many of the world's industrial sectors, including forestry, mining, textiles, construction, and agriculture. The consumption of petrochemicals is growing at a rate of approximately 10% annually. We have become so accustomed to these products in our daily lives that we don't even notice our increasing dependence on them.

How the Industry Began

The petrochemical industry has grown hand in hand with the refining industry. The first large-scale advance for both came in about 1912 when cracking made its first large-scale appearance, and both refiners and petrochemical manufacturers have benefited from each new advance in cracking methods. Cracking not only improved both the quality of gasoline and its quantity produced from each barrel of feedstock, it also vastly increased the output of *alkenes.* These are highly reactive unsaturated hydrocarbons that form the basic material for the petrochemical industry.

Petroleum is composed mainly of compounds of the elements carbon and hydrogen or, more simply, hydrocarbons. But nitrogen and sulfur are also found in crude oil. All four of these are extremely valuable in the manufacture of chemicals. Since these chemicals are derived from petroleum, we call them petrochemicals.

Chemistry Review

Today, our scientists and physicists are trying to break matter into smaller and smaller particles. The discovery of nuclear fission was made possible through an accumulation of knowledge of the structure of matter which began with Becquerel's 1896 detection of radioactivity. Later experiments led to the publication in 1939 of results showing that the nucleus of a uranium atom can be split into two parts by the absorption of a neutron. Further experiments led to the atomic bomb and the development of nuclear power, and the resulting increase in knowledge about atomic structure has led to the development and manufacture of many products from hydrocarbons.

The structure of the atom is assumed to have a dense, positively charged nucleus made up of closely packed protons and neutrons. Each proton carries an elemental positive charge of electricity. Each neutron is an electrically neutral particle with a mass slightly greater than the proton. Protons and neutrons are also referred to as nucleons. Each electron orbiting around the nucleus has a mass much less than either the proton or neutron, but it is a charged particle (Figure 13–4).

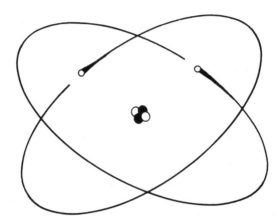

Figure 13–4 Diagrammatic representation of an atom of helium. The nucleus consists of two protons and two neutrons, and accordingly has a mass number of four. There are two electrons (negative charges) to balance the positive charges on the two protons. Since there are two protons in the nucleus, this atom is number two in the table of elements.

The distance between the nucleus and the orbiting electrons is approximately 10^5 times the dimension of the nucleus, which explains why X-rays are able to pass through dense materials.

In the un-ionized state, the number of protons is balanced by an equal number of electrons. An atom becomes ionized by gaining or losing one or more electrons. A gain of electrons yields negative ions and a loss results in positive ions.

When an atom has more than two electrons, their orbits are located in a series of separate and distinct shells, each capable of containing a specific number of electrons. Normally, an inner shell fills to its maximum number of electrons before electrons begin to form in the next shell. The number of electrons in the outer shell determines certain properties of the element.

Physicists are continually surprising the scientific world by breaking matter into small and smaller particles. Fortunately, the smallest particle considered in petroleum refining is the atom. Examples are carbon, hydrogen, sulfur, or oxygen, whose chemical symbols are C, H, S, and O.

The characteristics of matter depend on the types of atoms that it is composed of and the way in which these atoms are attached to each other in groups called molecules. There are rules by which atoms can be arranged into molecules. The most important of these rules have to do with valences and bonds.

The valence of an atom of any element is equal to the number of hydrogen atoms (or their equivalent) that the atom can combine with. Each type of atom (element) has an affinity for other elements according to its atomic structure. For example, carbon atoms would always like to attach themselves to four other atoms. Hydrogen atoms would like to attach to only one other atom.

The connection between two atoms is called a bond.

The atom has both size and mass. Mass is the number of protons plus the number of neutrons in a nucleus.

Chemists and physicists list the mass of atoms and fundamental particles in *atomic mass units (amu)*. This is a relative scale in which the nuclide, carbon-12 is assigned the exact mass of 12 amu. One amu is the equivalent of approximately 1.66×10^{-24} grams. A gram-atom of an element is a quality having a mass in grams numerically equal to the atomic weight of the element.

Atomic weight is the relative weight of a neutral atom of an element, based on an atomic weight of 16 for the neutral oxygen atom. A neutral atom is one in which the number of positive charges in the nucleus is equal to the number of electrons that surround the nucleus. On this basis, hydrogen has an atomic weight of 1.008, and carbon 12.010.

An *ion* is an electrically unbalanced form of an atom or group of atoms. An atom is electrically neutral until it loses an electron from its outermost shell, leaving the portion that remains behind with an extra unmatched positive charge. This unit is known as a positively charged ion. If the outermost shell gains an electron, the ion has an extra negative charge and is known as a negative-charged ion (Figures 13–5 and 13–6).

There is no atom with more than eight electrons in its outermost shell. The rules of electrons and orbits are important because the number of electrons in the outermost orbit largely determines the chemical properties of the element. Atoms strive for stability by filling up their outermost orbit. They can gain, lose, or share electrons with another atom in order to make a complete orbit of two or eight electrons.

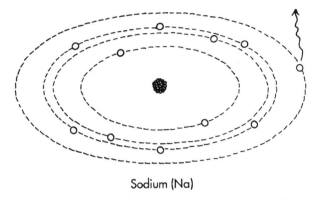

Sodium (Na)

Figure 13-5 Formation of the sodium Na⁺ results when the sodium atom loses the only electron in its outermost shell.

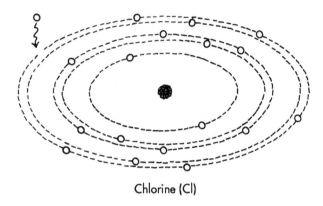

Chlorine (Cl)

Figure 13-6 Formation of the chlorine ion Cl⁻ results when a chlorine atom gains an electron in its outermost shell, to make the total eight.

Molecules

A *molecule* is the smallest particle into which an element or compound may be divided and still retain the chemical properties of the element or compound in mass. A molecule is a group of atoms bound together in a definite way. There are two types of molecules possible and they are identified according to the kinds of atoms they contain.

The formula of a molecule lists the atoms present and tells how many there are of each kind in the molecule.

There are rules for writing formulas, such as:

1. Each kind of atom present is represented by its symbol.

Modern Petroleum

2. The number of atoms of a given kind is indicated by placing a subscript number to the right of the symbol.
3. The sequence in which the different elements are listed is mostly arbitrary.

A molecule is called monatomic if it consists of one atom. Molecules that consist of only two atoms, either alike or unalike, are called diatomic.

Molecules containing three atoms are often called *triatomic*, and are called *polyatomic* when containing three or more atoms. CO_2 could be called a triatomic or polyatomic, because it contains three atoms (one of carbon and two of oxygen).

The molecular formula does not give any information as to how the atoms are arranged in relation to one another. Sometimes a formula called a structural formula, which shows how the atoms are connected, is useful. An example of the structural formula for a common diatomic molecule is: H-H. For polyatomic molecules, an example is ammonia, (NH_3) H-N-H.
$$\overline{H}$$

A chemical equation is a way of representing what happens by using formulas for each of the substances involved in a reaction. A chemical reaction occurs when one or more substances, called the reactants, are transformed into one or more different substances that are called the products.

One popular method of writing an equation for reactants is to place the formulas of the reactants on the left with a plus sign between each of them and the formula of the product on the right. To show that reactants lead to products, an arrow is drawn from left to right.

Element

A substance with molecules that contain only one kind of atom is referred to as an elemental substance or an *element*. There are some 92 known atoms (elements) that occur in natural form, or that are a product of nature, and an additional group of about 12 that are man made. This means that there are more than 100 different kinds of atoms recognized today. Many of our natural atoms such as gold and silver, have been known for thousands of years.

Among the elements there are some substances built up solely of individual atoms that are not connected in such a manner as to form molecules. Some examples of these are six noble gases, helium (He), neon (Ne), argon (Ar), krypton (DR), radon (Rn), and xenon (Xe). The metallic elements which consist of single atoms that are packed together closely, but without the formation of any individual groups can accurately be identified as molecules. Examples of these are iron (Fe), tin (Sn), silver (Ag), and gold (Au) (Table 13–2).

All of the elements have names, but symbols are normally used to identify them. Most of the symbols are made up of the first letter of the English name together with one other letter from the name. The first letter of the symbol is capitalized but the second one is not. Some consist of the capitalized first letter only. The symbols may be divided into four main groups:

1. Some symbols are derived from the old Latin names or other non-English names, such as: gold = aurum = Au
2. Symbols derived from the first letter only of the English name of the element, such as: carbon = C
3. Symbols derived from the first two letters of the English name, such as: helium = He
4. Symbols derived from the first letter of the English name and one other letter which is not the second letter, such as: cadmium = Cd.

The number of atoms of a given kind, if there are more than one, is indicated by placing a subscript number immediately to the right of the symbol. The symbol alone without any following subscript represents one atom.

Compounds

A substance in which the molecules contain different kinds of atoms is referred to as a *compound* substance or compound. Water is the most familiar compound substance in our universe (Figure 13–7).

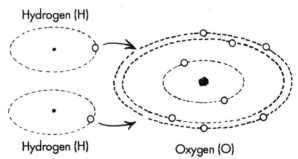

Hydrogen (H)

Hydrogen (H)   Oxygen (O)

Figure 13–7 Two hydrogen atoms and one oxygen atom join to form water, H_2O by a covalent bond. In this bond, the hydrogen electrons do double duty in a sense, filling the two empty places in the outer shell of oxygen, yet remaining at their normal distance from their hydrogen nuclei. The result is the formation of a molecule of water, the smallest unit that displays the properties of that compound.

Element	Symbol	Atomic Number	Element	Symbol	Atomic Number	Element	Symbol	Atomic Number
Actinium	Ac	89	Gold	Au	79	Promethium	Pm	61
Aluminum	Al	13	Hafnium	Hf	72	Protactinium	Pa	91
Americium	Am	95	Helium	He	2	Radium	Ra	88
Antimony	Sb	51	Holmium	Ho	67	Radon	Rn	86
Argon	A	18	Hydrogen	H	1	Rhenium	Re	75
Arsenic	As	33	Indium	In	49	Rhodium	Rh	45
Astatine	At	85	Iodine	I	53	Rubidium	Rb	37
Barium	Ba	56	Iridium	Ir	77	Ruthenium	Ru	44
Berkelium	Bk	97	Iron	Fe	26	Samarium	Sm	62
Beryllium	Be	4	Kypton	Kr	36	Scandium	Sc	21
Bismuth	Bi	83	Lanthanum	La	57	Selenium	Se	34
Boron	B	5	Lead	Pb	82	Silicon	Si	14
Bromine	Br	35	Lithium	Li	3	Silver	Ag	47
Cadmium	Cd	48	Lutetium	Lu	71	Sodium	Na	11
Calcium	Ca	20	Magnesium	Mg	12	Strontium	Sr	38
Californium	Cf	98	Manganese	Mn	25	Sulfur	S	16
Carbon	C	6	Mendelevium	Me	101	Tantalum	Ta	73
Cerium	Ce	58	Mercury	Hg	80	Technetium	Tc	43
Cesium	Cs	55	Molybdenum	Mo	42	Tellurium	Te	52
Chlorine	Cl	17	Neodymium	Nd	60	Terbium	Tb	65
Chromium	Cr	24	Neon	Ne	10	Thallium	Tl	81
Cobalt	Co	27	Neptunium	Np	93	Thorium	Th	90
Columbium	Cb	41	Nickel	Ni	28	Thulium	Tm	69
(or Niobium	Nb)		Niobium	Nb	41	Tin	Sn	50
Copper	Cu	29	(or Columbium	Cb)		Titanium	Ti	22
Curium	Cm	96	Nitrogen	N	7	Tungsten		
Dysprosium	Dy	66	Nobelium	No	102	(or Wolfram)	W	74
Einsteinium	En	99	Osmium	Os	76	Uranium	U	92
Erbium	Er	68	Oxygen	O	8	Vanadium	V	23
Europium	Eu	63	Palladium	Pd	46	Wolfram	W	74
Fermium	Fm	100	Phosphorus	P	15	(or Tungsten)		
Fluorine	F	9	Platinum	Pt	78	Xenon	Xe	54
Francium	Fr	87	Plutonium	Pu	94	Ytterbium	Yb	70
Gadolinium	Gd	64	Polonium	Po	84	Yttrium	Y	39
Gallium	Ga	31	Potassium	K	19	Zinc	Zn	30
Germanium	Ge	32	Praseodymium	Pr	59	Zirconium	Zr	40

Table 13–2 Alphabetical List of the Elements

When we attempt to identify a given substance as an element or a compound, it becomes clear that the purity of the substance is of great importance. A pure substance is one in which all of the molecules are the same. This means that its composition is uniform and it has certain constant and characteristic properties. In practice, the purity of any sample of a substance is determined by how close its properties are to those of the actually pure substance. We realize that no sample is truly pure, and what we try to achieve is a sample in which the level of impurities is so low that they will have little or no measurable effect upon its characteristic properties.

Most of the elements occur in nature as compounds. For this reason, the discovery of elements has depended on the ability of scientists and chemists to decompose compounds and to purify them to show that these substances could not be decomposed further. Much of this work has been done in recent times, since prior to 1700 only about 13 elements had been recognized.

The Structure of Hydrocarbons

Hydrogen and carbon both have their own distinctive atomic structures. It is by rearranging their individual molecular structures that chemists and scientists are able to produce many industrial materials. Hydrogen has the simplest atomic structure of any element. It has no neutrons in its nucleus, but instead has a single electron moving around it at a very low energy level. There is room for only one additional electron. An atom becomes chemically stable when its outermost energy level is filled. Since hydrogen has only one energy level with a single vacancy, it becomes stable when one more electron is acquired.

By contrast, the carbon atom is much more intricate. There are six neutrons and six protons in the nucleus. Two electrons move in the innermost and lowest energy level and four in the next. This second energy level has room for eight electrons, so four more must be added for stability. Hydrogen and carbon then achieve stability and form hydrocarbons sharing electron pairs. These shared electrons jointly occupy the outer levels of both atoms. Such sharing is called covalent bonding (Figure 13–8).

Large hydrocarbon molecules such as those found in petroleum are formed by the bonding of many hydrogen and carbon atoms. Thus the properties of hydrocarbons depend upon the number and arrangement of the hydrogen and carbon atoms in their molecules. Since compounds with similar structures have similar properties, the chemists can aim their research at group rather than individual characteristics. Hydrocarbon is a compound that contains only

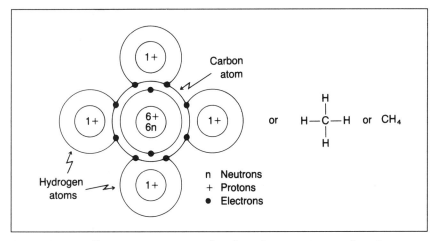

Figure 13–8 Different representations of methane (courtesy PennWell Books, *Petrochemicals for the Nontechnical Person,* Burdick and Leffler)

hydrogen and carbon atoms. It may be saturated (no double or triple bonds), unsaturated (containing double and/or triple bonds), aromatic (containing benzene or benzene-like rings) or cyclic (saturated or unsaturated but containing a ring of carbon atoms) (Figure 13–9).

Alkanes

Compounds consisting solely of carbon and hydrogen atoms bonded together by single bonds only, are called saturated hydrocarbons. Such compounds are also known as the *alkane* series of the paraffin series. The alkanes are the principal compounds present in natural gas and in petroleum. Usually, under normal conditions, low molecular weight compounds are gases.

One important group derived from petroleum is the alkane or *paraffin homologenes series* (Figure 13–10). It includes the gases methane (CH_4), ethane (C_2H_6), propane (C_3H_8), and butane (C_4H_{10}). Also included, but not shown are the liquids pentane (C_5H_{12}), and hexane (C_6H_{14}), as well as the solid, heptadecane ($C_{17}H_{36}$). The so-called normal alkanes are those in which the carbon atoms are strung in a single chain. Other hydrocarbons whose carbon atoms branch out are called *branched-chain.*

Isomerism

The alkanes are compounds with a common chemical formula, but because their structure is different they have different melting and combustion temperatures. Such compounds are called *isomers.* Isomers have the same chemical formula and molecular weight (the same set of atoms), but with different structures. These different

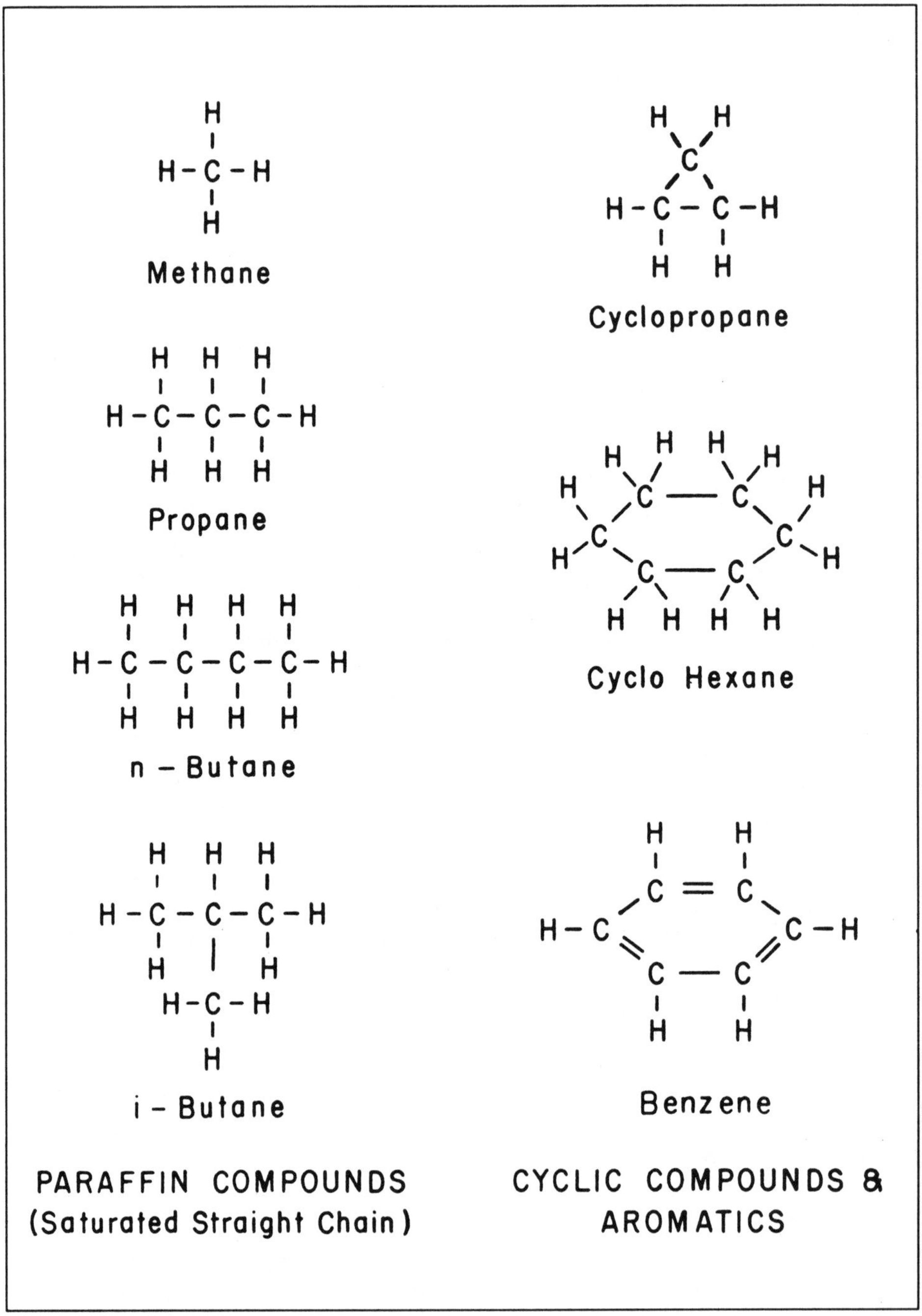

Methane

Propane

n – Butane

i – Butane

PARAFFIN COMPOUNDS
(Saturated Straight Chain)

Cyclopropane

Cyclo Hexane

Benzene

CYCLIC COMPOUNDS &
AROMATICS

Figure 13-9 Hydrocarbon gas molecule structures.

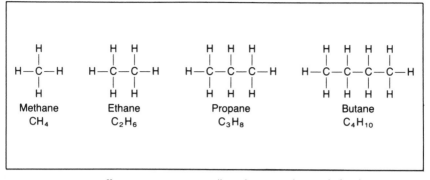

Figure 13–10 Paraffins (courtesy PennWell Books, *Petrochemicals for the Nontechnical Person*, Burdick and Leffler).

structures lead to different physical and chemical properties. With each additional carbon atom, the number of possible isomers increases. For example, pentane (C_5H_{36}) with 17 carbon atoms has three isomers. Heptadecane ($C_{17}H_{36}$) with 17 carbon atoms has several thousand. Isomerism plays an important role in making new products out of petroleum.

Unsaturated Hydrocarbons

Petroleum also can be processed to produce other types of hydrocarbons. These are the alkenes *(olefins)*, such as ethene or ethylene, in which the molecule has a double bond consisting of two pairs of shared electrons. Another is the *alkadienes* (diolefins) such as butadiene. In some compounds, such as acetylene, three electrons from each carbon atom are mutually shared, thus producing a six electron bond called a triple bond.

These three are much more reactive than the alkanes because of the carbon-to-carbon double and triple bonds that permit additional reactions. The bonds open easily and become attachment points for new atoms of other elements. Since there is room for these additions, these series are called unsaturated. Literally thousands of derivatives can be formed this way, and over the years ethylene, acetylene, and butadiene have become the building blocks that have led to dozens of useful products and spurred the continuing growth of the petrochemical industry.

Cyclic Hydrocarbons

Another type of hydrocarbons is the *cyclics*, so called because instead of their carbon atoms being linked in either straight or branched chains they have a ring of carbon atoms in their structure (Figure 13–11). Benzene and its derivatives form the group with the

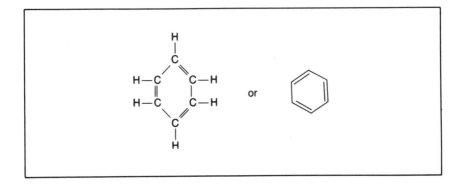

Figure 13-11 Benzene (C_6H_6), a cyclic hydrocarbon (courtesy PennWell Books, *Petrochemicals for the Nontechnical Person*, Burdick and Leffler).

greatest economic value. Collectively they are known as the aromatics since they were originally found in aromatic gums or oils. (The others are known as *aliphatics*.) When these compounds were first manufactured they were made from coal tar, but over the years ways were found to make them from petroleum.

Polymerization

Some compounds can be built up from smaller molecules and others can be made by breaking down larger ones. As noted above, the double and triple bonds of unsaturated hydrocarbons will open up under suitable conditions and join with other molecules of the same compound. This combining of molecules is known as *polymerization*.

The highly reactive unsaturates, and especially the olefins which are byproducts of cracking, are the starting point for the manufacture of many organic compounds. Prior to World War II, only 45% of these were provided by the petrochemical industry. Today, the majority of the total U.S. production of organic chemicals comes from petroleum. It is believed that many, many more compounds can be synthesized from the hydrocarbons in natural gas and crude oil.

Ethylene Production

One of the general types of petrochemical processing directly related to petroleum is ethylene production, which is a branch of the more diversified petrochemical industry (Figures 13-12 and 13-13).

Modern Petroleum

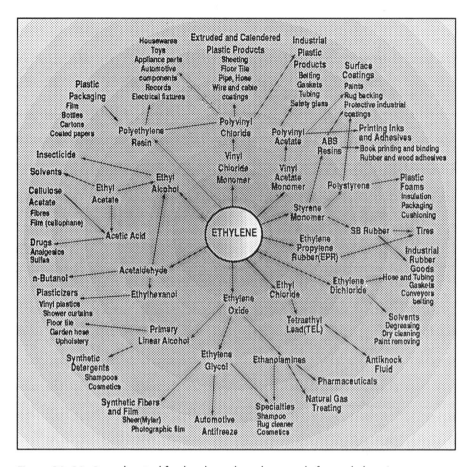

Figure 13-12 Petrochemical feedstocks and products made from ethylene (courtesy Imperial Oil Ltd.).

The ethylene or olefins plant can use ethane, an ethane and propane mix, propane, butane, naphtha, or gas oil as feedstock. These plants feeding the heavier materials are more complicated; therefore, we will examine only the ethane/propane cracking process to demonstrate the fundamentals of the process.

As shown in Figure 13-14, ethane and propane can be fed separately or as a mixture to the cracking furnaces where a short residence time followed by a sudden quench yields a high volume of ethylene. The feed is not completely reacted, so downstream in the product fractionator the remaining ethane and propane are split out and recycled to the feed. Ethane generally is recycled to extension, but some of the propane goes with the propylene.

The plants that crack the heavier liquids—naphtha and gas oil—create ethane on a once-through basis, so those olefin plants often have a furnace designed to handle the recycled ethane. As a result

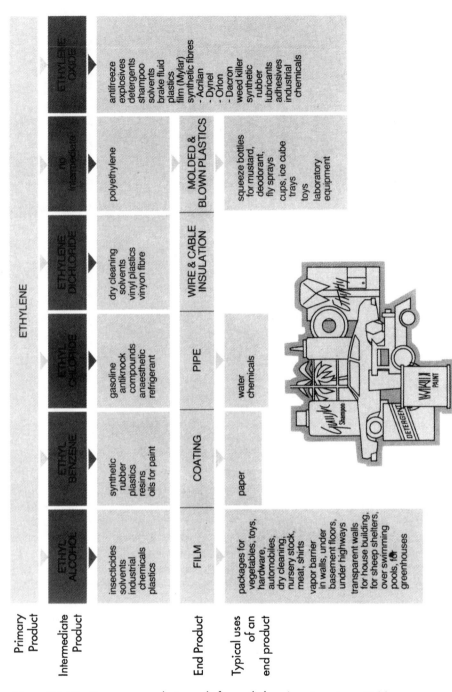

Primary Product: ETHYLENE

Intermediate Product:

ETHYL ALCOHOL	ETHYL BENZENE	ETHYL CHLORIDE	ETHYLENE DICHLORIDE	no intermediate	ETHYLENE OXIDE
insecticides solvents industrial chemicals plastics	synthetic rubber plastics resins oils for paint	gasoline antiknock compounds anaesthetic refrigerant	dry cleaning solvents vinyl plastics vinyon fibre	polyethylene	antifreeze explosives detergents shampoo solvents brake fluid plastics film (Mylar) synthetic fibres - Acrilan - Dynel - Orlon - Dacron weed killer synthetic rubber lubricants adhesives industrial chemicals

End Product:

FILM	COATING	PIPE	WIRE & CABLE INSULATION	MOLDED & BLOWN PLASTICS	

Typical uses of an end product:

packages for vegetables, toys, hardware, automobiles, dry cleaning, nursery stock, meat, shirts vapor barrier in walls, under basement floors, under highways transparent walls for house building, for sheep shelters, over swimming pools, for greenhouses	paper	water chemicals		squeeze bottles for mustard, deodorant, fly sprays cups, ice cube trays toys laboratory equipment	

Figure 13–13 Consumer products made from ethylene (courtesy Anne McNamara, Petroleum Resources Communication Foundation).

Modern Petroleum

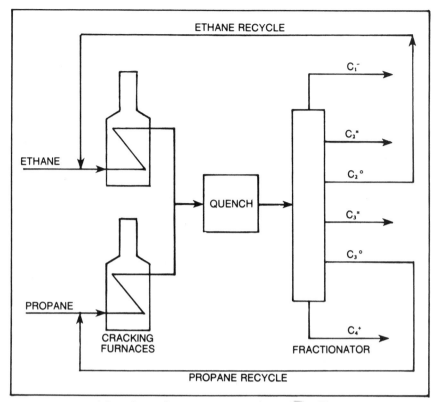

Figure 13-14 Olefins plant, ethane-propane cracker (after Leffler).

of the cracking reaction that produces the olefins, some butadiene is also formed. It is used in making plastics and rubber compounds.

Solvent Recovery of Aromatics

The second general type of petrochemical processing directly related to petroleum is BTX recovery. The aromatics compounds (*benzene, toluene,* and *xylene*) have many chemical applications. These three compounds appear in relatively large quantities in catalytic reformates, so these materials are recovered by treating the reformate with a solvent that preferentially dissolves them. The solvent with the dissolved aromatics can then be readily separated from the rest of the compounds.

The dissolved benzene, toluene, and xylenes (or BTX) can be easily recovered by a simple distillation of the solvent. Then the BTX materials may be separated into individual components by further distillation.

Figure 13–15 shows how the solvent-recovery process works. A heart cut reformate is the feed. This is obtained by separating off materials lighter than benzene and materials heavier than xylenes. Feed and solvent enter a mixer column where solvent flows downward, contacting the rising feed stream. The solvent absorbs BTX as it proceeds down the column.

The overhead *raffinate* contains almost no BTX. It is treated for solvent removal (which is recycled) and returned to the reformer. The bottoms extract-laden solvent moves to the solvent-extract distillation separator. Recycled solvent is taken overhead, and an extract containing product BTX is ready for fractionation. Some of the solvents used are *sulfolene, phenol, acetonitrile,* and liquid SO_2.

Alcohol

Alcohol is a hydrocarbon in which one or more of the original hydrogens has been replaced by a *hydroxyl* group. For many years, one of the most common alcohols, ethyl or grain alcohol, was made by the yeast fermentation of starches and sugars. But because the demand for ethanol as a solvent increased drastically, now 90% is produced from ethylene.

The same is true for methanol, which was formerly distilled from wood and is now made from *methane.* Common rubbing alcohol, *isopropyl,* is produced from propane.

Some alcohols contain two more of the functional hydroxyl groups. When two hydrogen atoms are replaced by hydroxyl groups, the compound becomes a *glycol.* Ethylene glycol made from ethane is used as an antifreeze and to synthesize Dacron.

Glycerin, which used to be a byproduct of the soap industry, is now made from petroleum and is used in the explosive nitroglycerin.

OH is the symbol of the alcohol family. By removing a hydrogen and adding a hydroxyl group to an ethane molecule, we can form ethyl alcohol, C_2H_5OH. Methyl alcohol is methane linked up with the hydroxyl group, CH_3OH. For an unusual reason, a carbon atom will attach itself to no more than one hydroxyl radical, although each carbon atom in a molecule can have its own hydroxyl radical.

Single and Multiple Bonds

Carbon and hydrogen can be attached with other carbons and hydrogens. When hydrogen is attached with another hydrogen, it forms H_2 and the electron needs of each of the atoms are satisfied. When carbon is attached to another carbon, each carbon atom still has three unsatisfied electrons. One possibility is to fill them out with hydrogen atoms, which results in the formation of the compound ethane.

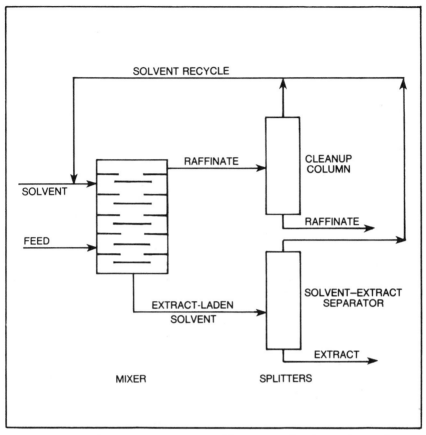

Figure 13-15 Solvent recovery process yields aromatics for chemical use (after Leffler).

The tendency of carbon to connect with four other atoms helps to point out at least partially why so many carbon compounds exist. There are many other ways that atoms can be attached with carbon. Carbon can be attached to itself via a single bond with three other atoms. It can also attach to itself with double or triple bonds to satisfy its valence requirements of four. Two carbon atoms are attached together with single, double, or triple bonds and are filled out with hydrogens, forming three different compounds.

This multiple-bond configuration is unusual in that in petrochemical processes the greater the number of multiple bonds, the more reactive the compound might become. Acetylene is more likely to react with other chemicals than ethylene, which is far more reactive than ethane.

Oxidation

Oxidation is the reaction of an atom or a molecule with oxygen. The oxygen may come from air or from another compound that readily gives up its oxygen, for example, hydrogen peroxide, H^2O^2. If the reaction reaches complete oxidation of an organic compound, the result is oxygens connected to every carbon, forming carbon dioxide and water. Combustion or burning is an example of complete oxidation. Partial oxidation is usually more desirable in petrochemicals.

Detergent

A detergent is a cleaning agent consisting of molecules with two distinct functions. The hydrocarbon end is fat soluble and the other end is electrically charged and water soluble. These distribute themselves where the oil and water meet, reducing the surface tension of both so that dirt particles are easily suspended to form an emulsion that is washed away with rinse water. Petroleum-derived detergents are much more effective than soap.

Polymers

In *polymerization*, two or more smaller molecules, usually from unsaturated compounds, are combined to form one large molecule. The reaction conditions can be controlled to produce *polymers* of varying size. As the size and weight of the molecules increase, the melting point of the polymers rises and the solubility decreases. This way, polymers with many different physical properties can be made. Fibers are made from polymers that can be stretched out to form a filament. Those that have rubber-like qualities are *elastomers* or synthetic rubber, and others are plastics. A few of the products of polymerization would include polyethylene plastics, nylon, and neoprene synthetic rubber.

Olefin Plants

The olefin plant is a major part of the petroleum industry. It is the producer of the industry's basic building blocks—ethylene, propylene, butylenes, butadiene, and benzene. The olefin plants have become so important that the newer ones are larger than many medium-size refineries. Their capacities have grown from millions to billions of pounds per year.

Benzene

Benzene can be derived from coal and other materials, but the major source of benzene is from the refinery catalytic reforming process. The main purpose of this process is to make quality gasoline components out of low-octane naphtha.The feed to a catalytic reformer is a mixture of paraffins and cyclic compounds in the C_6-C_9 range, usually called naphtha. The cat reforming process changes the chemical composition of the naphtha by causing paraffins to be converted to *isoparaffins* and to *naphthenes*. Then the naphthenes are changed to aromatics including benzene. These reactions are favorable because the end products, isoparaffins, naphthenes, and aromatics, have higher octane numbers than the compounds from which they are created.

The three important uses of benzene include the manufacturing processes for styrene, cyclohexane, and cumene.Polymers and many other plastics are produced from styrene. Synthetic fibers are a product of cyclohexane, and cumene leads to phenol, which usually ends up in resins and construction adhesives.

Toluene

Toluene is manufactured by cat reforming process, coke production, and olefin-plant operations and by recovering the small amount found naturally in crude oil. It is a byproduct like benzene.

Toluene has a high octane characteristic and it became a very important part of the manufacturing and blending of aviation gasoline during World War II.Later, it became an ingredient in manufacturing as an industrial solvent and an ingredient in manufacturing toluene diisocyanate, the forerunner to polyurethane foams. Other derivatives include phenol, benzyl alcohol, and benzoic acid.

Butadiene

Butadiene from olefin plants is coproduced with the other olefins, a product that just happens. Olefin-plant butadiene is about 75% of the total butadiene production in the United States. Other butadiene supplies are made by catalytically dehydrogenating, which is removing the hydrogen from butane or butylene.

Butadiene has two double bonds that make it very reactive and it easily forms polymers, reacting with itself to form polybutadiene. It is used as a feedstock for synthetic rubber, elastomers, and fibers.

Styrene

Styrene is made from two other basic building blocks, ethylene and benzene. Most all styrene now comes from dehydrogenating, which is removing two hydrogen atoms from the ethyl group attached to the benzene ring. *Dehydrogenation* is accomplished by cracking in a mixture with steam.

Plastics and synthetic rubber are the major uses for styrene. Styrene derivatives are found in toys, construction pipe, foam, boats, latex paints, tires, luggage, furniture and many more products modern man feels he must have.

Chapter References

Bill D. Berger and Kenneth E. Anderson, *Refinery Operations*, (Tulsa: PennWell Books, 1979).

Bill D. Berger and Kenneth E. Anderson, *Modern Petroleum—A Basic Primer of the Industry*, 2nd Ed. (Tulsa: PennWell Books, 1981).

Donald L. Burdick and William L. Leffler, *Petrochemicals for the Nontechnical Person*, (Tulsa: PennWell Books, 1983).

14

Equipment

From the very beginning of our search for oil we depend upon instruments to locate geographic areas that are likely to contain reservoir rock. In the pioneer days of the oil industry, no methods had been devised for locating reservoirs other than to search the earth's surface for seeps. It wasn't until the theory that oil and gas accumulate along anticlines came about that the sciences of geology and geophysics began to play an increasing role in the oil industry. Looking for surface seeps was not enough to locate deep formations, and new methods were needed to give the explorationists an idea of what lay beneath the surface. The first instruments were crude. Today however, technology has advanced to the point where now satellites are being used.

The oil and gas industry is so large and complex that it is almost impossible to cover all of the instruments, controls, sensors, valves, gauges, and many more pieces of equipment that are required to search for, drill, produce, transport, and deliver oil and gas. There are certain valves and controls that are considered basic to the industry, and the use of these will be explained. Today, practically any device that monitors, or controls regulators and valves from a remote location can be adapted to space-age technology to operate via satellite from points all over the world. Cost and feasibility are usually the factors that keep operations simple and effective.

Beginning with the search for oil, we will cover the more common instruments used throughout the industry.

Gravimeter

The gravimeter indicates the density of rock formations, measures the gravitational pull of buried rocks, and provides information about their depth and nature. It is a very sensitive and an inexpensive instrument to use.

Magnetometer

Magnetometers detect very slight fluctuations in the earth's magnetic field that show the presence of sedimentary rock. Magnetism is affected by the kind and depth of rock. This instrument measures the magnetic pull of underlying layers of rocks to identify basement rocks, especially those containing large concentrations of magnetite. The magnetometer can be attached to an airplane that will fly over an area to be explored. Also, magnetic fluctuations occurring over vast areas of the Earth have been mapped by satellite.

Seismograph

This instrument was first used to detect shockwaves from earthquakes and was later adapted to geological surveys. The seismograph, or seismometer, measures the transit times of sound waves generated by an explosion. As the waves from the explosion or thump from a vibration, travel downward, they are reflected from the various layers below back to the prearranged geophones on the surface. A geophone is a flat, circular device that picks up seismic reflected waves and transmits them to a recording instrument in a truck for evaluation.

Remote Sensing

This involves using heat-sensitive (infrared) color photography or television to reveal ground water, salt-water intrusion, mineral deposits, faults, and other features not visible to the naked eye. This infrared sensing equipment is carried aboard aircraft or satellites and the information is fed into special computers that compile the data and make printouts for use.

Laser Rangefinder

The *laser rangefinder* is sometimes used to locate areas for seismic work or for locating a site to drill for oil or gas. It is a portable rangefinder using a battery-powered ruby laser in combination with an optical telescope for aiming a laser beam, and a photomultiplier for picking up the laser beam reflected from the target.

Drilling Controls

The driller is able to see all of the necessary instruments, gauges and controls for the various functions of drilling. Within his reach are levers, throttling valves, and other control devices which can change engine speed, fluid pressure, weight on the bit, and make emergency shutdowns. The *weight indicator* is one of the most important drilling instruments used and observed by the driller. It has a large dial with a pointer that shows the weight of the drill string at all times. The rate of penetration is maintained principally by releasing the draw works brake, thereby lowering the drillpipe which applies weight on the bit. The proper combination of weight on the bit, rotary speed, and fluid circulation results in the most favorable rate of penetration. An *indicating pressure gauge* is usually installed in the mud line which extends up the derrick to the rotary hose which is attached to the swivel. A remote-reading dial showing the circulating fluid pressure is usually located on the same panel with the weight indicator. An accurate indicator of rotary torque will help the driller to determine the condition of the bit, or to assess borehole conditions. Engine or rotary speed indications can be observed from tachometers on the driller's console which indicate the revolutions-per-minute. For electrically driven rotaries, an ammeter shows the amount of current needed to turn the rotary at the best speed. Fuel-air intake manifold pressure gauges will indicate the power output of gas or gasoline engines. Diesel engines are provided with *pyrometer instruments* to measure cylinder temperature to indicate engine power.

Pit level indicating instruments, and fluid flow sensors with remote reading devices at the driller's position give immediate indications of drilling fluid gains or losses. Some of these devices are fitted with indicating alarm lights.

Most drilling rigs are provided with recording devices which automatically record the time of day versus the weight of the drill-string suspended in the derrick, pump pressure, and other important data as well as the time of day the observation was made. Hole depth is usually noted on the strip chart at the start of each day, and then the instrument will mark off each foot as the hole is drilled. Another indicating device will show total depth as the hole is deepened.

Hydraulic controls for the blowout preventers are usually placed on the rig floor near the driller's position to enable rapid closure of the preventers in case the well shows signs of a blowout. An alternate set of controls is often located at a remote location away from the rig in case fluid does blow out and for some reason the driller cannot operate his controls. A fluid level or volume instrument for the mud pits is a significant recording device to indicate any approaching hazards of a blowout. These devices are combined with an indicating instrument for a record showing time of day and the level of the fluid in the mud pits. Often a record is made of the indicated volume of fluid returning from the well and its density or weight (Figures 14–1 and 14–2).

The density (weight) of drilling muds is usually measured with a mud balance or scale. By using the density of fresh water as a standard, it is possible to compare it with the sample mud density,

Figure 14–1 Blowout preventer stack at wellhead (courtesy API).

and get a reading. Normally, mud density is reported in pounds per gallon.

Mud viscosity is difficult to measure. However, routine on-site field measurements are made with a *marsh funnel* (Figure 14–3).

Control of the density of drilling mud is an excellent way for the driller to prevent formation fluids from rising to the surface prematurely. When some of the drilling fluid seeps out into the formation causing mud pressure drop, a blowout may occur in certain formations. To help prevent overpressure problems, blowout preventers (BOP) are installed at the top of the wellhead. If a blowout does occur, three sets of rams can be used to close in the well. These seal off

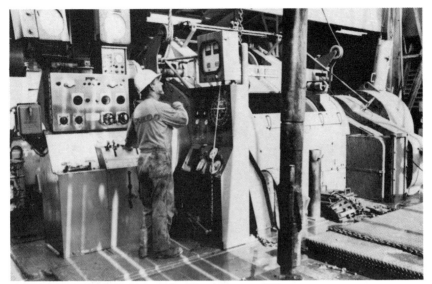

Figure 14-2 The driller's control station (courtesy API).

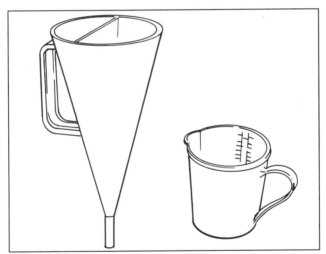

Figure 14-3 A marsh funnel.

annular spaces and can often prevent the fluids from venting to the surface (Figure 14–4).

Underwater blowout preventers, wellheads, and other related equipment are manufactured especially for offshore drilling and production use. The continuous motion of a drillship or floating structure requires that wellheads and blowout preventers be installed on the ocean floor.

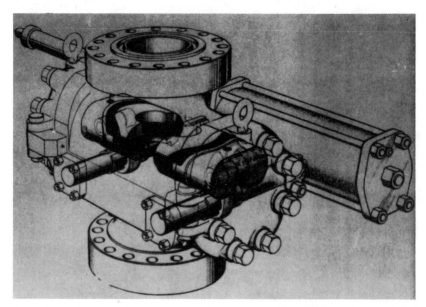

Figure 14-4 Cameron F ram preventer (courtesy PennWell Books, *Workover Well Control*, Adams).

Drilling an oil or gas well is not restricted to only making a hole, as there are many other services required that use different instruments and tools. One is coring, and core analyses are taken at various depths. An assembly ranging from 25 to 60 ft in length is attached to the drillpipe and run to the bottom of the hole. As it rotates, it cuts a cylindrical core which is retrieved in the tube part of the assembly and brought to the surface by tripping out of the hole. Other types are the side-wall samplers, one of which is a percussion-type tool from which hollow projectiles are fired into the wall at the desired depth and then retrieved with the core sample inside. The other is a rotary sidewall coring tool that cuts $^{15}/_{16}$ in. by $1\frac{3}{4}$ in. samples from the formation and retrieves them. It is operated by wireline and a high voltage cable in an uncased borehole. The tool is about 17 ft in length and can obtain 30 cores at each lowering (Figure 14-5).

Cuttings caught at the shale shaker are checked both visually and by ultraviolet light to detect the presence of oil, and also by use of a gas detection instrument. Samples of drilling mud are sometimes taken to a portable laboratory located at the rig site. The unit contains sensitive instruments for detecting gas in the mud.

A formation testing device can be run on a wireline when the drillpipe is out of the hole to obtain a sample of the formation fluids. The

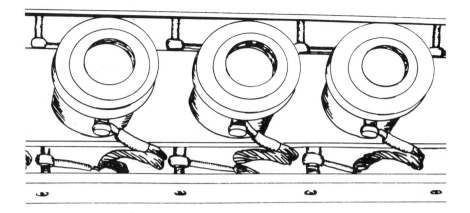

Figure 14–5 Sidewell coring device (courtesy Schlumberger Well Surveying Corporation).

sample quality may be only a gallon of fluid and several cubic feet of gas, if any is present. Pressure information is recorded on the surface as the test proceeds.

A drill stem test (DST) may be run to obtain a sample of the gas, oil, or water and pressures involved by permitting the formation fluids to flow into the drillpipe. Trained personnel and specialized equipment are used for this operation.

There are special tools that are used in offshore and also increasingly in onshore drilling that are called MWD (measurement while drilling). This system is a collection of tools that are built up as groups of modules on the primary steering tool. It normally provides gamma-ray, resistivity, and directional data acquired while drilling. This data is transmitted back to the surface by pulses through the mud, which move at 4,100 ft/sec and are decoded by a computer on the surface. The changes in pulsation are interpreted by the computer as various formations are being drilled through. In some cases a down-hole computer is used and the information is recovered when the drillpipe is tripped out.

There are many causes for the drill bit to wander from the vertical, and there are ways of measuring the "drift" and correcting it. The drift survey instrument is positioned above the bit to record drift angle. The record is made when a paper disk is punched by a pendulum-balanced stylus inside the instrument. The driller can place a stabilizer above the first collar on the string that tends to naturally swing back to the vertical.

Controlled directional drilling requires careful planning and special tools. Directional survey instrument assemblies are lowered on

Equipment **385**

wirelines inside the drillpipe, that give both the angle of deviation from vertical and the direction toward which the hole is directed.

Drillpipe wears, and small cracks may appear that could cause failure and result in lost pipe in the hole. The drillpipe is usually taken out of service periodically and inspected with electronic equipment that will locate flaws and cracks. New instruments have been developed that will fit on the rig floor and inspect the drillpipe as it is tripped out. This new instrument employs eddy currents for the detection of magnetic values that are associated with cracks in the pipe.

Logging Instruments

Electric logging is probably the most common of several wireline measurement methods used to assess the potential of rock formations. After the hole has been drilled, and sometimes at various intervals during drilling, instruments are lowered into the well on an electric cable (wireline) (Figure 14–6). Some of the earlier instruments used only sent an electrical current into the formation, and then measured the returning current to calculate the resistance of various formations and fluids in order to make evaluation studies. Also, the spontaneous potential (naturally occurring voltage of potential) measurement was usually included with the electric logs. Other wireline logging instruments include the neutron tool for neutron logging and the gamma ray log. The neutron log is often a standard counterpart of the gamma ray. The original neutron tool was an early development. It bombarded the formation with neutrons from a chemical source in the logging tool, and engineers could measure the response of the formation as a function of the number of hydrogen atoms present. The gamma-ray tool measures the naturally occurring radiation that emanates from radioactive potassium, thorium, and uranium. Potassium and thorium are closely associated with shale, while uranium may be found in sands.

The acoustic log is an instrument that uses sound waves and measures the reaction of the rock.

Some of the latest instruments used are computer-generated logs in which the various parameters may be displayed in a number of different ways, including colored 3-D maps.

Line Pipe and Fittings

During the past several centuries, various materials have been used for making pipe to transport water and gas. Some of these projects were monumental in size and scope and represented great

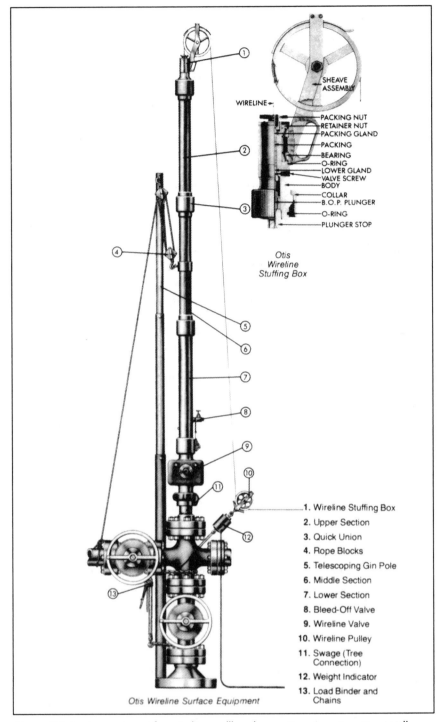

SHEAVE
ASSEMBLY

WIRELINE

PACKING NUT
RETAINER NUT
PACKING GLAND
PACKING
BEARING
O-RING
LOWER GLAND
VALVE SCREW
BODY
COLLAR
B.O.P. PLUNGER
O-RING
PLUNGER STOP

Otis
Wireline
Stuffing Box

1. Wireline Stuffing Box
2. Upper Section
3. Quick Union
4. Rope Blocks
5. Telescoping Gin Pole
6. Middle Section
7. Lower Section
8. Bleed-Off Valve
9. Wireline Valve
10. Wireline Pulley
11. Swage (Tree Connection)
12. Weight Indicator
13. Load Binder and Chains

Otis Wireline Surface Equipment

Figure 14–6 Arrangement for wireline wellhead operations (courtesy PennWell Books, *Workover Well Control*, Adams).

Equipment

technological advances at their time. However, the volumes transported and the distances involved were minuscule in comparison with today's petroleum industry. The pipe materials—bamboo, hollowed wooden logs, lead, and copper—were inferior to today's steel and could not withstand much pressure.

The use of wrought iron and steel for making pipe is a fairly recent development. The manufacturing of iron and steel tubular goods only began in full swing after Drake's discovery of commercial quantities of oil in 1859. The drilling, production, transportation, storage, refining, and marketing of petroleum and its products have led to the development of new classes of tubular goods to meet industry demands. Processes for manufacturing wrought iron and steel suitable for the petroleum industry were developed just prior to, and during, the early stages of the oil age. The puddling process to produce wrought iron was introduced in 1784, and made available the first ferrous metal suitable for forge welding.

In 1855, the Bessemer process for making steel was developed, and the open-hearth process in 1861. In the beginning of both processes, the output was geared to the production of rails, plates, and structural steel. It was not until 1887 that steel pipe was made in commercial qualities. Today, almost all pipe used in the oil and gas industry is made of steel. It is made mostly be open-hearth, electric furnace, and basic-oxygen processes.

To keep up with current demand it is now necessary to transport large quantities of oil at higher pressures through single pipelines. This has resulted in steady requests for larger-sized pipe with increased yield strength. Because of this, API standards now include some special extra-strength pipe that will withstand pressures up to 100,000 psi.

Steel is the only available material that has the strength and other qualities required for general use in pipelines, refineries, and gas handling that is obtainable to meet the demand. Steel is subject to corrosion in varying degrees and requires some protection from it, as discussed in Chapter 13. Some severe situations may require use of pipe and fittings made of material other than steel, such as asbestos-cement, extruded plastic, and glass fiber-reinforced plastic that are suitable only for low-pressure applications.

Connections and Fittings

Connections are one of the main types of pipeline fittings. They are used to connect various pipes and vessels and provide a liquid-tight system that must be maintained in the industry. Some examples are collars, elbows, tees, flanges, and unions. Fittings other than couplings such as plugs, reducers, and valves are also used in the industry.

Valves

Valves are the second main type of pipeline fittings. They are special fittings in that they are not static parts of the pipeline system but are the means of providing control of the fluid being handled. The most common form of control is that of stopping flow completely or restricting it to a determined flow with a valve. Many types of valves have been developed to provide fluid flow and only some of them will be discussed in this chapter. The bottom line is that without valves, the petroleum industry could not function as it does today (Figures 14–7, 14–8, and 14–9).

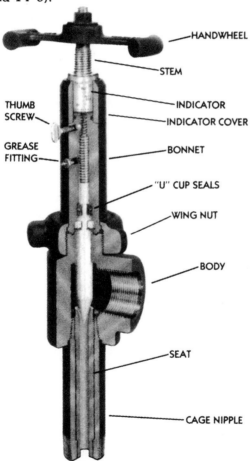

Figure 14–7 Barton adjustable choke cutaway (screwed X screwed) used for flow control of deeper wells that operate under great pressure (courtesy Barton Valve Company, Inc.).

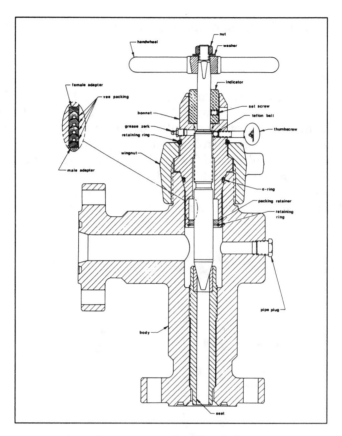

Figure 14-8 2,000 through 10,000 psi adjustable chokes (courtesy Barton Valve Company, Inc.).

Gate Valves

The *gate valve* is a comparatively simple piece of equipment that is very common in all phases of the oil and gas industry. The term gate is more appropriate for the closing element because a disc is defined as a nearly flat, circular plate although some gates now have other shapes. There are several types of gate valves available (Figures 14-10, 14-11, and 14-12).

Plug Valves

Plug valves are similar in design to the stopcock in that the plug usually has an elongated port that may be smaller in cross-sectional area than the inside diameter of the pipe that it is connected to. The valve is operated by giving the plug a quarter turn. It may be operated by a lever fitted to the stem, by a gear system, or by mechanical

Figure 14-9 Barton API 6D high-temperature geothermal valves on a geothermal well (courtesy Barton Valve Company, Inc.).

operators that are power driven. Where fast operation and reliability is desired the plug valve is gaining favorable application. Many plug valves are designed with lubricating grooves in the plug to make the turning easier and improve the sealing effect. Normally plug valves in the pipeline industry range in sizes up to 12 in. but are not used where scrapers or pigs are run through the pipeline.

Ball Valves

Ball valves are a fairly recent adaption of the plug valve principle for use in the pipeline industry whereby valves of nearly any size may be made with full-sized, circular conduits.

Butterfly Valves

Butterfly valves have a disc mounted on a stem across its diameter inside a conduit approximately the same size as the pipe. They are used to control fluid flow at low pressures.

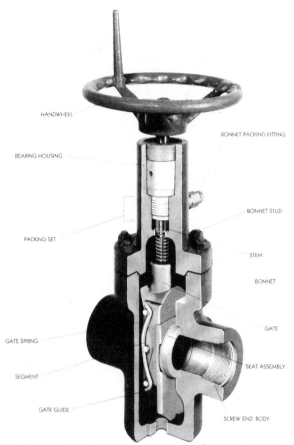

HANDWHEEL

BONNET PACKING FITTING

BEARING HOUSING

BONNET STUD

PACKING SET

STEM

BONNET

GATE

GATE SPRING

SEGMENT

SEAT ASSEMBLY

GATE GUIDE

SCREW END BODY

Figure 14-10 Barton API 6A 2,000#, 3,000#, and 5,000# E.G. (expanding gate) gate valve (courtesy Barton Valve Company, Inc.).

Check Valves

Check valves are designed to permit full flow in one direction but to close automatically to prevent reverse direction flow. They use the force of gravity or back flow of fluid to close a disc, thus closing flow in the reverse direction. Check valves are used in gathering systems and in station piping more than in any other locations. Some valves are designed to permit scrapers to pass through them and are put in service in other piping systems as well.

Relief Valves

The function of *relief valves* is to remain closed when system pressures are normal and to open automatically and permit fluids to flow from the system when pressures rise above predetermined levels. They are often called safety valves because the release of pressure may prevent damage. They are set to open at appropriate pressure levels and release fluids into a safe discharge system. They usually reclose automatically when the excess pressure has been relieved, and then are kept closed by spring pressure or weight loading as long as normal pressure is maintained. Some relief valves are designed so that their operation may be remotely indicated or monitored by communications and control systems.

Globe valves

Globe valves are commonly used in auxiliary piping in oilfield gathering systems where pressures are low. They are useful for shutoff and throttling, such as in fuel supply lines to engines and motors. They offer high resistance to flow and are very reliable.

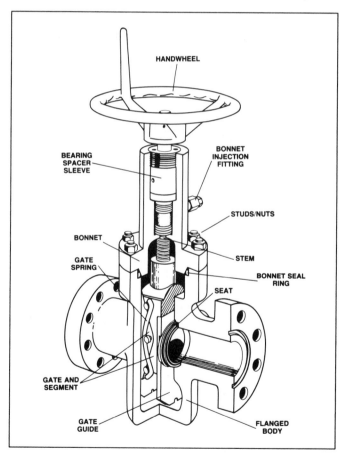

Figure 14–11 High-pressure E.G. gate valves with flanged body (courtesy Barton Valve Company, Inc.).

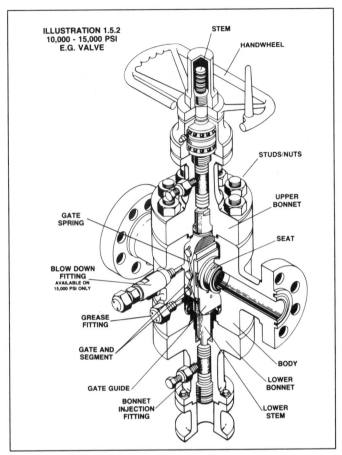

Figure 14-12 API 6A 10,000# and 15,000# dual-bonnet E.G. gate valve (courtesy Barton Valve Company, Inc.).

Lock Stopcock

Lock stopcocks are used when there is need of assurance that a valve has remained in the appropriate open or closed position during a specific operation or time period. They are often used in the custody transfer of crude oil from producers' tanks to pipeline gathering systems. Often they are also used to seal possible bypass piping around measuring devices such as meters.

Tank Shutoff Valves

Tank shutoff valves operate on a float principle and are often used to prevent air from entering pipeline systems. They are usually installed adjacent to a tank from which oil is drawn. They close automatically when the oil level drops to preset level above the outlet pipe.

Modern Petroleum

Refining and Gas Handling Instruments

The various types of instruments and equipment used to control and measure liquids and gases during refining and natural gas handling would take hundreds of pages to describe. Thus, only some of the basic instruments and controls will be discussed here: those that are designed to maintain or measure the liquid or gas at a specific pressure, temperature, flow-rate, volume, or liquid level. Liquids and gas are often collectively referred to as controlled medium.

Temperature Measurement

The thermometer is an instrument used for determining the temperature of a body or space. Temperature measurement is common in refining and processing; it is one of the first and oldest measurements associated with the industry. There are many types of thermometers used, and the common thermometers consist essentially of a confined substance such as mercury, the volume of which changes with a change in temperature. An example is the common weather thermometer, in which a column of contained liquid rises or falls as it measures atmospheric temperature. It is obvious that weather thermometers are not practical for measuring liquids or gases in processing, because of the very high temperatures involved and other physical factors.

Generally, operating temperatures in process units and gas compressors are measured with *thermocouples, electrical-resistance thermometers, optical pyrometers, radiation pyrometers*, and other specially designed instruments.

All comprehension of temperature is relative and should be stated in terms of a known, accepted scale. Temperature can be measured by several scales as shown in Table 14–1 and Figure 14–13. The Fahrenheit scale (°F) is commonly used but the use of the Celsius scale (°C) is gaining popularity. In processing two absolute zero scales are used. The absolute scale using Celsius degrees as the temperature interval is called Kelvin (K), in which absolute zero temperature is –273.16°C (rounded off to –273°C). The absolute scale using Fahrenheit degrees is called the Rankine (R) scale, in which absolute zero temperature is –459.69°F (rounded off to –460°F). Absolute zero is a hypothetical temperature characterized by the complete absence of heat.

Heat balance, heat exchange, and many other problems involving temperature are important in all phases of both plant operations and

To Convert	Use Formula
Fahrenheit to C	$C = \frac{5}{9} \times (F - 32)$ or $C = (F - 32) \div 1.8$
Celsius to	$FF = \frac{9}{5} \times C + 32$ or $F = 1.8 \times C + 32$
Celsius to K	$K = C + 273$
Kelvin to C	$C = K - 273$
Fahrenheit to R	$R = F + 460$
Rankine to F	$F = R - 460$
Kelvin to R	$R = \frac{9}{5} \times K$ or $R = 1.8 \times K$
Rankine to K	$R = \frac{5}{9} \times K$ or $R = K \div 1.8$

Table 14–1 Conversions

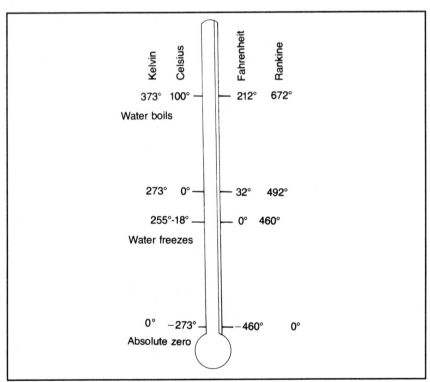

Figure 14–13 Temperature relationships.

gas compression. Means of indicating, recording, and controlling temperatures are necessary for the proper operation of equipment. Heat-affected properties of substances, such as thermal expansion, radiation, and electrical effects, are used by commercial temperature-measuring instruments. These instruments vary in their precision, which is affected by the property used and the substance tested as well as the design of the instrument.

Thermocouple

The *thermocouple* principle is based on the relation between different metals insulated from each other and joined at the ends to form a simple continuous electrical circuit (Figure 14–14). If one of the junctions is maintained at a temperature higher than that of the other, an electromotive force (EMF) is set up that will produce a flow of current through the circuit. The magnitude of the net EMF depends on the difference between the temperature of the two junctions and the materials used for the conductors. The EMF is measured with a galvanometer and corresponds to temperature measurement.

Thermocouples have a high degree of accuracy, are relatively inexpensive, are very durable, and are versatile in application. They are convenient for recording temperature readings from one or many remote points and can be incorporated into control and regulating equipment.

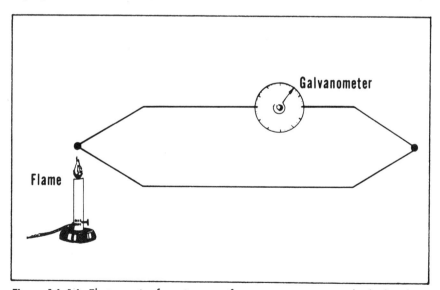

Figure 14–14 Electromotive force is set up if temperature at one end is higher than temperature of the other end (after Babcock and Wilcox).

Fusion Pyrometers

Fusion, or the change of state from solid to liquid, occurs at a fixed temperature for a pure chemical element or compound, such as ice to water at temperatures above 32°F. The melting points of various materials are therefore suitable fixed points for temperature scales.

A suitable material is formed into small pyramids about 2 in. in height. These pyrometric cones have a known melting point at established temperatures and are sometimes used as a method of measuring very high temperatures in refractory heating furnaces. They are suitable for the temperature range from 1,000 to 3,000°F. Fusion pyrometers are also made in other forms such as crayons, paints, and pellets.

Vapor-Pressure Thermometer

The changes of vapor pressure of a specific liquid with temperature is utilized in the *vapor-pressure thermometer*. The working range of a given instrument is limited to several hundred degrees and usually lies between – 20° and 200°F.

Expansion-Type Thermometer

Most substances expand when heated, and usually the amount of expansion is almost proportional to the change in temperature. This effect is utilized in various types of thermometers using gases, solids, or liquids. Nitrogen gas-filled thermometers are suitable for a temperature range of – 200 to 1,000°F.

Radiation-Type Thermometer

All solid bodies emit radiation. The amount is very small at low temperatures and large at high temperatures. For example, the temperature of hot iron can be estimated visually by its color. When iron is dark red, its temperature is about 1,000°F and it is about 2,200°F when white. Two types of temperature-measuring instruments are based on the radiating properties of materials: the optical pyrometer and the radiation pyrometer.

Thermometer Well

The commercial thermometer inserted in a *thermometer well* is widely used in the petroleum industry. It consists of a bulb filled with a liquid such as mercury, pentane, toluene, alcohol or other such liquid in a glass stem that is properly marked to scale. The thermometer well is permanently mounted in the piping system. It is inserted in the well and packing is installed in the annular space around the stem in order to eliminate variations due to ambient temperature. These thermometers are limited to relatively small changes of temperatures.

Pressure Measurement

Pressure is a common term in the oil and gas industry. It can be any force from less than atmospheric (vacuum) to thousands of pounds above atmospheric pressure. Pressure measurement is possibly the most often-made measurement in refineries and process plants. There are usually more pressure gauges used in a process plant than any other instrument. Pressure is a good quick indication of the work done by pumps and compressors. Also, it is the most important measure of the status of operating pressure vessels.

The pressure of each vessel and tower is shown on the instrument control board of well-designed process plants and compressor stations. Because of its effect on boiling and condensation points and because it responds more rapidly to changes in these values, pressure measurement is sometimes used instead of temperature to monitor and control these processes. For example, heavy oils are boiled off (distilled) at a feed inlet temperature of about 700°F (371°C) in a vacuum pipe still with approximately 1 psia (pounds per square inch absolute) vacuum. If the vacuum were not used, the same oil would boil off at a feed inlet temperature of about 1,100°F (593°C).

Pressure is a measure of the force exerted by a fluid due to its molecular activity. The molecules of a fluid are in continuous motion, colliding with one another and the walls of the surrounding container. This continuous impact of molecules creates a force (pressure) against the container walls.

Liquids are fluids that have definite volumes independent of the shape of the container under conditions of constant temperature and pressure. Liquids will assume the shape of the container and fill a part of it that is equal in volume to the amount of liquid. For practical purposes, liquids are considered noncompressible because extreme pressures are required to get a very small reduction in volume.

Gases are fluids that are compressible; gases will vary in volume to fill the vessel containing them. The volume of a given mass of gas will change to fill the container. Gases cannot be contained in open-topped containers as liquids can.

Bourdon Tube

The pressure gauge (or gas) is probably the earliest instrument used in refining operations. It is a direct-connected, locally mounted pressure indicator using a *Bourdon tube* measuring element (Figure 14–15). The measuring element is named for Bourdon, a French engineer who invented the device in 1849. The Bourdon tube is an elastic device that functions as a pressure-measuring element because its tip movement is proportional to the pressure being measured. It is sealed at one end (the tip) and equipped with a pressure connection on the other end.

Any pressure inside the C-shaped tube exceeding the pressure on the outside will tend to straighten the tube. The movement of the sealed end of the 250° arc is used to position a pointer, a recording pen, or other indicating equipment. The amount of movement of the sealed end is formed by the tube. Usually, the movement is small and must be amplified with proper linkage if a broad scale is required. Calibrations are possible for measuring pressures up to thousands of pounds.

A spiral Bourdon tube is made by winding a partially flattened metal tube into a spiral having several turns instead of a C shape. The spiral form produces more tip movement but does not change the

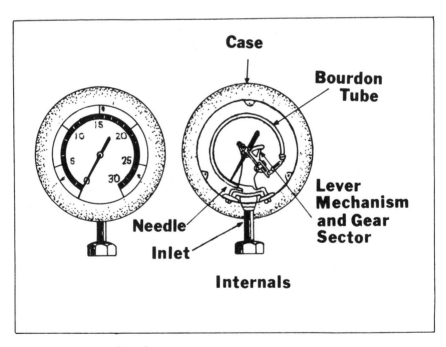

Figure 14–15 Bourdon-tube gauge.

operating principle of the Bourdon tube. The tip movement of a spiral equals the sum of the tip movements of all its individual C-bend arcs. Another design uses the tube wound in the form of a helix.

Diaphragms

A *diaphragm element* is used in measuring low pressures up to a maximum of a few pounds. Like the Bourdon tube, it works because of the elastic deformation characteristics. It is extremely sensitive to small pressure changes and can provide reading superior to the Bourdon tube within the range of 0 to 10 psi. It can also be used to operate an automatic control device.

There are generally two types of diaphragms used; one is made of a thin sheet metal (Figure 14–16) and the other is made of nonmetallic materials. The nonmetallic diaphragms are used to measure extremely low pressures.

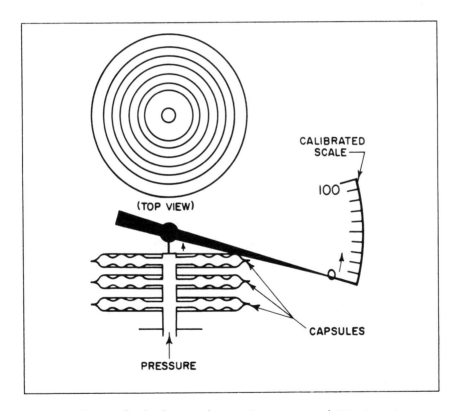

Figure 14–16 Metallic diaphragm schematic (courtesy Ametek-U.S. Gauge).

Bellows-Actuated Instruments

A *bellows unit* is similar to the diaphragm (it works on the elastic deformation principle) except that the corrugations extend in series rather than expand outward. When the pressure inside the bellows increases, the metallic discs thicken and the length of the bellows increases. This increase in length is the sum of the expansion of all the discs and is a measure of the pressure inside the bellows. Their usual range is 0 to 100 psi (Figure 14–17).

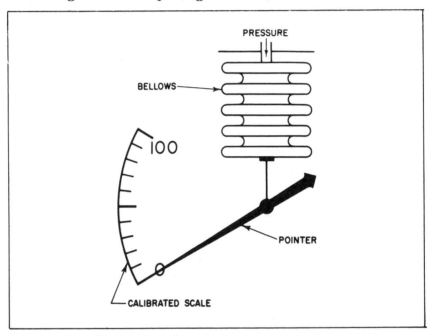

Figure 14–17 Bellows element schematic (courtesy Ametek-U.S. Gauge).

Bellows-measuring elements have a variety of uses in measuring pressures. They can be used singly or in pairs to measure absolute or differential pressure. When two bellows are used, one pressure is admitted to each bellows. By mechanically connected the two, the difference in their expansion or contraction is a measure of the differential pressure. The most widely used bellows-actuated instrument for measuring differential pressure is the *flowmeter.*

U-Tube Manometers

The *manometer* is a fairly simple instrument used for measuring pressures from zero through a few pounds per square inch. This device is a U-shaped tube made of glass or other transparent

Modern Petroleum

material. One end of the manometer is connected to the pressure to be measured, and the other end is open to the atmosphere. A scale with markings is attached behind the tube, and the tube is filled about half full with fluid. The pressure will force the fluid to move in the tube until the weight of the displaced fluid is equal to the force of the exerting pressure. The difference between the height of the fluid in the two sides of the U-tube is the pressure measurement. Because of their fragile construction and limited range, they are seldom used except in cases where another type of instrument would be is impractical.

Electrical Pressure-Measuring Instruments

Electrical pressure-measuring elements convert the volumetric changes caused by pressure into mechanical movement. The most common instrument of this type is the strain gauge. Basically, the strain gauge consists of a small wire arranged so a change in the pressure stretches the wire. Its electrical resistance is changed as the wire stretches, and this change can be measured and converted to indicate pressure change.

Liquid Level Measurement

Sight Glass

The *sight glass* is used in certain tanks and vessels for a direct reading of liquid level. It consists of a transparent column, usually glass, attached to the tank or vessel by suitable leakproof fittings. There is usually a scale attached to the column by which the liquid level can be observed when the valve(s) is opened. If the vessel is at atmospheric pressure, the top end of the column can be vented.

Float-Type Units

Liquid level as measured by *float-type units* is determined from the buoyancy of a floating body partially immersed in the liquid. Buoyancy is the upward force exerted on the floating body by the liquid. The float is usually a hollow metal ball that is mechanically connected to the measuring instrument. If necessary, suitable leakproof connections are installed between the float and instrument.

Liquid level has no absolute value and is always relative to a reference point such as the tangent line of a vessel and the bottom of the vessel. Liquid level is easily understood and can be directly observed.

Tank Gauging

Manual *tank gauging* has been practiced for nearly a century and is still used through the oil industry, especially for gauging vertical cylindrical production and storage tanks. When there might be some doubt in the performance of an automatic gauge or for double checking, the steel gauging tape and bob are used.

Recently, field tests have been made with radar tank gauging equipment with promising results. It lends itself for applications to vertical, cylindrical, atmospheric storage tanks that contain asphalt, acid, wax, and heavy, viscous products. It can also be used for gauging high-temperature liquids.

Radar gauging offers several advantages, such as trouble-free operation and the reduction of manpower requirements. Radar technology is not new but its adaption to gauging is (Figure 14–18).

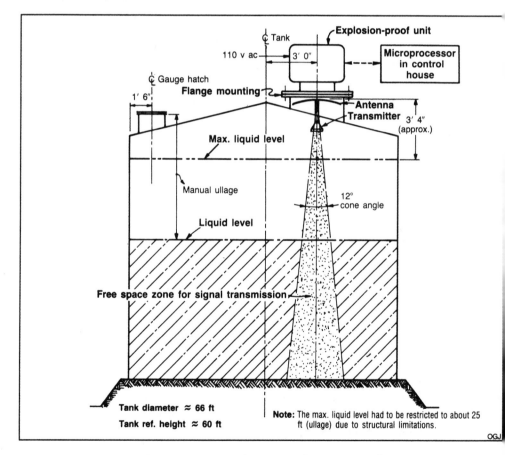

Figure 14–18 Radar gauging system (courtesy *Oil & Gas Journal*).

Flow Control

One of the important factors that determines the physical setup of a refinery or gas plant is the rate of fluid flow. After determining flow rate, the next problem is providing the equipment that is best suited to do the job. Some control functions can be accomplished manually; however, all control functions can be better accomplished automatically. Automatic control systems should have a measuring element, controlling element, and a final control valve.

The controlling element produces a signal when any variable occurs. The signal can be produced by various pieces of equipment. Some employ an electrical sensor used with a conventional measuring element. These sensors convert the force or movement of the element to a measurable characteristic of an electrical circuit. Movement of the element can shift a contact across a resistance coil. It can also change the position of a coil in a magnetic field, or increase or decrease the gap between the conductors in a capacitor. Other controlling elements can utilize pneumatic sensors.

Gas Regulators

There are several varieties of mechanical devices designed for controlled reduction of gas pressure. Any gas regulator matches the flow of gas through the regulator to the demand for gas downstream, and it should maintain the system pressure within certain limits. Regardless of their size, all gas regulators have three basic components in common, type of service, or complexity: a *regulating element*, a *sensing element*, and a *loading element*.

Regulating Element

The regulating element creates a restriction in the gas line that is decreased or increased in size to vary the gas flow rate. It is usually a valve, although it can be an expandable sleeve or any device that will modulate the flow.

Loading Element

The loading element applies the needed force to the restricting element to cause it to vary. It can be a spring, diaphragm, weight, or piston (Figure 14–19).

Sensing Elements

Sensing elements will vary depending upon the type and classification of the regulator. Normally, regulators are classified as direct (self) operated or relay (pilot or remote) operated. The most common

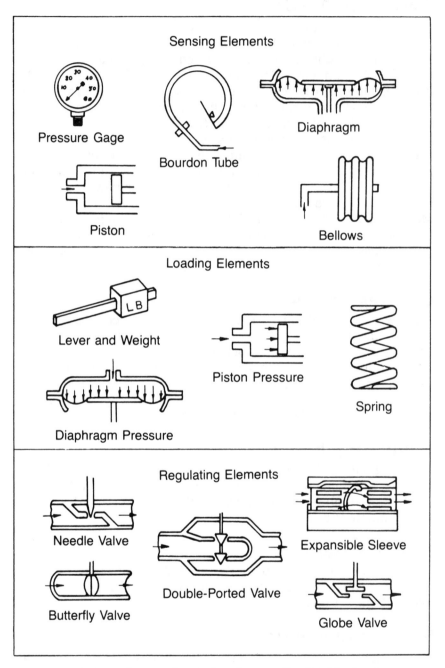

Sensing Elements

Pressure Gage

Bourdon Tube

Diaphragm

Piston

Bellows

Loading Elements

Lever and Weight

Piston Pressure

Spring

Diaphragm Pressure

Regulating Elements

Needle Valve

Expansible Sleeve

Double-Ported Valve

Butterfly Valve

Globe Valve

Figure 14-19 Basic elements of a regulator (courtesy PennWell Books, *Gas Handling and Field Processing*, Vol. 3, Berger and Anderson).

sensing element in a direct-operated regulator is the diaphragm. All of the energy required to position the restricting element comes from the gas stream through the direct action of the diaphragm (Figure 14–20).

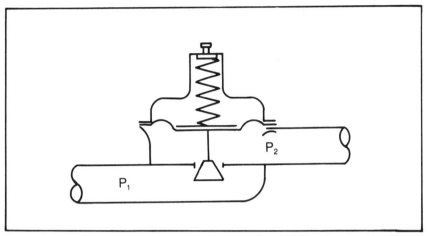

Figure 14–20 Direct-operated regulator (courtesy PennWell Books, *Gas Handling and Field Processing, Vol. 3*, Berger and Anderson).

When the direct-operated regulator is operating under steady-state conditions, the plug valve is in equilibrium, with the pressure force exactly balancing the spring force. The pressure force is developed by the sensed pressure acting against the diaphragm area, and the spring force is developed from the compression in the spring.

Relay-Operated Regulators

At locations where the inlet pressure varies too much to use a direct operated regulator, one of the many types of *relay-operated regulators* should be used. In relay-operated regulators, not all the energy required to position the regulating element is obtained from the gas stream through direct action of the sensing and loading elements. Instead, additional energy is obtained by the action of a variable loading force. This force is controlled by an auxiliary sensing element, but it is not transmitted directly by the element. The sensing element acts as a relay to increase or decrease the loading force. A common type could be the pilot amplifier regulator shown in Figure 14–21.

Instrument-Controlled Regulator

Some installations require relay-operated, instrument-controlled, or remote-controlled regulators. One type of *instrument-controlled*

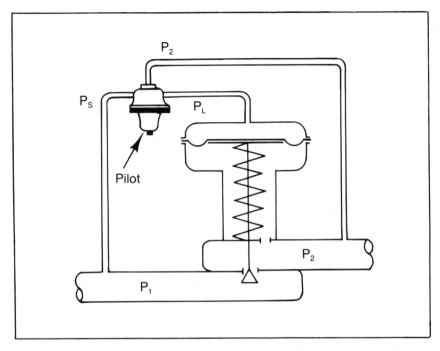

Figure 14–21 Pilot-operated regulator (courtesy PennWell Books, *Gas Handling and Field Processing, Vol. 3,* Berger and Anderson).

regulator has two loading elements, and the sensing element is not one of the basic working parts of the regulator. One of the loading elements is a compressed spring underneath the diaphragm, and the other is pressure on top of the diaphragm, which is controlled by the Bourdon tube and flapper.

The loading spring supplies a positive force to overcome the weight of the diaphragm and valve to move them from the open to the closed position, and to seat the valve positively. The small spring-adjustable air supply regulator is set for an outlet air pressure high enough to produce loading pressures on the main diaphragm that can create enough force to open the valve against the spring loading force (Figure 14–22).

An increase in the downstream pressure tends to straighten the Bourdon tube. This tube movement pivots the flapper away from the air bleed hole, and the bleed rate increases. As the air bleed rate increases, the loading pressure falls and the spring-loading element forces the valve toward the closed position.

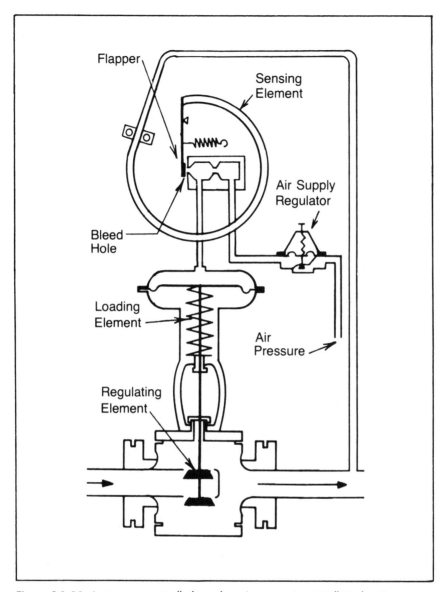

Figure 14–22 Instrument-controlled regulator (courtesy PennWell Books, *Gas Handling and Field Processing, Vol. 3,* Berger and Anderson).

Piston Regulator

In a *piston regulator,* the diaphragm of a conventional regulator is replaced by a specially designed piston. The piston regulator contains a pilot regulator, which responds to the loading pressure applied to the piston. Any increase in the downstream load decreases the downstream pressure and unbalances the forces on the piston. The net downward force then further opens the regulator valve. If the downstream pressure falls below the setting of the pilot regulator, the pilot valve opens wider, and the loading pressure under the piston causes the main regulator valve to open wider. Any decrease in the downstream load causes the reverse to occur.

Expansible-Tube Regulator

An expansible sleeve fitted over a hollow, slotted metal cylinder forms the regulating element of an *expansible-tube regulator*. Gas passing through the regulator flows out of the cylinder through the upstream row of slots, around the partition in the middle of the regulator back into the cylinder through the downstream flow of slots and out of the regulator. These flow passages are open only when the expansible sleeve is forced off the cylinder against the body of the regulator. The length of the sleeve that remains against the cylinder, and hence the area of the downstream flow passage is controlled by the pilot regulator through its control of the pressure in the jacket space between the body of the regulator and the outside of the rubber sleeve. This pressure is less than the inlet pressure because of the pressure drop across the orifice which depends on the rate of flow of gas through the orifice and out the pilot regulator valve.

During normal operation, the force exerted on the sleeve by the jacket pressure, plus the sleeve tension, is equal to the force exerted by the pressure underneath the sleeve at the point where it stretches from the regulator body to the cylinder. When the downstream pressure falls below the control point, the pilot regulator opens wider, the rate of flow through the orifice increases, and the jacket pressure decreases. This unbalances the forces across the rubber sleeve, and it rolls open wider until the increased flow through the regulator reestablishes the desired downstream pressure and a new balance point.

Motor Valves

Remote-controlled motor valves are often used for special applications such as final control of a larger piece of equipment or a system. The essential parts consist of a controlled pressure connection, diaphragm, loading spring, and restricting valve. The restricting valve can be either single or double ported.

The motor valve can be designed as either normally open or normally closed. Either way is equally effective. The loading spring attempts to hold the restricting valve either completely open or completely closed. Controlled pressure is applied to the top of the diaphragm, overcoming the resistance of the loading spring and increasing or decreasing the restricting valve, depending on the arrangement of the valve seats.

Pneumatic Control Instruments

Pneumatic control instruments are used in areas where remotely monitoring and operating control valves are desirable. These systems require instrument air supply. In general, the pneumatic control includes various components of the control loop, such as transmitters, receivers, and controllers.

The transmitter takes a particular span of measurement and converts it to a pneumatic signal between 3 and 15 psi. The 3 to 15 psi signal of the transmitter goes to a receiver unit, where it enters the receiver bellows and moves a pen or indicator a distance on the scale that is proportional to the particular air valve between 3 and 15 psi. This receiver unit has a valve preset point and the incoming air pressure is compared to it. Any difference in the pneumatic valves in the measurement and set bellows at the controller is the actuating signal to the controller. The output of the controller goes directly to the valve motor of the final control element and positions the restricting valve (Figure 14–23).

Shutoff Valves

Shutoff valves automatically shut off the supply of gas to a system when the pressure rises above a specified value. There are various

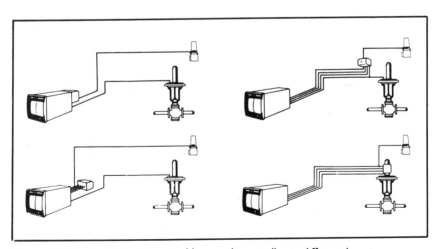

Figure 14–23 Variations in control loop with controller in different locations (courtesy Foxboro).

types in use through the gas industry. In one industrial type, a latch normally holds the valve open. The latch is connected to a diaphragm that is exposed to the pressure to be controlled. If the pressure becomes too high, movement of the diaphragm releases the latch and a spring closes the valve. Once closed, the valve must be opened and reset manually (Figure 14–24).

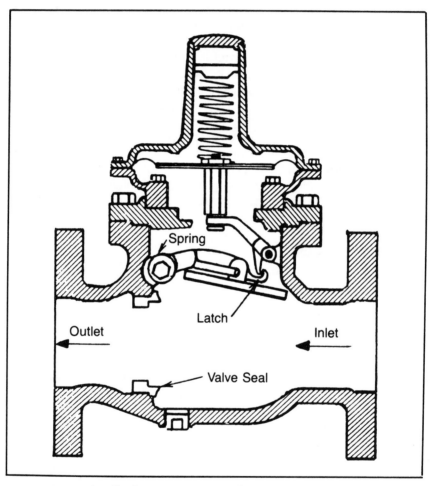

Figure 14–24 Shutoff valve (courtesy PennWell Books, *Gas Handling and Field Processing, Vol. 3,* Berger and Anderson).

Orifice Meter

Gas flowing through a pipe can be measured by placing a constriction (an *orifice plate*) in the line to cause a pressure drop as in the gas it flows through the orifice plate. This pressure drop is known as *pressure differential*. There is a direct relationship between the amount of this pressure drop and the rate of flow. Since gas is a highly compressible fluid, its density will vary greatly with the pressure existing at the orifice plate. So when flowing gas is measured with an *orifice meter*, it is necessary to measure both the differential pressure and the flowing pressure, which is called the *static pressure*. There are many other measurements and factors to be taken into account in the accurate measurement of gas with an orifice meter. They are usually stationary and located where they are easily accessible at all times. Orifice meters are often used near wellheads. (Figure 14–25).

Gas volumes can be measured with orifice meters, *positive-displacement meters, turbine meters, venturi meter, rotameters,* and other specially designed equipment. One common type that is familiar to most people is the typical domestic meter used by gas utility companies.

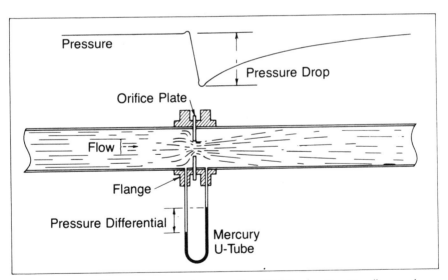

Figure 14–25 Differential pressure (courtesy PennWell Books, *Gas Handling and Field Processing, Vol. 3*, Berger and Anderson).

Chapter References

Neal Adams, *Workover Well Control*, (Tulsa: PennWell Books, 1981).

Bill D. Berger and Kenneth E. Anderson, *Gas Handling and Field Processing*, (Tulsa: PennWell Books, 1980).

S. Sivaraman, "Field Tests Prove Radar Tank Gauge Accuracy," *The Oil & Gas Journal*, April 23, 1990.

A Primer of Oilwell Drilling, (Austin: University of Texas Petroleum Extension Service, 1976), Fifth Printing.

15

Corrosion

The oil and gas industry is one of the largest users of iron and steel (metallic) equipment in the world. Metallic equipment includes drilling rigs, refinery components, steel tanks, well casings, ships, pipelines, and the thousands of other structures, devices, and containers needed to produce and deliver petroleum *products*. For example, there are more than a quarter-million miles of buried and submerged pipelines in the United States alone. Throughout the world are hundreds of thousands more miles, many of which are far beneath the sea and which often are connected to offshore production and gathering structures.

These pipelines carry a wide variety of raw materials and finished products, some of which are corrosive and have a harmful effect on the internal walls of the pipe. This internal *corrosion* and because buried and submerged lines are constantly under attack from external *corrosion* creates an area of great concern and massive cost to the industry. Heavy fines can be levied against a company or person responsible for creating a hazardous situation that poses a danger to life or the environment. Thus a rupture or break in any petroleum pipeline is capable of causing damage far beyond the cost of repairing the pipe or the loss of the material being transported.

Today's environmental concerns are focusing more and more attention on the oil and gas industry, and there are more and more demands for producers and transporters to take drastic steps to protect their pipelines in order to prevent oil and product spills and natural gas leaks.

The Cost of Corrosion

The total cost of the damage done to metallic structures throughout the worldwide petroleum industry is staggering, and can easily run into the billions of dollars. Today, great efforts are being made to train engineers and technicians in ways of mitigating corrosion. And new and more effective means are being developed and used to combat this costly problem.

Protecting metallic structures in not cheap. However, it is only a fraction of the cost of the damage corrosion can cause to unprotected equipment. This protection, referred to as *cathodic protection*, will be discussed in detail in this chapter. Given the high cost of a new, coated and wrapped line installation, the addition of cathodic protection is very a small factor considering the benefits it provides.

Some of the earlier pipelines were laid as bare pipe, and the cost of recovering these lines and replacing them with coated and wrapped pipe is enormous. An alternative to reclaiming and recovering is to use the *hot spot* method of corrosion control. Generally, this is done by making a survey along the entire length of the line to determine the areas or "hot spots" where corrosion is mostly likely to occur. This might represent only 10 to 15% of the line, and once the hot spots are located, cathodic protection can be added to them to reduce the risk of leaks.

Economics will be the guide in determining which of the two common systems of cathodic protection will be used: *sacrificial anodes* or the *impressed current (rectifier)* system.

What Is Corrosion?

Corrosion is the deterioration of a metal object.Usually, we commonly refer to it as the rusting of an object. Sometimes it is referred to as oxidation. Other definitions include chemical attack, electrolysis, or an electrical phenomenon. However, even though each of these is partially correct, none of them tell us very much. The process of corrosion that affects the petroleum industry is basically electrochemical in nature and requires the presence of some form of oxygen. This is especially true in the case of buried pipelines.

The Corrosion Mechanism

A corrosion cell will not function unless certain conditions are met. These include:

(A) There must be an *anode* and a *cathode* (definitions to follow).

(B) There must be an electrical potential (voltage) between the anode and cathode. There are various conditions that can cause this potential on a buried metallic structure.

(C) There must be a metallic path that electrically connects the anode and cathode. This could be the metallic structure itself, such as a continuous length of pipeline.

(D) The anode and cathode must be immersed in an electrically-conductive *electrolyte* which is *ionized.*

To be ionized, some of the water (H_2O) molecules are broken down into positively charged hydrogen ions (H^+), and negatively charged hydroxyl ions (OH^-). In most cases, the water or soil moisture that surrounds pipelines fulfills this condition.

When subjected to an ionized electrolyte, metals will be consumed at the anode, as an electrical current flows between the two poles. At the same time, the cathode will be protected from a loss of metal.

Anodes and Cathodes

The Anode

An anode is often described as the positive electrode (also called a terminal or pole) of a battery or other electrochemical device. A cathode is the negative electrode (terminal, or pole) of such a device. Usually, an anode will be of a different metal than the cathode, such as a magnesium anode and an iron cathode, and there will be a voltage potential or difference between the two poles.

There are many combinations, such as both the anode and cathode being made of iron, or there may be an iron anode and a copper cathode. The term anode is also used to refer to a device known as a *sacrificial anode assembly.* This is a bag, usually containing a magnesium or zinc ingot and other chemicals which is connected by a wire to an underground metallic system such as the buried pipe which delivers natural gas to your home. Or anode may be used to describe the ground bed of a electrical impressed current (rectifier) system. The important thing to remember is, that whatever its use, or what it is called, the anode protects the cathode from destruction by corrosion.

The Cathode

While under certain conditions the cathode can be made of the same metal as the anode, it is usually made from a different metal. Electrical current, caused by an electrochemical reaction between the buried structure and the electrolyte in which it is immersed, is picked up by the structure at the cathode. And, as this happens the cathode is protected against corrosion. *Polarization* or a buildup of hydrogen film, takes place on the cathode as the current is picked up. This hydrogen film coats the cathode surface, acts as an insulator, and reduces the flow of current which causes corrosion.

Corrosion Cells

Corrosion cells differ in the elements that provide the electrical potential or driving force that makes them work. A good understanding of these elements is essential to the design engineer who is planning the construction of buried metallic structures: pipelines, steel piers, or gasoline storage tanks at service stations or other underground installations. The engineer should use this knowledge to incorporate corrosion control features into the basic project design.

Dissimilar Metal Cells

Dissimilar metal corrosion cells are the simplest kind, and are becoming well-known to design engineers. These are called *Galvanic cells* after the Italian, Luigi Galvani, who discovered that electrochemical cells, consisting of an electrolyte, an anode, and a cathode, could be used as we use a battery today to produce useful amounts of electrical current. An iron-copper Galvanic cell is illustrated in Figure 15–1. The iron is anodic and corrodes while the copper cathodic is protected while the cell is active. An iron-copper cell is capable of producing a voltage of .75 to 1 V, depending upon the purity of the elements. In such a cell, iron will corrode at a rapid rate. The amount of metal that will be removed is directly proportional to the amount of current flow. One ampere of direct current discharging into the electrolyte can remove approximately 20 lbs of steel in one year. Metals other than steel will be removed at different rates, some rapidly, some slowly.

Seldom do corrosion currents in buried or submerged pipelines approach one ampere at one location. In most cases, the currents found and measured are recorded as thousandths of an ampere (*milliamps, millieamperes,* or *ma*). Although the current flowing is only in milliamps, it could cause severe damage. A single milliamp flow during one year that is concentrated to a few small points of discharge could cause seven, ¼-in. holes in a 2-in. standard wall thickness steel pipe. Usually however, the current discharge is distributed over wider areas, so that the rate of penetration is not necessarily so rapid.

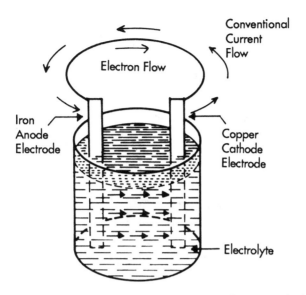

Conventional
Current
Flow

Electron Flow

Iron
Anode
Electrode

Copper
Cathode
Electrode

Electrolyte

Figure 15–1   A copper-iron cell will produce a voltage of ¾ to 1 volt, depending on the purity of the elements. Iron will corrode rapidly in such a cell.

In an iron-magnesium cell, the iron will be cathodic, and thus will be protected by the magnesium anode. Such a cell (as shown in Figure 15–2) immediately suggests a method of preventing corrosion of metallic structures that are immersed in an electrolyte of soil moisture or seawater. The voltage produced by an iron-magnesium cell is relatively high, and the magnesium is considered an effective anode for certain cathodic protection installations. This cell is the basis for the *sacrificial anodes* which will be discussed later in this chapter.

Electromotive Series for Various Metallic Elements

Each metal has its characteristic potential with respect to each of the other metals. These potential differences can be measured in terms of their electromotive forces on voltages under certain standard conditions. Table 15–1 lists the typical potential normally observed in natural soils and water measured with respect to a copper sulfate reference electrode. When a metallic path electrically connects together any of the metals immersed in an electrolyte, any metal listed in the table as being higher will be anodic (which means it will corrode) relative to any metal with a lower listing. That is, the anode sacrifices itself to protect the metal listed lower in the table. The potential of the cell would be the difference in the listed voltages.

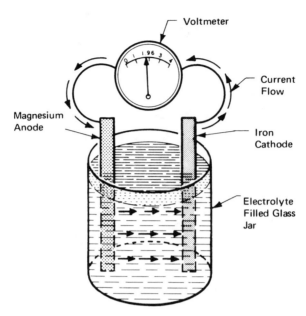

Figure 15-2 In an iron-magnesium cell, the iron will be cathodic and will be protected by the magnesium. The voltage produced by such a cell is relatively high, and the magnesium is considered an effective anode for certain cathodic protection installations.

METAL	VOLTS*	
Commercially pure magnesium	– 1.75	Anodic
Magnesium alloy (6% Al, 3% Zn, 0.15% Mn)	– 1.6	
Zinc	– 1.1	
Aluminum alloy (5% zinc)	– 1.05	
Commercially pure aluminum	– 0.8	
Mild steel (clean and shiny)	– 0.5 to – 0.8	
Mild steel (rusted)	– 0.2 to – 0.5	
Cast iron (not graphitized)	– 0.5	
Lead	– 0.5	
Mild steel in concrete	– 0.2	
Copper, brass, bronze	– 0.2	
High silicon cast iron	– 0.2	
Mill scale on steel	– 0.2	
Carbon, graphite, coke	+ 0.3	Cathodic

*Typical potential normally observed in natural soils and water, measured with respect to copper sulfate reference electrode.

Table 15-1 Potentials

Notice that commercially pure magnesium has a potential voltage of –1.75 V and copper, brass, and bronze have a potential voltage of –0.2 V. Lists of potentials in volts of high purity metals compared with a hydrogen reference electrode will be different. For example, magnesium has a potential voltage of –2.40 V, and copper of +0.34 to +0.50 V. These are given under laboratory conditions and are measured with laboratory equipment. However, laboratory conditions never occur along a buried or submerged pipeline. In the field, the electrolyte is of varying concentration and chemical composition. The metals are not pure, and their surfaces are not chemically clean. Based upon study of field data, Table 15–1 was compiled to list a practical series where the voltage values given represent average to realistic readings.

Dissimilar Soils

Differences in the electrolyte in contact with various parts of an underground metal structure can result in the formation of a *corrosion cell.* Causes of corrosion often coincide as to location, and the action at some anodes may be attributable to a combination of several conditions. Areas that are anodic because of oxygen deficiencies are likely to also have higher concentrations of dissolved salts in the soil water electrolyte. The separation of anodic and cathodic portions of buried pipelines are not clear-cut, and corrosive soils are not uniform. Portions of pipelines buried in them would not likely be corroded over their entire surfaces, and some parts would be unaffected. These unaffected areas would result from their being isolated from the environment. Many complete small or local cells can exist within areas that are generally classified as anodic or cathodic. Naturally occurring differences in electrolyte concentrations and composition account for most of the corrosion cells on buried pipelines. The pipe usually rests on undisturbed soil at the bottom of the ditch, and around the sides and on top of the pipe is relatively loose backfill that has been placed in the ditch during pipe burial. Because the backfill is more permeable to oxygen diffusing down from the surface, a cell is formed. The anode is the bottom surface of the pipe, and the cathode is its top and sides. The soil becomes the electrolyte, and the connecting electrical circuit is provided by the pipe itself. This is why corrosion is almost always more extensive on the bottom of the pipe than on the sides and the top. This is especially true of uncoated or poorly coated pipelines.

When a pipeline passes under pavement, such as a street or highway, the portion under the paving has less access to oxygen than does those portions of the line buried under dirt alone. In such a situation, a cell is formed with the pipe under the pavement becoming the anode and the pipe outside the roadway the cathode. Again, the soil is the electrolyte, and the metal pipe completes the electrical circuit. Although the entire length of the pipe under the pavement is anodic, most of the corrosive attack will take place not far from the edge. Because the path through the electrolyte is shortest to this point, most of the current will take this path of least resistance.

The casing of an oil or gas well is normally connected to a network of surface piping such as flow lines, which is buried just beneath the Earth's surface. This can form a cell with the steel well casing becoming the anode and the flow line, which has greater access to oxygen, the cathode. Soil moisture provides the electrolyte, and the flow lines and casing make up the connecting circuit. The corrosive action of this cell can be controlled by insulating the surface piping from the casing by means of a special insulating union or other fitting designed for this purpose.

New Pipe and Old Pipe

A condition very closely related to dissimilar metal corrosion arises when new steel pipe is intermixed with old steel pipe (Figure 15–3). This has happened in smaller distribution systems when a new, uncoated length of pipe is installed to replace a corroded length of old pipe. Logically, one would expect the new pipe to last longer than

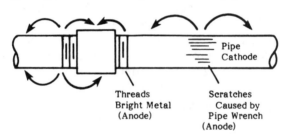

Threads
Bright Metal
(Anode)

Scratches
Caused by
Pipe Wrench
(Anode)

Corrosion Caused by Dissimilarity of
Surface Conditions

Figure 15–3 New steel pipe becomes anodic to old steel pipe when exposed to the same corrosion conditions. A condition closely related to dissimilar metal corrosion.

the old pipe, but this is not always the case. Table 15–1 shows this is simply an application of the typical potential of mild steel (clean and shiny), – 0.5 to – 0.8 V, being anodic to mild steel (rusted), – 0.2 to – 0.5 V, which would be cathodic. When replacing or repairing buried threaded pipe even small scratches caused by a pipe wrench can cause that portion to become anodic to the other sections of the pipe (Figure 15–4).

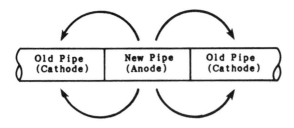

Figure 15–4 Corrosion caused by dissimilarity of surface conditions.

Although not a metal, mill scale on hot rolled steel acts as a dissimilar metal in contact with the pipe steel. As shown in Table 15–1, pipe steel will be anodic to mill scale on steel.

Stray Current

Stray current is current that flows through paths other than the intended circuit. One example would be where a buried pipeline passes near a commercial area where large DC (direct current) welding machines are being used. The welding machines would be grounded to earth, and the positive leads *(stingers)* would be working near or on the pipeline. During welding, some of the current could leave the metal being welded, enter the earth and flow through the pipeline back to the negative ground where the current would leave the pipeline. The point where this current would leave the pipeline would be anodic. Other areas where DC machines are in use could also cause trouble.

Another source of stray currents occurs when two buried pipelines cross each other and one has cathodic protection and the other does not. This is becoming more common as more and more pipelines are being laid. Assume that Line A has cathodic protection and that Line B is bare and unprotected. At several points, remote from their crossing point, current will be picked up on Line B from cathodically protected Line A. This stray current will flow along Line B back to the crossing point, and as it leaves Line B to return through the soil to Line A, an anodic spot will be created. One way to correct such a situation would be to install a metallic bond between the two lines at their crossing point. Normally this would consist of a piece of *resistance wire*, carefully measured and calculated to regulate the correct amount of current flowing between them so that Line B will not be damaged, and the cathodic protection of Line A will not suffer too greatly. Such a connection is called a *mutual bond.*

Conventional Current Flow

By definition, *current* is the rate of transfer of electricity from one point to another. Current is usually a movement of electrons, but may also be a movement of positive ions, negative ions, or holes. As you may recall, electrical current is measured in *amperes, milliamperes* (one-thousandth of an ampere), and *microamperes* (one-millionth of an ampere). The conventional concept of current flow is from the plus (+) or positive terminal to the minus (–) or negative terminal. Although popular among craftsmen for many years, it can be confusing because this conventional representation of current flow is opposite in direction to the flow of electrons described in corrosion problems. However, remembering the following will make the terms of conventional current flow understood:

> 1) Conventional current flow from (+) to (–) will be
> from the cathode terminal to the anode terminal in
> the metallic connecting circuit.
> 2) Conventional current flow from (+) to (–) will be
> from the anode electrode to the cathode electrode
> submerged in the electrolyte.
> 3. Metal is consumed where it leaves the anode in the
> electrolyte.
> 4) Metal receiving current from the surrounding
> electrolyte does not corrode, except certain materials
> such as aluminum and lead which can corrode if
> they receive excessive amounts of current.

Corrosion Control Methods

There are many conditions that can cause corrosion, such as those already discussed which dealt with underground pipelines and metallic structures. No two conditions are exactly alike, because of the wide variety of materials involved and the soil composition and moisture content. Other forms of corrosion can attack structures exposed to the Earth's atmosphere. Usually however, good protective coatings are sufficient for these, although there are instances where other methods must be used.

For underground and submerged structures, the principal methods for mitigating corrosion are:

> 1) Coatings and wrappings.
> 2) Insulated joints and isolation.
> 3) Cathodic protection.

There are also more specialized corrective methods available which can be applicable to unusual and unique situations.

Coatings and Wrappings

Coatings normally are intended to form a continuous film of an electrically resistive insulating material over the metallic surface to be protected. The function of a coating is to isolate the metal from direct contact with the surrounding electrolyte and to impose such a high electrical resistance in the anode to cathode current that there will be no significant corrosion current flowing from the anode to the cathode. If it were possible to coat and wrap the pipeline with a material that was absolutely waterproof and free from holes (holidays), external corrosion would be stopped. At this time, there is no perfect coating that has these two properties that will remain on the pipe permanently. An approach to perfect coating is very expensive, so it is obvious that imperfect coatings must be used. Experience has shown that a combination of a reasonably good coating and plus cathodic protection will retard corrosion. A good coating and wrapping job should have many special properties, including a high electrical resistance, resistance to water, and an ability to withstand mechanical damage, which is the main purpose of the wrapping.

There are many varieties of coatings used on pipelines, and each has its own merits and restrictions. Some of the more common ones used today include vinyls, epoxies, chlorinated rubbers, alkyds, phenolics, asphalts, and coal tars. The methods used to apply these coatings will also vary. Some are applied by hand, while others are applied by machine, either in the pipeyard or over the ditch. And in some cases extruded polyethylene or polypropylene coatings are applied at the mill where the pipe is manufactured.

The thickness of the coating will also vary depending on the material used and the method of application. A typical application including the primer and hot enamel is about 0.10 to 0.15 in. thick, plus the thickness of the wrapper. Thermoplastic resins and tapes are in the 0.010 to 0.030 range.

Protective coatings and wrappings are expensive and can represent from 5 to 15% of the total pipeline construction cost. But, when compared with having to replace part or all of the pipeline, their cost is really very economical.

Insulated Joints and Isolation

Insulated joints and flanges are used to break the metallic electrical connection between the anode and cathode and thereby prevent the flow of current between the two. Insulated flanges are used to isolate a section of pipeline that is cathodically protected from a section that is not to be protected or to isolate a pipeline from a pump station or other facility.

Insulated joints can be used at the junction of two dissimilar metals to retard corrosion between them, but without cathodic protection, localized corrosion cells could act on the surface of the submerged structure.

Cathodic Protection

Cathodic protection is an effective means of retarding corrosion on underground or submerged metallic pipes and structures. It is also an effective means of retarding corrosion on steel tank bottoms that rest on soil, the hulls of steel ships, and many other metallic objects and devices.

As previously mentioned, there are two basic types of cathodic protective systems: the galvanic anode type and the rectifier type. Each has its own characteristics which makes it more effective for a given situation. Each is effective on both coated or bare underground or underwater structures. And, both are more efficient on well-coated pipelines. Each type uses DC electricity from its anode or its anode ground bed to protect the cathode. The galvanic type has a very limited current output, whereas the rectifier type has a very high current output capability as well as methods of controlling current output. Both types make the underground or submerged structure become a cathode, hence the term cathodic protection.

Galvanic Anodes

Galvanic electrodes require no outside source of electrical power for operation. As the metals in the anodes corrode or oxidize back to their more stable forms, electric power is generated. For this reason, they are often referred to as sacrificial anodes. These are molded pieces of magnesium, zinc, or aluminum, buried in the soil with insulated solid conductors connected to the pipeline. In the petroleum industry, galvanic anodes are becoming more commonly known as *magnesium anodes* because they are the most widely used (Figure 15–5). Referring again to Table 15–1, we see that magnesium is more anodic (–1.75) V than mild steel (–0.5 to –0.8) V, and a galvanic cell exists when the two metals are submerged in an electrolyte with a metallic path connecting the two. If a DC voltmeter was connected in the metallic circuit that connects the anode to the cathode, a negative reading would be observed. This reading could be in the range of –1.25 to –.95 if the system is working properly. The

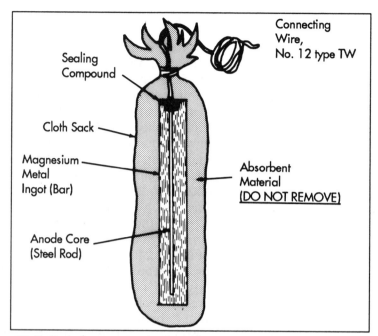

Figure 15-5 Typical magnesium (Mg) anode.

reason for the negative reading on the voltmeter is because the magnesium is more negative with respect to mild steel (Table 15–1).

The current delivered by a magnesium anode conforms to all DC electrical circuits where the current (voltage and amperage) varies directly with the potential difference between the source (battery) and the structure, and inversely to the total circuit resistance, including the wire, battery, pipe, etc. Ohm's Law (E=IR) can be used to determine the magnesium current output. However, this is difficult because of frequent changes in the soil moisture content.

One mile of 8-in. pipeline represents 11,915 ft² of metal to be protected. Assuming that each anode (Type 1B, 17 lb) will deliver 100 ma of current in average soil conditions, and that each square foot of pipe surface will require 1 ma of current flow for protection, 1 mi would require 119 anodes. Assuming 500 amp-hr of useful current for each pound of magnesium consumed, each anode would last approximately 10 years. If each anode cost $15.00 installed, the total cost would be $1,785.00 for 10 years' protection.

Electrical current requirements per square foot of metal surface to be protected vary greatly. The current values required to protect bare steel in natural soils and waters have been found to range from 0.5 to 5 ma/ft². For coated steel, the values range between .01 and .5 ma.

The external bottoms of steel storage tanks, either grouped in batteries or in more spread-out arrangements such as those typically found at tank farms are often subject to severe corrosion attacks. Here, a system of magnesium anodes can provide a very economical solution to the problem. In the case of an oil tank farm, it could be advantageous to install individual anodes close around the periphery of each tank, rather than to spot concentrated beds of anodes between the tanks. In such applications it is usually difficult to determine just how many anodes should be installed in order to achieve proper cathodic protection. First, the total number of square feet of metal to be protected must be determined, then an estimate made of the current requirements. This is usually accomplished by assuming that 2 ma/ft^2 will be needed. Next to be determined is the amount of useful current that each anode will deliver. This can usually be obtained from existing tables or from anode suppliers.

Three to six months after installation, field tests should be made by taking potential readings at various points in the soil around the tank bottoms. When the tank-to-soil potential, using a copper-copper sulfate *half-cell* as a reference electrode, is –0.90 V at the edge of the tanks, sufficient protection is being provided. In an average situation, the drop in potential from the outer edge of the tank to the center should be approximately 1 millivolt (mv). Taking potential readings with a high-resistance voltmeter and a copper-copper sulfate (Cu-CuSO$_4$) half-cell as a reference electrode will be discussed in greater detail later in this chapter.

Not all galvanic anodes are supplied in bags, such as those normally used with pipelines. Some are bare metal with connecting mechanisms attached and molded in various configurations. They may be made of magnesium, zinc, or any other suitable metal. Galvanic anodes are being used, and are performing satisfactorily for the cathodic protection of wharves, docks, drilling barges and platforms, marine equipment, ship's hulls, pipelines, and other underwater structures.

Miscellaneous uses for sacrificial anodes in the cathodic protection of metallic structures and installations are almost endless as the anodes have applications wherever these structures are surrounded by soil or water functioning as an electrolyte.

However, despite their usefulness, sacrificial anodes do have some disadvantages, including

1) They have a limited useful life and should be replaced when it is exceeded.
2) They have a limited voltage and amperage output.
3) They are not ideal for use in protecting long, large-diameter pipelines.

Modern Petroleum

Rectifiers for Cathodic Protection

In many instances, rectifier-type cathodic protection systems can be used in place of galvanic anodes.

Rectifiers used in cathodic protection require an outside source of electrical power. A typical system will take alternating current (AC) power from a conventional power line, pass it through a rectifier to convert it into direct current (DC), and then feed it into anodes located along the pipeline (Figure 15–6). At remote sites, solar panels can effectively be used to supply power.

Rectifier systems are most often used on larger buried pipelines because of their larger area of influence. They are capable of very high voltage and ampere output. It is not uncommon to protect 20 mi of large-diameter pipeline with each rectifier.

Rectifier systems normally use carbon graphite or high-silicon cast-iron anodes that are buried to form a ground bed. When carbon graphite anodes are used, they are placed in the soil and surrounded

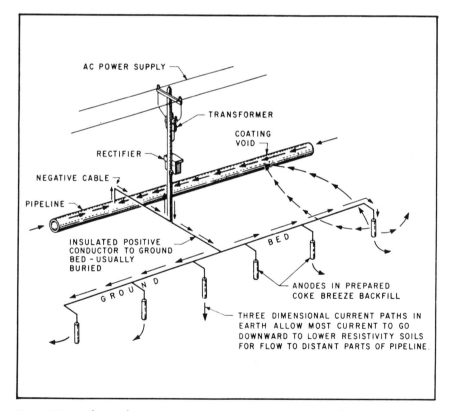

Figure 15–6 Electrical corrosion protection system (courtesy Petroleum Extension Service).

Corrosion

by a coke breeze backfill. The number of anodes used in each ground bed will vary with the situation at hand. Preliminary tests are usually run with a engine-driven portable welder as a power source, together with temporary metal ground bed material, to calculate the DC volts and amperes required to protect a given section of pipeline. The size and capacity of the rectifier, and the size and number of the anodes needed can be calculated from the data obtained from the tests. An average figure might be 20 impressed current anodes per bed, each with a calculated life of 20 years.

Cathodic Protection Measurements

After cathodic protection is installed, it is necessary to take measurements to insure that the system is working properly. It is desirable to have the pipeline or other metallic structure protected forever. Unfortunately, this does not happen. Coatings deteriorate, protective devices are consumed or wear out, and "foreign" structures that are not included in the protected system can create electrical short circuits to the protected device. To ensure that a protective system is continuing to function properly, its effectiveness must be periodically checked.

This is done by measuring the *pipe-to-soil* (or *structure-to-soil*) voltage. In the case of buried pipelines, the negative terminal of a high-resistance voltmeter is connected to the pipe through a test wire or riser. The positive terminal of the voltmeter is connected to the earth through a copper-copper sulfate ($Cu-CuSO_4$) half cell. The half-cell is a device which usually has a copper rod immersed in a copper sulfate solution that is contained in a glass or plastic tube with a porous plug at the bottom of the tube and an electrical terminal at the top of the copper rod (Figure 15–7). Types of half-cells other than the one described are also available. A half-cell is used because it makes a better electrical contact with the earth than a wire and it gives a constant voltage. The half-cell should be placed directly over the pipeline when possible. Readings can be taken along the length of the pipeline as long as the voltmeter terminal can be connected to it (Figure 15–8). The high resistance voltmeter and half-cell can also be used to locate anode areas on a pipeline, and to find any hot spots and also to find interference from foreign metallic structures.

Note that all pipe-to-soil voltages are negative when the cathodic protection system is working. Voltage readings of –0.85 or more negative than –0.85, indicate that the pipeline is protected at the point of the reading: –0.85 V is the criterion normally accepted for protecting coated pipe. Voltage readings more positive than –0.85 V indicate either partial or no protection. It should also be noted that

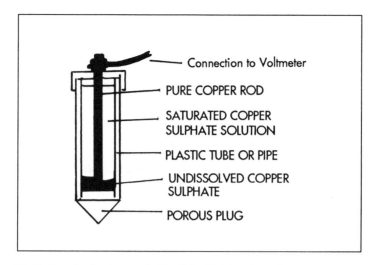

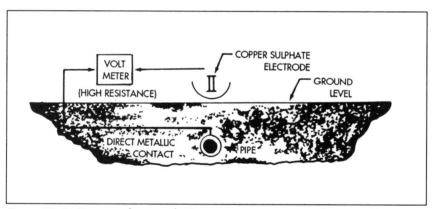

Figure 15-7 Standard reference half-cell Cu-CuSO₄ electrode.

Figure 15-8 Pipe-to-soil potential, 1) investigate corrosive conditions, and 2) evaluate the extent of cathodic protection.

a 0.30 volt negative shift will indicate that bare pipe is protected. For example, if the pipe-to-soil reading of a bare, unprotected pipeline reads –0.45 V, and cathodic protection is installed and the voltage reading becomes –0.75 V, it indicates that the pipe is being protected.

Cathodic protection and protective outer coatings retard the external corrosion of pipelines but do not protect the internal surface. Internal corrosion is not a real problem for many crude oil pipelines. However, it can be a problem for some crude lines and lines carrying natural gas or petroleum products. In these cases, internal protection is achieved by sandblasting and cleaning the interior walls and then coating them with a thin protective film.

Today, the cathodic protection of buried pipelines has become more than a desire to save money formerly lost to the forces of corrosion; in many areas it is increasingly becoming a requirement of government regulatory agencies.

Chapter References

Guidance Manual for Operators of Small Gas Systems, (Washington D.C.: U.S. Department of Transportation, Research and Special Programs Administration, 1985).

Safety Requirements for Gas Pipeline Systems Conducted for Materials Transportation Bureau and State Agencies, 2nd. Ed. (Washington D.C.: U.S. Department of Transportation, Transportation Safety Institute, Third Printing, January, 1987).

16

Natural Gas

Natural gas is one of the most popular forms of energy in use today. It is colorless, odorless, and it burns with an even temperature. And, as it burns, natural gas leaves no harmful residue that would harm our environment. More and more reserves of natural gas are being found each year, and more and more uses for it and benefits from these uses are being discovered. Yet the burning of natural gas for heat is not new. In fact, the Chinese were doing so more than 2,000 years ago.

Natural gas not only heats our homes and cooks our meals, it also powers our school buses, garbage trucks, and other fleet vehicles at a fuel cost of about one-fourth that of gasoline or diesel fuel. Its many useful by-products include carbon black, an additive to synthetic rubber that enables us to get 50,000 mi or more from a set of radial tires, and a host of consumer products ranging from drugs and medicines to cosmetics and man-made fibers for clothing. Although technically a "fossil fuel," its major constituent—methane—can be generated from biological wastes from plants and animals. Incidentally, natural gas is odorless. That familiar "gas" smell you may have whiffed is actually mercaptan, an odorant added to the gas as a safety measure. More uses for this "fuel for today" are discussed in the chapters on petrochemicals and fuels for the future.

Natural gas is a homogenous fluid of low density and low viscosity. It is classified as a fluid, since both liquids and gases are fluids. Unlike liquids, however, gases have neither definite shape nor definite volume; e.g, a gas will expand to fill its container.

Natural gas is a form of energy that is basically a hydrocarbon mixture that contains some impurities. It is a naturally occurring mixture that has its own composition and can change that composition as its underground reservoir (where natural gas is always found) is depleted.

Natural gas may be found by itself in a formation, but it is often found with oil in an oil producing zone. In this case, gas under pressure provides the prime driving force needed to remove the oil from a reservoir. This is especially evident when a well is first drilled and the gas pressure is high. Natural gas from the reservoir can also be used in secondary recovery efforts. Compressors are used to repressure the gas which is then reinjected into the formation though special wells to again act as a driving force in bringing the oil to the surface.

Natural gas from oil wells is either free, *unassociated gas* (that is, separated from the oil), or it is *associated gas* (dissolved in the oil) in a *gas cap* lying over a layer of crude oil in the reservoir. Associated gas that is dissolved in the oil, usually under very high pressure, is also called *entrained gas*. Natural gas is also found in a natural mixture of gas and liquid hydrocarbons called *condensate*. The relationship between the oil and gas depends on how heavily the liquid oil is saturated with dissolved gas. The oil and gas remain together in solution as long as the temperature is low and the pressure is high. Since both the oil and gas occupy space in the reservoir while they are in solution, this space must be taken into account when the volume of oil is calculated. When the oil is pumped, or flows to the surface, and the pressure is lowered in a separator, the gas comes out of solution. A good analogy to this is a can of soda pop. Carbon dioxide gas, which gives the soda its "fizz," is dissolved under pressure in the liquid. If you place the unopened can in the warm sun for a while, and then shake it violently, you have imitated the conditions that exist in an underground reservoir by increasing the temperature and pressure. If you suddenly pull the tab, you reduce the pressure, and the rapidly expanding gas spews the sticky pop all over you and your clothes as it escapes from the container.

If less natural gas is present in a reservoir than the oil can absorb, the oil is undersaturated. If the opposite is true, then the oil is supersaturated. When *free gas* is present in the reservoir, it will rise to the top of the reservoir and form a gas cap. In such cases, the oil below it will be saturated with dissolved gas in solution. Saturated oil is lower in viscosity, and will move easily through the formation to the well casing.

The normal hydrocarbon gases found in natural gas are methane, ethane, propane, butanes, pentanes, and small amounts of heptanes, octanes, hexanes, and heavier gases. The propane and the heavier fractions are removed and later processed for their value as gasoline-blending stock and chemical plant raw feedstock. Methane and ethane are the most abundant mixtures found in natural gas that have a value as fuel. The more methane a given sample of natural gas has, the greater its quality. The usual components of a typical natural gas sample taken at the wellhead are shown in Table 16–1.

Typical Natural Gas Components	
Hydrocarbon	**Amount, %**
Methane	70 — 98
Ethane	1 — 10
Propane	trace — 5
Butane	trace — 2
Pentane	trace — 1
Hexane	trace — $1/2$
Heptane +	none — trace
Non-hydrocarbon	
Nitrogen	trace — 15
Carbon dioxide	trace — 1
Hydrogen sulfide	trace occasionally
Helium	trace — 5

Table 16–1 Typical Natural Gas Components

Types of Natural Gas

The four general types of natural gas are wet, dry, sweet, and sour. Wet gas contains some of the heavier hydrocarbon molecules, which are vital for processing operations, and water vapor while it is in the reservoir. When this gas reaches the surface, some of the hydrocarbon molecules form a liquid. Dry gas indicates that the fluid does not contain enough of the heavier molecules to form a liquid at surface conditions. Sweet gas has a very lower concentration of sulfur compounds, particularly H_2S. Sour gas contains excessive sulfur compounds, which have an offensive odor and which are harmful to breathe, and exposure to it can cause death.

The Origin of Natural Gas

The origin of petroleum or natural gas has been the subject of many heated discussions over the years. Many theories about natural gas exist, but it has been impossible to determine the exact place or materials from which the gas in any particular reservoir originated. The two most popular theories are the organic and the inorganic, which were discussed in Chapter 2.

The Physical Properties of Natural Gas

Because each production stream of natural gas may vary in composition, and in the relative amounts of compounds present, physical properties may vary widely between streams. In order to predict the behavior of a particular stream for processing purposes, it is important to know its physical properties. The most accurate method of determining them is by analyzing samples of the gas. This requires the use of a specially designed gas bottle for collecting the samples, which must be sent to a properly equipped laboratory for analysis. The following is the laboratory analysis of an actual natural gas sample collected at the wellhead of a producing oil well in north central Oklahoma:

Constituent	Amount
Methane	62.9%
Ethane	14.6%
Propane	10.3%
N-Butane	8.2%
I-Butane	1.2%
N-Pentane	.61%
I-Pentane	.54%
Hexane +	.69%
Nitrogen	.52%
Carbon Dioxide	.39%

And, each cubic foot of the sample contained 1,376 BTUs.

This last is very important for determining the heating value of natural gas. The *British Thermal Unit* or *BTU* or *Btu*, is the most common industrial measure of heat in the United States. A BTU is the amount of heat necessary to raise the temperature of 1 lb of water 1° F. In laboratories, heat is measured in units called *calories*. Each calorie is the heat energy needed to raise the temperature of one gram of water from 14.5°C to 15.5°C.

The above laboratory measurement of 1,376 BTUs per cubic foot at the wellhead represents considerably more heating value than that of the natural gas delivered to homes for residential use which is typically only 1,000 BTU per cubic foot. Why such a discrepancy can be important will be discussed later in this chapter.

Once the composition of the gas is known, the different physical properties of each pure component can be determined. This informa-

Modern Petroleum

tion is of value in reaching a decision regarding the value of the gas and deciding what type of processing it must undergo.

Physical properties are usually termed either *intensive* or *extensive*. *Density, specific volume,* and *compressibility factor* are independent of the quantity of material present and are intensive properties.

Volume and *mass*, whose values are determined by the total quantity of matter present, are extensive properties.

Density, gas specific gravity, critical temperature, critical pressure, bubblepoint, dew point, specific heat, latent heat of vaporization, boiling point, molecular weight, and *phase* are the physical properties of natural gas that have an important role in gas processing.

Density is the weight (or mass) per unit volume. The density of natural gas is usually expressed as the weight in pounds per cubic foot (lb/ft³). Normally, the volume will be expressed at the standard condition of 60°F and 14.7 psia. Natural gas is lighter than air, which has a normal density of 0.0763 lb/ft³.

The *specific gravity* is the ratio of a gas density to the density of air at the same conditions of temperature and pressure. Specific gravity is an important factor in gas measurement.

Critical temperature is the temperature above which a hydrocarbon gas cannot be liquefied regardless of how much pressure is applied. The critical temperature for methane is above –116.6°F. This means that when the temperature of a methane gas rises above that point, it will not liquefy regardless of how much pressure it is subjected to.

Critical pressure is the pressure in pounds per square inch (psi) required to liquefy a gas at its critical temperature. For methane at –116.6°F, a pressure of 667.8 psi would be needed for liquefaction.

Bubblepoint is the pressure at which gas, held in solution in crude oil, breaks out of solution as free gas and forms a small bubble; saturated pressure.

Dew point is the temperature at a given pressure at which the first drop of liquid forms in the gas system. Or it is often referred to as the point at which condensation takes place.

Specific heat is the number of BTUs needed to raise the temperature of one pound of material 1°F.

Latent heat of vaporization is the heat necessary to change a liquid into a gas.

The *boiling point* is that point when a liquid will boil whenever the vapor pressure of a liquid is equal to the pressure being exerted on it. Since the vapor pressure of a liquid changes with temperature, any liquid has many different boiling points, depending upon the pressure being exerted on the liquid.

The temperature at which the hydrocarbon mixture begins to vaporize at a given pressure is the bubble point. It is the temperature at a given pressure at which the first bubble would form.

The structure of a gas and its *molecular weight* are related. The formula of a gas indicates the relative numbers and kinds of atoms that have united to form the gas molecule. An example is the formula for methane, CH_4. This indicates that carbon and hydrogen are present in the compound in a 1:4 ratio. By taking the atomic weight of carbon, 12.010, and adding it four times to the atomic weight of hydrogen, 1.0080, a molecular weight of 16.042 can be obtained.

A *phase* can be defined as any homogenous and distinct part of a system that can be physically separated from other parts of the system by distinct boundaries. For example, ice, liquid water, and water vapor are three phases. Gas processing is concerned with two phases: gas and liquid. Therefore, for our purposes here, we can consider the terms *gas* and *vapor* to be synonymous.

Types of Wells

Natural gas comes from three types of wells: oil wells, gas wells, and condensate wells (Figure 16–1). In a traditional oil field, one can see various types of pumping units at the wellheads as well as separators and stock tanks which are usually located nearby. These wells were drilled primarily for their crude oil production. However, some of them will also produce enough natural gas to make its processing and handling profitable.

Gas Wells

Gas wells are usually drilled for the natural gas they produce. Note that the term *gas well* usually means a natural gas well. However, wells are drilled specifically for other gases such as carbon dioxide. Natural gas wells are normally drilled in a known gas producing field, such as the Hugoton Field in Kansas or the Anadarko Basin in Oklahoma. Gas wells are obviously different than oil wells. One of the differences is at the wellhead, where only pipe connections, valves, and gauges are visible. These configurations are commonly referred to as the *christmas tree*. No pumping unit is required, because natural gas being lighter than air naturally flows from the reservoir upwards through the pipe to the closed gathering and treating vessels above ground.

Some gas wells do produce amounts of crude oil and require stock tanks to collect it when it reaches the surface. However, large quantities of natural gas are found unassociated with oil and at deeper levels. Oil is rarely found at depths greater than 10,000 to

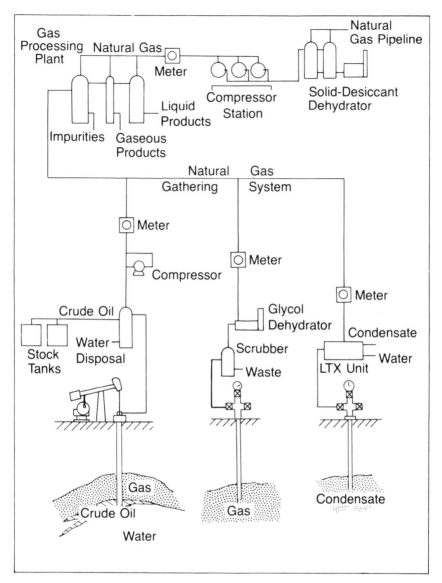

Figure 16–1 Concept of a natural gas system (courtesy PennWell Books, *Gas Handling and Field Processing, Vol. 3,* Berger and Anderson).

16,000 ft, while natural gas producing zones can range down to 30,000 ft. Our knowledge and experience in finding gas at such depths in large quantities is still limited. In recent years, the cost of deep well drilling and the cost of developing equipment capable of functioning successfully at such depths has restricted our search for deep production.

It has only been in the latter part of the twentieth century that such ventures have been attempted. Yet it may well be that there are vast areas in all parts of the world where drilling at great depths may result in high success rates and the resulting production of large quantities of natural gas. At the present time, great quantities of natural gas have been discovered in deep zones, mainly in Oklahoma. In the deep Anadarko Basin of western Oklahoma, the success rate for the deep wells drilled in recent years is believed to be 70%, the highest in oil and gas history. Robert A. Hefner III, one of the pioneers of the Anadarko, and one of its most successful producers, is one of those who believes natural gas may be found in great quantities and great depths in many parts of the world. And the current success of the Arkhoma Basin in southeastern Oklahoma lends great credence to his theories of natural gas accumulation.

Condensate Wells

Condensate wells are usually flowing wells with the christmas tree connected to the casing at the wellhead. These wells produce both condensate (a liquid form) and natural gas (the gaseous form). Condensate is a liquid hydrocarbon that lies in a range between gas and oil. Normally, it is separated from the gas by cooling and other methods, and then stored in vessels for future use.

Over the years, condensate has gone by many names of which *casinghead gas, casinghead gasoline, white gas,* and *drip gas* are but a few. Back in the days of simple automobile and farm tractor engines, it was not uncommon for anyone having access to a well producing condensate to obtain enough "drip" to fill his tank. The results were not always predictable since the fuel value of the condensate varied widely. At times it might not even ignite, and at other times it might cause thundering backfires and clouds of foul smelling smoke. And, in any event, its use on public highways was a violation of fuel tax laws.

More recently, under the name "white oil," it has been in the news as a subject of controversy and litigation in the Texas Panhandle caused by a dispute over whether it was gas or oil and who actually owned it.

Hydrates

Hydrates are created by a reaction of natural gas with water. They are solid or semisolid compounds that form ice-like crystals. The pressure and the actual composition of natural gas determines the temperature at which hydrates will form upon cooling. There must be water or water vapor present in the natural gas for hydrates to form. The greater the gas pressure, the higher the temperature at which the hydrate will form. They can form at temperatures just above 32°F (0°C), and it is not unusual for them to form under certain conditions at 68°F (20°).

Hydrates have a specific gravity of about 0.96 to 0.98, and float on water but not on hydrocarbon liquids. They are about 90% water, and the other 10% may be composed of methane, ethane, propane, normal butane, isobutane, carbon dioxide, and hydrogen sulfide. Heavier hydrocarbons will not form hydrates during cooling when water is present. Methane and ethane are both the most common constituents of natural gas and are also the most common to combine with water to form hydrates. Hydrate formation is usually accelerated by unusual turbulence in the gas stream.

The gas usually loses heat and can hold less water vapor after it leaves the formation and flows through the wellhead, separator, and other equipment. Some of the water vapor condenses as the gas is cooled through the separator and drops out, but some of it will remain in the gas stream. Under certain conditions hydrate formation can take place with the remaining water vapor left in the gas. The more water vapor removed from the gas the better, especially when it is to be transported through a pipeline. There are specifications that require that natural gas that is to be transported through a pipeline transmission system have a water vapor content of 7 lbs or less per 2 MSCF (million standard cubic feet).

Hydrate formation in a gas stream may create a hazard because its accumulation may plug lines and valves so completely that no flow can occur. When this happens, there is a danger of great pressure building up behind the hydrate plug which can result in a rupture of the pipe. Thus hydrate formation in valves and lines should be prevented whenever possible.

There are three general methods for preventing hydrate buildups. One is to keep the gas temperature above hydration formation levels. Another is to inject an inhibitor into the gas stream to lower the freezing point of the water. The third is to remove more water vapor from the gas stream by means of a dehydration unit.

Not all hydrate formations are in valves and lines or are considered to be hazardous. Some methane hydrates have been discovered in many cold areas of the ocean floor and also in permafrost regions.

Field Processing

Usually, all natural gas taken from a wellhead requires some processing. At some locations gas from the wells is piped to a central plant for processing. Generally, however, it is more practical to install field equipment near the wellhead to remove impurities and harmful hydrates. The actual field equipment used will vary widely between locations, since the composition of natural gas can vary widely from field to field, from well to well, and sometimes the composition of gas from a single well can change, often in a short period of time, as downhole conditions change.

Most field processing equipment is constructed to operate under and withstand the pressures inherent in gas processing. They are usually built of steel, and all seams are welded. Where high pressures are anticipated, the equipment is built to strict pressure vessel specifications pressure relief valves and other suitable safety devices provided. Where severe corrosion may be encountered, the internal parts are either made of corrosion-resistant material, or corrosion protection is provided.

It is not possible to cover all types of field processing equipment here, since such a wide variety is in use today. However the more common types include *conventional separators, low-temperature separators (LTX), heaters, dehydrators,* and *scrubbers.*

Conventional Separators

The simplest form of an oil and gas separator is a closed tank in which the force of gravity separates the gas from the oil. These vessels are usually located near the wellhead of producing oil wells.

Low-Temperature Separators (LTX)

Low-temperature separation (LTX) is often used to handle the production of high-pressure gas wells that produce low liquid volumes of very light crude oil or a condensate. This process uses the effect of expanding high-pressure gas across a special choke to obtain a cooling effect. As the inlet gas is cooled, water and liquid hydrocarbons condensate. If hydrates are formed, they are melted by a heat exchange method. The dry gas, condensate, and free water are then discharged from the vessel through a control system (Figure 16–2).

The LTX system functions below hydrate formation temperatures. The gas and liquid separation is usually achieved at 0° to 10°F. A higher percentage of water vapor also is accomplished, thus resulting in gas dehydration.

If the production stream is hot enough to melt any hydrates that may form in the bottom of the low-temperature separator, no heater is required between the wellhead and the LTX. The stream passes through coils immersed in liquid in the lower portion of the separator. Next, the production stream passes through a heat exchanger

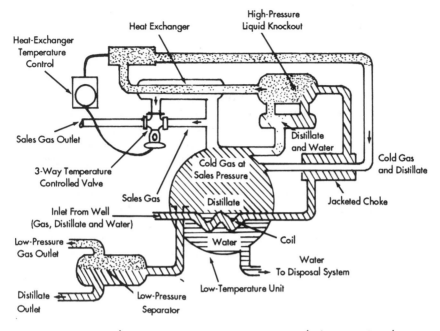

Figure 16–2 How a low-temperature separator system works (courtesy Petroleum Extension Service).

where it is slightly cooled. The stream then exits through a high-pressure liquid knockout where any liquids are separated out and dumped into the low-temperature separator.

After leaving the high-pressure liquid knockout, the stream, still at almost wellhead pressure, flows to an adjustable choke. As it leaves the choke, it is expanded into the low-temperature separator located above the liquid. This expansion causes the cooling effect that is necessary to lower the temperature in the separating chamber. The dry gas stream leaves the separator and flows through a heat exchanger that is heated by the high-pressure gas stream that has retained some of its wellhead temperature. Flow of the gas through the heat exchanger is controlled by a three-way temperature valve. Finally, the gas exits the heat exchanger and passes into the gas piping system.

Both the condensate and the water that has been removed accumulate in the bottom of the separator. The control system dumps the waste water and pipes it to a disposal system. The condensate flows into a low-pressure vessel where any remaining gas is separated and removed through the low-pressure outlet. The *distillate* is then piped out through the distillate outlet and into the storage facility.

The LTX system cannot be used at crude oil reservoirs where less gas is available and heavier liquids are present.

Heaters

At certain locations, heaters are installed to heat the gas stream to a temperature high enough to control hydrate formation. Heaters are relatively simple and require little maintenance or attention. They use natural gas for fuel and are less expensive than dehydration units. They are not as efficient as dehydrators, and often heating must be repeated at each point in the system where hydrate formation is a possibility.

The two general types of heaters are the flow-line heater and the indirect heater. The flow-line heater operates by directly heating the pipes with a gas flame in an enclosed chamber. Indirect heating can be done by passing the gas lines through tanks of water heated by fired heaters. Indirect heaters are more commonly used than flow-line heaters (Figure 16–3).

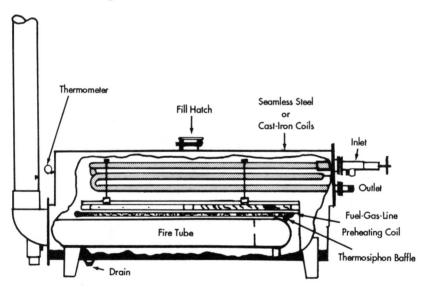

Figure 16–3 Cutaway view of an indirect heater (courtesy Petroleum Extension Service).

Dehydration

Dehydration means removing water from a substance. Here, the substance is natural gas. The processes used to dehydrate, or remove the water from natural gas, may be either *absorption* or *adsorption*. Absorption means the water vapor is sucked up, or taken out, by an agent such as *glycol*, and this requires a reaction. Adsorption means the water vapor is collected in condensed form on the surface, and requires no chemical reaction.

Glycol Dehydration

One general type of dehydrator is the liquid desiccant dehydrator which uses glycol as the absorption agent since glycol has an affinity for (gathers or collects) water (Figure 16–4). The glycol used is usually either a solution of diethylene glycol (DEG), or triethylene glycol (TEG). TEG is a superior material because it is more easily regenerated to about 99% concentration, has a higher decomposition temperature (about 40°F), and is subject to lower vaporization losses. For our purposes, we will refer to the absorption agent simply as glycol because the method of operation is the same for either TEG or DEG.

As shown in the flow diagram in Figure 16–4, the glycol is brought in contact with the wet natural gas in a double compartment called the contactor. Inside the contractor are several mesh baffles which serve to create maximum gas bubble surface areas which expose the wet gas to a high concentration of lean absorbent supplied by a

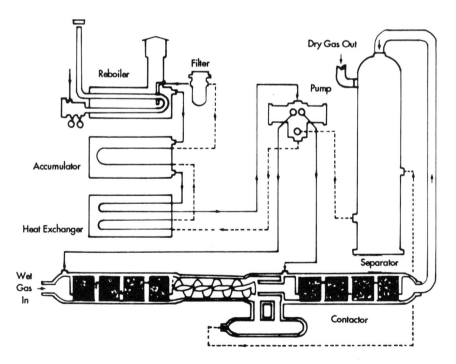

Figure 16–4 Flow diagram of horizontal contractor (courtesy BWT-Moore).

multipoint injection system. This exposure causes the wet gas to give up its water vapor to the "water hungry" glycol. As the glycol particles collect water, they become heavier and drop into multiple saturated glycol removal points located below the contactor.

The saturated absorbent leaves the reservoir at the bottom of the contactor, and enters the separator from where it is pumped to the heat exchanger. It then passes through a filter on its way from the heat exchanger to the reboiler.

Glycol has a boiling point above 400°F, and since water boils at 212°F, purifying the glycol by removing the water is relatively simple. By heating the water-wet glycol in the reboiler to a point just below the 400°F boiling point of the glycol, but well above the 212°F boiling point of the water, the water will be boiled out. The now water-free glycol leaves the reboiler, enters the accumulator, and then passes through the heat exchanger where it is cooled before being pumped back into the contactor to begin a new cycle.

Meanwhile, the gas leaves the contactor through an outlet located at the other end, and flows into the top of the gas-liquid separator. Finally, the dry gas exits the separator through a side outlet ready for pipeline transportation.

Solid-Desiccant Dehydration

The adsorption method of dehydration consists of a solid-desiccant dehydrator using activated alumina or a silica gel type of granular material. Water is retained on the surfaces of particles of solid material as wet gas is passed over and around them. Solid-desiccant dehydrators are more effective than glycol dehydrators and are better suited for large volumes of gas at very high pressures. They are usually installed in a natural gas pipeline transmission system on the downstream side of a compressor station.

The solid-desiccant dehydration unit consists of two or more adsorption towers that contain a solid, granular, gas adsorbent material, controls, piping, manifolds, and switching valves. There is also a high-temperature heater located in one of the towers to produce hot regeneration gas for drying out the wetted solid desiccant. Other essential equipment includes a regeneration gas separator to remove water from the regeneration-gas stream, and a regeneration-gas cooler for condensing water from the hot regeneration gas (Figure 16–5). The basic operation of the solid-desiccant dehydration unit takes place in the tower which is in service during the operating cycle (the towers normally operate on 8-hour cycles). The wet gas enters the tower near the top and flows downward through the desiccant during the adsorption cycle. If the tower has

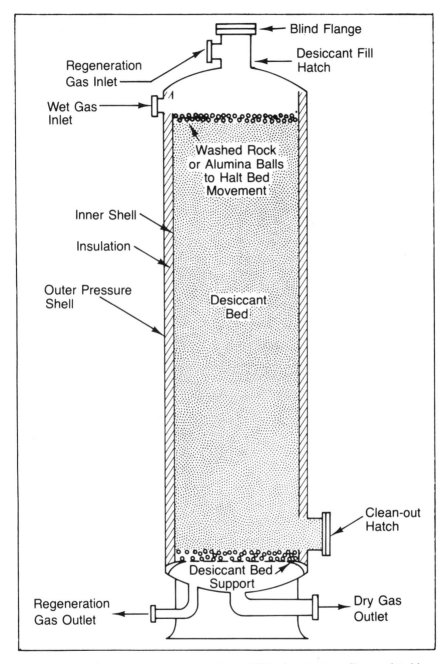

Figure 16–5 Absorber tower (courtesy PennWell Books, *Gas Handling and Field Processing, Vol. 3*, Berger and Anderson).

just been put into service, the water vapor is immediately adsorbed into the top layers of the desiccant bed. The other adsorbable gas components, ethane, propane, and butane, are adsorbed at different rates as they pass downward through the bed. As the cycle proceeds, the heavier components displace the lighter components.

The top layer of the bed becomes saturated with water, and the wet gas moves down to the next layer where water displaces the hydrocarbon components and forces them further downward. During this process, there will be sections or zones, called the mass-transfer zones, where a component transfers its mass from the gas stream to the surface of the solid desiccant. When this mass transfer zone reaches the bottom of the tower, it must be switched over to the regeneration cycle and another tower switched to the adsorb cycle.

A portion of the wet gas stream is taken from near the inlet section of the unit and sent through a heater where it is heated to approximately 450°F. The water is boiled off, and the hot regeneration gas is forced through the tower that has been switched into the regeneration cycle. As the hot gas passes through the desiccant, the liquids will be boiled out or vaporized, leaving a dry desiccant bed ready for reuse after being cooled.

The dry gas that comes out of the bottom of the tower that is in service is ready to be transported by pipeline.

Scrubbers

The natural gas flowing from a gas well normally contains small particles of sand from the formation, impurities, and free water. A scrubber located near the wellhead is generally used to remove these unwanted contaminants from the gas. It is usually a vertical steel vessel with internal screens and baffles and a collection area for impurities. The scrubber has an outlet for clean gas and provisions for dumping the impurities.

Compressors

Practically all natural gas that is transported through pipelines will have to pass through one or more compressors to increase its pressure. Some natural gas streams have very low pressure at the wellhead that must be increased by a compressor before the gas can enter a pipeline carrying gas at a higher pressure. Also, there is a pressure drop any time the gas must pass through a piece of equipment such as scrubbers, separators, heaters, dehydrators, or other devices that are necessary to make it clean and dry enough to transport. Long pipelines will have a pressure drop along their entire length.

Many gas compressors in use in the industry are *reciprocating,* which use gas engines as their prime movers. These have a long record of successful operation. Other types in use include *centrifugal, radial,* and *blowers.*

Reciprocating compressors compress the gas by means of a piston moving up and down in a cylinder, on somewhat the same principle as an automobile or lawn mower engine. They operate at low speeds (150–600 rpm) and create very high pressures. Their disadvantages are that they require a great deal of space and contain many valves and other internal moving parts that require maintenance and periodic costly replacement.

Centrifugal compressors can be powered by gas engines, electric motors, or gas turbine engines. Their disadvantage is that the compression ratio is less.

A single-stage reciprocating compressor compresses gas from a lower pressure to a higher pressure through one cycle, or rotation, of the unit. Two or more of these units can be placed in series with a resulting increase in pressure as the gas passes from one unit to the next. A multistage unit is the combining of two or more single stage units. Each stage of a multistage compressor is an individual, basic compressor in itself. Generally, the gas must be cooled as it exits one stage before it enters the next, as well as when it leaves the compressor unit.

Regulators

In addition to all the equipment described thus far, another factor is necessary for the successful delivery of natural gas from formation to final destination. That is the regulation and control of its pressure as it passes through the system. Control of gas system pressures is usually accomplished with pressure reducing equipment called regulators. It is the job of the gas regulator to match the flow of gas through the regulator to the demand for the gas downstream. Regulators accomplish this by means of a regulating element, a sensing element, and a loading element.

Gas Measurement

The volume of gas is measured to determine how much is being consumed or sold. Gas sales, royalty payments, and taxes are based on measured volumes. The measurement of gas requires that the unit of volume be defined. This means that the unit, pressure base, temperature base, and specific gravity must be determined or specified. The supercompressibility factor is often used when high pressures are being measured and rates are being determined. Usually a cubic foot of gas is the amount required to fill a cubic foot of space at 60°F and 14.73 psia. Also, the base temperature is usually converted to absolute zero (520°R).

Gas may be measured with orifice meters, positive displacement meters, turbine meters, venturi meters, flow nozzles, critical flow provers, elbow meters, and rotameters. The orifice meter is widely used in field installations gathering systems and in pipeline systems at the present time. It is described in Chapter 14.

The acceptance of gas turbine meters in Europe and their introduction into the United States accounts for their growing use in the past few years.

Domestic meters are not designed for gas measurements in the field or in high-pressure systems.domestic meters are the last piece of equipment owned and operated by the gas companies that the gas passes through on its way to residential and business users. It measures the cubic feet of natural gas delivered to the customer. They are usually read, and the amount of gas used, measured in cubic feet, billed for on a monthly basis. Residential gas is very low-pressure gas, usually only 6 to 8 oz/ft^3. It is clean, dry, and at approximately 1,000 BTU/ft^3, has less heating value than it did when it left the wellhead, but is still one of today's greatest energy bargains (Figure 16–6).

Figure 16–6 Meter reader Gail Hernandez records gas usage from a residential gas meter (courtesy Yale Gas Company, © Gary Gibson).

Chapter References

Bill D. Berger and Kenneth E. Anderson, *Gas Handling and Field Processing*, (Tulsa: PennWell Books, 1980).

17

Fuel for Tomorrow

With the world's demand presently reaching an estimated 25 billion bbl of crude oil a year, it is time to take a good look at our petroleum reserves and alternate sources of energy. According to various industry sources, we are accumulating new discoveries at the rate of only 10 to 15 billion bbl annually, which indicates that our present known reserves are shrinking rapidly. Much of the industrialized world's imports come from the Middle East, where an estimated 60% to 70% of the known oil reserves are located.

Crude oil demand is increasing at an alarming rate. The fastest growing area of petroleum demand is the Far East. Expanding economies in Japan and other countries in the Pacific Rim are using more and more oil and there is no decrease in sight. The spread of democracy throughout Eastern Europe has brought about another potential for increased demand for oil once these countries gain control of their economies.

Dwindling reserves, growing demand, a desire for clean air, and the fear of global warming have sparked new interest in finding alternate fuels and sources of energy for the future. Fortunately, we do have some choices. However, there will be many changes involved and many tradeoffs will become necessary.

Alcohol

Methanol

Methanol is an alcohol fuel made from either coal or natural gas that is compatible with gasoline for blending and shipping. It has a higher octane rating than gasoline and emits fewer ozone-causing compounds. There are less emissions of benzene, and some other toxics, depending upon the type of feedstock from which is it refined.

In the United States, several hundred methanol-fueled vehicles are now being tested in service. Ford Motor Company recently began testing a fleet of 30 flexible fuel vehicles that are capable of operating on gasoline, methanol, ethanol, or any combination of these.

Among methanol's disadvantages are that it is more expensive than gasoline. It is corrosive. And it does emit formaldehyde, thus it requires the development of better emissions control systems. However, a number of nations do use methanol as a motor fuel. Currently it is available at a handful of filling stations in the United States in the form of a blend with gasoline.

Ethanol

Ethanol is an alcohol that has been called grain alcohol because for thousands of years it was manufactured by the yeast fermentation of starches and sugars. Or in more familiar terms, it is the alcohol found in beer, wine, and spirits made from fermented corn, grapes, rice, or other vegetation. It can also be made from fermented municipal solid waste.

Today, ethanol is being considered for wide use as a an alternate motor fuel. It has a high octane rating and emits less hydrocarbons, benzene, and carbon monoxide than does gasoline. Race cars at Indianapolis have been running on alcohol fuel for years where it has been found to be better and safer than gasoline.

Ethanol is presently very expensive to produce and it has a lower volatility, which makes a vehicle harder to start in cold weather. It is also more corrosive than gasoline.

Many feel that alcohol fuels will help reduce our dependence on imported oil and at the same time improve our sagging farm incomes. However, growing crops just to produce fuel-grade alcohol is not feasible from either an economic or an energy standpoint. Brazil learned this during the last decade. With limited supplies of crude oil, that nation tried to switch to an alcohol-based fuel system. However, the process worked only so long as the government contributed vast sums of borrowed money to subsidize the program.

Part of the problem lies in our present system of agriculture which calls for massive amounts of diesel fuel and gasoline to cultivate, plant, and harvest crops. In between planting and harvesting, our farmers apply tens of thousands of tons of petrochemical-based fertilizers and other chemicals. That in effect places us in the questionable position of using crude oil derivatives to grow crops from which we hope to make fuel to replace the fuel we spent growing the crop in the first place.

Modern Petroleum

Propane

Propane has been used to power vehicles, especially farm trucks and tractors for many years as have fleets of police cars, garbage trucks, and utility vehicles that have been converted to use both propane and gasoline. This liquefied petroleum gas has higher octane content than gasoline and is less expensive. Emissions of hydrocarbons, carbon monoxide and nitrogen oxide are significantly reduced. Owners of propane-powered vehicles also report longer engine life and lower maintenance costs.

Among its disadvantages are that propane is heavier than air and in the event of a fuel line or tank rupture, the escaping gas sinks to the ground where it may constitute a fire hazard. This has led to the banning of propane-powered vehicles from using tunnels. As a vehicle fuel, it must be contained in a pressurized tank, and there are a limited number of fueling stations available to fill the tank.

Compressed Natural Gas

Compressed natural gas (CNG) is a clean burning fuel that reduces many of the pollutants that create smog. Based on test figures, it reduces emissions of reactive (non-methane) hydrocarbons by up to 48%, nitrogen oxides about 33%, and carbon monoxide by up to 47%. CNG requires less engine maintenance and needs no refining. On a per mile basis, CNG-fueled vehicles cost about 30% less to operate than gasoline powered vehicles.

CNG vehicles require pressure tanks that are filled with natural gas drawn from utility pipelines which is fed into compressors that compress the gas to 3,000 psi. Refueling using the fast-fill method for filling two tanks takes about 5 minutes. Using the slow-fill method presently takes about 10 hours of unattended time. The fast-fill method requires the use of a commercial "service station." The slow-fill method uses a small, low pressure unit which can be attached to the vehicle owner's home gas line. Such devices are currently available from at least three manufacturers in the United States.

Most gasoline-powered automobiles and trucks can be modified to operate on CNG at a cost at this time of about $2,500. A typical car with two pressure tanks can travel approximately 200 miles before refueling or switching to gasoline.

School buses and other fleet vehicles that can be refueled overnight from domestic supplies are presently being used to burn CNG. Its popularity is growing, and several states, including Texas and California, are mandating that their vehicle fleets be converted. The number of public fueling stations is increasing; Oklahoma recently announced plans for CNG facilities to be placed at interstate highway locations. Research is also underway to design new highly efficient CNG engines and fueling equipment.

Absorbed Natural Gas

Research is currently underway at the Institute of Gas Technology on the use of *Absorbed Natural Gas (ANG)* as an industrial fuel. This is a process whereby natural gas is absorbed by a metal and used for fuel in a solid state rather than in a gaseous state. The advantages in storage and use of such a fuel are obvious.

Coalbed Methane

As coal was being formed by the chemical and thermal alterations of organic debris, a series of by-products were also generated, including water and methane. Approximately 5,000 ft^3 of methane was generated per ton of coal. The methane produced greatly exceeded the capacity of the coal to hold this gas. This excess gas migrated into the surrounding rock strata and into any traditional sand reservoirs that overlaid the deep coal beds; however, significant quantities of methane gas are held in coal seams (Figure 17–1).

The *coalbed methane* industry intends to produce the gas still remaining in the coal seams. Gas contents of 350 to 600 ft^3 T have

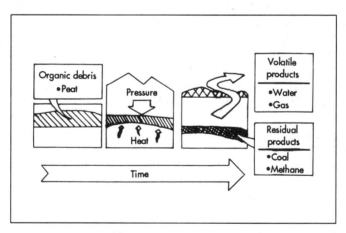

Figure 17–1 Key steps in coalification process (courtesy *Oil & Gas Journal*).

been measured in the United States for the higher ranked bituminous and anthracite coals. This makes coal an attractive gas reservoir. The rank and depth of the coal is closely related to the amount of methane stored. The higher the coal rank and the deeper the coal seam is, the greater its capacity to hold gas. Producing this absorbed methane is made possible by lowering the pressure, which usually involves removing the water and lowering the hydrostatic pressure on the coal seam.

Reducing the pressure by 50% may release less than 29% of the absorbed methane. Normally, the reservoir pressure needs to be reduced to near atmospheric to achieve efficient release of the stored methane.

As the methane is released from the coal it must diffuse through the coal matrix until it reaches the natural fracture network in coal. The gas then flows through these fractures until it reaches the wellbore. The diffusion through the coal matrix is controlled by the concentration of methane, the inherent diffusive properties of the matrix, and the distance to reach the fracture. Both water and gas move through the fracture in a two-phase flow to the wellbore (Figure 17–2).

Coals are easily damaged and difficult to stimulate properly to produce methane, which poses a problem for the driller and reservoir engineer. The proper drilling fluid must be chosen and the best method to complete the well must be implemented. The first wells were cavity-type open hole completions; however, the trend now is toward cased wells with perforations.

Coalbed methane wells need to be artificially stimulated to achieve commercial rates of gas flow, except in some high natural permeability areas. Stimulation is a challenge because coal is naturally fractured and somewhat plastic. High treatment pressures, limited proppant penetration, and complex fractures are common for effective stimulation.

Production of methane gas requires proper pumping equipment to dispose of the water and gas treatment equipment and compressors to handle the gas. The production challenges are different from those associated with producing crude oil and natural gas, but generally solvable.

There are 13 large coalbed methane reserves in the United States with nearly 2,000 wells presently producing gas from coal seams. Most of the active drilling is in the Black Warrior basin of Alabama and in the San Juan basin that extends from Colorado to New

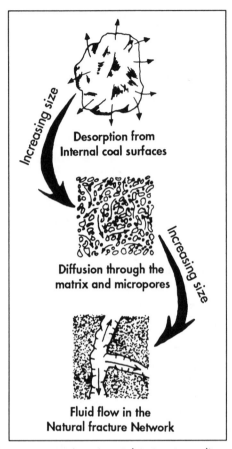

Figure 17–2 Methane transport (courtesy Oil & Gas Journal).

Mexico. High rates of coalbed methane are being produced in the San Juan basin where the coals are about 60 ft of net coal at a depth of 3,000 ft with a rate of 350 ft³ of gas per ton.

Several U.S. operators have recently announced major expansions in their coalbed methane plans.

Coal

While perhaps not *petroleum*, coal is a hydrocarbon just as is crude oil, and the United States and the rest of the world have vast reserves of coal.

Although coal can be burned directly to produce heat and thus electrical power, *gasification* and *liquefaction* can provide critically needed liquid fuels for cars, trucks, farm tractors, and other equipment in the event of an emergency or boycott. And at the point of end-use air, pollution from burning this substitute gas and oil will possibly prove to be less polluting than the present methods of burning coal.

A commercial-size plant similar to the SASOL plants in South Africa could produce 58,000 bbl of synthetic crude oil a day, using 40,000 T of coal. A commercial-size plant using a direct liquefaction process could produce approximately 100,000 bbl of synthetic crude a day. This would require 40,000 T of coal a day. In the United States, this quantity of coal would have to come from western strip mines and would necessitate the construction of slurry pipelines and expensive plants that would make the cost per barrel neither monetarily nor energy feasible at the present time. Like ethanol from farm crops, vast amounts of petroleum fuels would have to be "spent" to mine and transport the coal.

Other Petroleum Fuels

Petroleum Coke

Petroleum cracking processes are presently used as a different means to produce a higher yield of lighter hydrocarbons and solid residue suitable for fuel from their heavy residuals. Solid fuels from oil include delayed coke, fluid coke, and petroleum pitch. The delayed coking process uses residual oil heated and pumped to a reactor for coking. The coke is deposited as a solid mass and is then stripped either in the form of lumps or as granular material. Part of these cokes will be easy to pulverize and burn, while others might be difficult.

Fluid coke is produced by spraying hot residual feed into an externally heated seed coke in a fluid bed. It is then removed as small particles that can be pulverized and burned.

Petroleum pitch is produced as an alternate to the coking process and yields fuels of different characteristics. Its physical properties vary from soft and gummy to hard and friable. Their melting points will vary with their particular forms. The pitches with low melting points may be heated and burned like heavy oil. Those with higher melting points may be pulverized and burned.

Heavy Oil

Heavy oil is very dense, highly viscous crude oil that is similar to ordinary petroleum; however, it is so gummy that it must be heated or otherwise coaxed to flow before it can be pumped out of the formation. U.S. reserves are somewhere between 10 and 15 billion bbl. Total reserves worldwide are estimated to be more than 10 times that amount.

The most common method of producing heavy oil is to pump steam into the formation through injection wells. In this process, called *huff 'n' puff* in Canada, the steam thins the oil and drives it out through a production well. Variations of this steam injection technique and underground combustion method are in use and are being improved upon in demonstration projects.

When refined, heavy oil tends to yield more residual (heavy) fuel oil and less gasoline and other light products. As a result, expensive upgrading equipment is needed to convert some of the heavy products into the more desirable ones.

Most of the nation's heavy oil reserves are located in California where there is a great demand for crude oil products, and thus a ready market. Heavy crude oil is considerably less expensive to produce than such fuels as coal liquids and *shale oil.* In addition, the infrastructure for oil production, transportation, procession, and distribution already exists in our vast network of pipelines and refineries.

Shale Oil

Oil shale is a fine-grained rock that contains varying amounts of a solid organic material called *kerogen.* When heated to nearly 900°F the kerogen decomposes into hydrocarbons and carbonaceous residue. The cooled hydrocarbons condense into a liquid called shale oil. This can be refined into a useable fuel. In fact, there was a highly active shale oil industry in the eastern states prior to Drake's discovery well in Pennsylvania.

Almost two trillion barrels of oil are trapped in shale formations in a 16,000 mi^2 area that extends into Colorado, Utah, and Wyoming. While these oil shale deposits are potentially large, only 600 billion bbl (one-third) are recoverable. This estimate is based on "high quality" shale that yields over 25 gal per T.

Presently, there are two basic processes for extracting the oil. In one, the shale is mined and heated in retorts located above ground to extract the oil. In the other, a well is dug, explosives are set, and a huge underground cavern of rubble results. The shale is then heated underground and the oil is pumped from the bottom of the cavern. Each method has its own set of problems.

Between each heating cycle, the rubble remaining in the retort must be removed. While the kerogen is driven out by the heat, heavy metals such as cyanide and other pollutants remain. Thus if the rubble is dumped in the open and subjected to rain and snow, toxic runoffs can occur.

In situ treatment is expensive; however, this method leaves the rock rubble in the ground and eliminates much of the surface restoration problem.

Its problems are that when the solid shale rock is reduced to a size ideal for heating, its volume expands by 40%. This means that there is always more rubble in the underground cavern than there was solid rock. Also, in at least one test project, the entire hill in which the cavern was blasted was composed of shale. Thus when the rubble was ignited to drive out the kerogen, the hill also burned.

Thus there are numerous technical, environmental, and regulatory obstacles that must be resolved. First, present methodologies are still in the test stage. No one yet knows the full capabilities of shale development. Second, water is necessary for the process. In the arid West where oil shales abound, water is scarce. Finally, obtaining approval from state federal agencies is always a time consuming task. And, much of the shale oil reserves are located on federal land, which means large-scale environmental studies would have to be made.

Of the various proposed technologies to produce synthetic fuels, shale oil is among those always mentioned as being close to commercialization. Proponents claim the United States could produce more than 300,000 B/D of shale oil.

Tar Sands

The majority of the *tar sands* in the United States are located in Utah. However, the largest deposits on the North American continent are in Alberta, Canada. Here bituminous sands cover 12,000 mi^2 and contain 900 billion bbl of oil. Tar sand deposits in the United States are estimated at less than 200 billion bbl. Once extracted, the tar-like sands are mixed with hot water and steam to form a slurry. Then, processing breaks the material down into sand and *bitumen*. Once separated from the sand, the bitumen is cracked, and upgraded with hydrogen and catalysts (Figure 17–3).

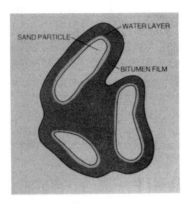

Figure 17–3 Composition of oil sands (courtesy Anne McNamara, Petroleum Resources Communication Foundation).

Producing oil from tar sands is somewhat similar to the processes used both for oil shale and heavy oil. It can either be surface mined and then treated to extract the bitumen or heated in situ and then pumped out of the ground (Figure 17–4). As with the shale rubble, the sand that is cleaned and removed from the bitumen expands in volume 40% when it is "fluffed up." However, unlike the shale rubble, the sand can be spread on the surface, treated with fertilizer, and used to grow plants and crops.

The development of oil from U.S. tar sands is somewhat constrained by environmental laws and regulatory delays, but the major obstacle is technological. Most U.S. deposits are relatively small, and in-situ technology is still relatively new. Until a method can be found to extract the oil more cheaply, developers will probably opt for other alternatives that are more economically feasible.

The Immediate Future

Natural gas and its derivatives are perhaps the fuels that offer the most promise for the immediate future. While not an "alternative fuel" in the strictest sense of the term, new uses for natural gas, plus its many attributes make it very attractive indeed. It is in plentiful supply from presently producing "conventional" basins, from vast

Modern Petroleum

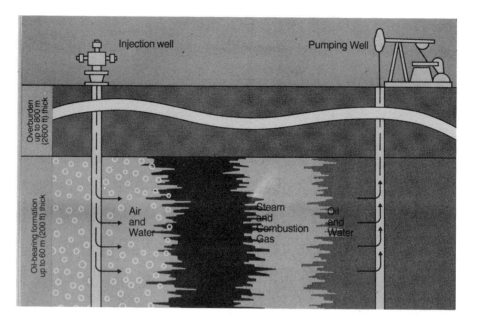

Figure 17–4 In-situ recovery process in the oil sands is based on principles similar to those used in the recovery of heavy oil. Heat thins the oil, enabling it to move toward producing well bores (courtesy Anne McNamara, Petroleum Resources Communication Foundation).

newly opened deposits such as the Arkhoma Basin and from deep areas such as the Anadarko. Plus, its presence is now being discovered in areas never before touched. It can also be produced as methane from coal seams and municipal waste disposal systems.

In most cases it can be transported more cheaply and easier than liquid or solid fuels. It burns relatively cleanly and already enjoys wide use. For years it has been used in "dual fuel" commercial installations which have the capability of burning either gas or fuel oil depending on the availability and cost of one or the other. Today, it is finding growing use in *cogeneration* plants. This is a plant that burns gas to make steam to generate electricity. After powering the turbine that turns the generator, the steam is recycled to either heat buildings or for commercial uses such as in automobile tire factories and other large consumers of heat energy.

Chapter References

"Alternate motor fuels due extensive testing", *Oil & Gas Journal*, (October 23, 1989).

"Coalbed methane resources of the U. S.," *Oil & Gas Journal* (October 9, 1989).

Our Petroleum Challenge: The New Era, The Petroleum Resources Communication Foundation, n/d.

"PG&E's Compressed Natural Gas Vehicle Program," Fact Sheet, Pacific Gas and Electric Company, San Francisco, CA, (March 1, 1990).

"Tough air-quality goals spur quest for transportation fuel changes.", *Oil & Gas Journal* (June 18, 1990).

18
Environmental Concerns

As the human population has increased, so have the demands for faster transportation, greater food production, and larger amounts of consumer products. Man's efforts to meet these demands have brought about an increase in pollution.

Today, the pollution of our environment is one of the greatest concerns facing the human race today, and the survival of all living things depends upon our ability to reduce and control it. Our atmosphere, our water, and our soil are all finite resources and once damaged by pollution may never be replaced.

Thus, the present modification of the environment from its once pristine state can be called a result of mankind's productivity, and this is viewed by some groups as an unnatural change.

Meanwhile, at the other end of the spectrum, there are those who consider all environmentalists "tree-huggers" and refuse to admit there are any environmental problems at all; that there will always be enough clean air, clean water, and uncontaminated soil to go around.

There are no easy answers. But cleaning up and protecting the environment and preventing a recurrence of the excesses of the past will call for reason and good judgment on the part of all. And we must develop new technologies and methodologies for dealing with our problems.

What Is the Environment?

Our environment is the aggregation of material factors and conditions surrounding all living organisms and their component parts. The human environment basically includes temperature (climate), food and water supply, and other people. The environment for the plants that furnish our food may be made up of the soil and its constituents, the climate, the availability of sunlight for photosynthesis, and various kinds of animals that might eat or damage the plants. Non-living factors such as soil, water, sunlight, the atmosphere, and radiation are forces that influence one another and the surrounding community of living things.

Humans, of course, are also living organisms, and although somewhat ruthless at times, they know the impossibility of turning back the clock of history. Given the magnitude of today's problems, hopefully they will have both the technical knowledge and the inclination to reverse the trend.

Everyone would like to reduce or eliminate pollution, but much of it is caused by the very things that benefit people. Solving the problem is as serious as it is complicated. For example, the automobiles that provide necessary transportation for millions of persons cause a major portion of the air pollution with their exhaust gases. Factories that manufacture essential goods and employ thousands of people discharge material that pollutes both the air and water. Farms produce food for human and animal consumption, but the use of chemical fertilizers and pesticides eventually harms the soil.

All throughout history people have polluted their surroundings. However, in the past this was not a great problem because most of the population lived in uncrowded rural areas where the pollutants were widely dispersed. But with the advent of the industrial revolution, people began concentrating in crowded cities, building large factories, and later, using automobiles. Hence pollution gradually became a serious problem. As population increased, and the world became more industrialized, pollution steadily worsened

The composition of various kinds of pollution and its concentration varies widely with location, climate, and topography. When climatic factors are favorable, some areas can be essentially pollution-free, but when these factors change to unfavorable, excessive pollution can bring about emergency situations that are hazardous to life. One such area is the Los Angeles basin, which is, by nature, a very efficient collector of concentrations of pollutants in the air. This is especially true when the basin is capped by an *inversion layer* during periods of windless warm days. Scientists and engineers can do little about the natural conditions of the basin. However, they have taken steps to control the thousands of substances which contribute to air pollution.

Air pollution is basically caused by artificial and natural sources that emit hundreds of millions of tons of gases and tiny particles of solid or liquid matter called *particulates* into the atmosphere each year.

The Atmosphere

Our atmosphere consists of gases, vapors, and suspended matter. It is predominantly a mixture of the two gases, oxygen and nitrogen, but several more complex gases are also present. Oxygen is the chemically active ingredient, and nitrogen plays the role of diluent. The atmosphere is a series of envelopes in the form of imperfect spherical shells of various materials that are bound to the earth by gravitational force, and extends some 370 to 930 mi into space. More than 75% of the atmosphere exists below about 15 mi altitude, where the components of the air remain relatively unchanged in their ratios, and are only slightly altered until they reach a height of approximately 30 mi (Table 18–1).

The composition of the lower layers of the atmosphere is estimated to be 79% nitrogen and 21% oxygen by volume. In the lower layer of the atmosphere, water vapor is an ingredient of air that varies seasonally and geographically and is a major factor where the air is adequate for certain chemical reactions. One of the lower layers of interest is the *troposphere*. This is a thermal atmospheric region extending from the Earth's surface to a height of about 8 mi. It is characterized by temperature decreasing with height, strong vertical wind motion, and appreciable water vapor content. It contains nearly all the clouds, storms, and pollutants.

The *stratosphere* lies above the troposphere and reaches to a height of about 19 mi above the Earth's surface and contains the *ozone* layer, which is of great current interest to scientists.

Constituent	Parts per million
Nitrogen	780,840.00
Oxygen	209,460.00
Argon	9,340.00
Carbon dioxide	330.00
Neon	18.18
Helium	5.24
Methane	2.00
Krypton	1.14
Hydrogen	0.50
Nitrous oxide	0.50
Xenon	0.09

Table 18–1 Composition of the Earth's Atmosphere

Complex interactions take place between the fluxes of electromagnetic radiation of various wavelengths, radiation from the sun, and concentrations of atmospheric gases. The Earth's magnetic field also governs particulate radiations. Short-wavelength radiations cause a variety of photochemical reactions. The most important of these is the creation of a layer of ozone that acts as an effective absorber of solar ultraviolet radiation, thus causing a warm layer at about 19 mi up in the atmosphere.

The atmosphere permits most of the sunlight to pass through and heat the Earth's surface. However, some of the heat energy, called infrared radiation, is sent back into the atmosphere. Certain gases in the atmosphere partially absorb the infrared radiation emitted by the Earth's surface and grow warm. They then re-emit infrared radiation partly upward and partly downward. These gases include carbon dioxide, ozone, and water vapor. The upward flux is essential to the radiation balance of the planet, and the downward flux adds to the heating of the surface of the earth and is of basic importance in the *greenhouse effect.*

Greenhouse Effect

The Earth's atmosphere acts much like the glass roof and walls of a conventional greenhouse, thus the name *greenhouse effect.*

Vast amounts of carbon dioxide, nitrous oxides, methane, and *chlorofluorocarbons* (a class of synthetic chemicals not normally found in the atmosphere), have added to this blanket of gases that allow sunlight in, but trap the resulting heat and prevent it from escaping. This is believed to be the reason for our global warming trend. No conclusive proof has linked this slow warming to the greenhouse effect, but some scientists believe the burning of coal and petroleum fluids have caused an increase of carbon dioxide in the atmosphere. Other scientists now think that carbon dioxide is only half the problem. The other half comes from chlorofluorocarbons, nitrogen oxides, and methane. Estimates are that from the late twentieth century to the late twenty-first century the amount of unwanted gases in the atmosphere could double, causing an increase of up to 9°F in the Earth's average temperature.

Chlorofluorocarbons (CFCs) are synthesized chemicals consisting of chlorine, fluorine, and carbon atoms that are nontoxic and inert. CFCs are excellent as coolants in refrigerators and propellant gases for spray cans and are inexpensive to manufacture. CFCs are good insulators and are also used as blowing agents in plastic-foam materials such as Styrofoam.

CFCs gained acceptance as a miracle chemical immediately after their introduction in the late 1920s. However, somewhat later they were discovered to have a harmful effect on our environment. Scientists claim that each molecule of CFC is about 20,000 times as efficient at trapping heat in the atmosphere as a molecule of carbon dioxide, thus increasing the danger of the greenhouse effect. Also, there is fear that the chlorine released when CFC molecules break up will destroy ozone molecules.

In the 1970s when scientists first warned that CFCs could harm the ozone, the United States responded by banning their use in spray cans. In their place, American manufacturers began using butane. However, the rest of the world continued to use CFCs in spray cans. In 1985 the threat became alarming when researchers reported a "hole" in the ozone layer over Antarctica. In response, 24 nations met in Montreal and agreed to reduce use of CFCs by 35% worldwide by 1990.

Global warming appears to be one of the major environmental issues of the 1990s because of the growing political and scientific consensus that the world can no longer dodge this issue. The drafting of an International Climate Treaty was held in 1990, and a formal adoption was expected in 1992 at a Global Environmental Summit in Rio de Janeiro.

The Clean Air Act

In recent years automobile manufacturers and oil refiners have begun to prepare for new emission rules. After several years of relaxing environmental policy under the Reagan administration during which, many believed, more attention was given to the stars than to the atmosphere, things have begun to change. Following the hot summer of 1988, fears of global warming resurfaced and made the environment a presidential campaign issue.

Suddenly, both environmental interest groups and scientists began to be heard by politicians. The *Exxon Valdez* ran aground on March 24, 1989, spilling an estimated 11 million gal of crude oil in Prince William Sound in Alaska, and the whole world took notice. Almost immediately polls started to show that the American public favored increased attention to our environment.

The resurgence of environmental awareness in the United States prompted Congress to pass the Clean Air Act Amendments of 1990.

The Clean Air Act requires the EPA to set limits for air pollutants. These pollutants include carbon monoxide, sulfur dioxide, nitrogen dioxide, lead, ozone, and particulate matter.

Since the passage and revision of the Clean Air Act, pollution levels have been reduced in most areas of the country. This reduction has qualified many areas to meet national ambient air quality standards for lead, nitrogen dioxide, sulfur dioxide, and particulate matter. Carbon monoxide and ozone standards have been more difficult to achieve, but smog has been noticeably reduced.

Reduction of carbon monoxide and ozone emissions from motor vehicles is one of the prime objectives of the revised Clean Air Act because they are believed to account for most of the carbon monoxide emissions and about 40% of the ozone emissions in urban areas. Motor vehicles are also a source of volatile organic compound emissions that either contribute to or cause the formation of hydrocarbons as both evaporative and exhaust emissions. These include benzene and oxides of nitrogen.

Oxides of nitrogen are targeted because they are one of the reactants in the formation of ozone. Their levels are affected more by engine combustion temperatures than they are by components in the fuel. Hydrocarbon emissions combine with oxides of nitrogen in the air and in the presence of sunlight to form ozone. Ozone is a lung irritant and may cause respiratory problems. Carbon monoxide emissions are a problem in areas where there are temperature inversions that trap pollutants in the surface atmosphere. Benzene is a known cancer-producing substance and is found in finished gasoline. It is also generated as a result of gasoline combustion in engines.

The 1990 Clean Air Act Amendments call for stricter control requirements in cities that have not yet met federal air quality standards (non-attainment areas). It also mandates cleaner gasoline and the use of clean-fueled vehicles in some cities. Called for is the establishment of an acid rain control system and the phase out of the production of chemicals that contribute to the depletion of the stratospheric ozone layer. Alternate fuels development is also encouraged. The 1990 Amendments set tailpipe emission levels for gasoline and diesel-fueled automobiles, trucks, and buses.

Automobile manufacturers have a major role in the fuels issues because any new fuels will have to work in concert with improved emission-control systems. The new fuels must not significantly change the fuel economy of current and future engine designs. Also, there should be a steady increase in fuel economy from all vehicles sold in the United States through the 1990s.

Compressed natural gas and liquefied petroleum gas are the two alternate fuels that have attracted interest in the clean-air debate. Methanol and gasohol will also probably find increased use. But

Modern Petroleum

time is required to make the changeover, and it is likely that conventional and reformulated gasoline will remain the dominant fuel in the United States for the next decade.

Refiners will be affected since their processes are considered as sources of air pollutants. Refineries that emit 10 T per year of any hazardous pollutant, or at least 25 T per year of a combination of hazardous air pollutants, will be targeted.

Corrective Measures Being Tested

Soil Pollution

Any liquid, solid, or gases that damages the thin layer of fertile soil that covers much of the Earth's land surface is considered to be a pollutant.Soil is essential for growing food crops to feed the multitude of people and animals on earth. The trees and vegetation that clean our air depend upon the Earth's soil for existence. Natural processes required thousands of years to form the soil that provides our sustenance.

Natural cycles, similar to those that absorb small amounts of waste in bodies of water, work to keep soil fertile. Dead organisms and plant and animal waste all accumulate in the soil. Bacteria and fungi decay these wastes by breaking them down into nitrates, phosphates, and other nutrients. Growing plants feed upon these nutrients, and when they die the cycle takes place again. Too much waste in a concentrated area overloads the bacteria and fungi decay process and upsets the natural cycle. And overly large amounts of chemical fertilizers may decrease the ability of bacteria to decay waste. Also, chemical pesticides may harm bacteria and other helpful organisms in the soil. An important development in agriculture is the use of biological controls instead of chemical pesticides.

Possibly the most visible forms of soil pollution are solid wastes. Such waste is a very serious problem because nearly all present methods used to dispose of the solid waste result in some type of damage to the environment. Burning produces smoke that cause air pollution. When dumped into bodies of water, waste contributes to different forms of water pollution.

Today, solid waste is disposed of in open pits or land fills, formerly called "dumps." But these sites provide homes for disease-carrying animals. These outside facilities also harm the attractiveness of areas. They also interfere with air travel, since many attract large numbers of scavenging birds, which can pose a hazard when the site is located near an airport. Disposing of waste in an outside facility does not eliminate liability for damage caused by the waste in cases where the facility is not in compliance with regulatory requirements.

Other methods of disposing of waste is the process of recycling, which reprocesses waste for reuse. Most solid wastes can be recycled and used over and over again. Aluminum cans are collected by individuals and organizations who sell them for cash to recyclers. Old newspapers are collected and turned into pulp to be used again for new paper. If a substance is reused, it is not a waste.

Another innovative method beginning to find acceptance in the United States and Italy, is the composting of *municipal solid wastes (MSW)*. Some 42 cities and counties in 14 states currently operate composting programs. Wood, paper, and other trash is processed through hammer mills and the resulting *silage* spread out in long windrows on the ground. These windrows are periodically turned to make sure the heat from the naturally occurring *thermophilic action* spreads evenly through the material. This heat, when it reaches temperatures of 140°F or more, destroys any pathogens that might be present, and converts the MSW into high-quality compost in demand by agriculture and horticulture industries.

Similar projects using sludge from municipal sewer systems instead of MSW are also in operation in a number of states.

To comply with new regulations, industries have begun to reduce the amount of waste they generate, recycle materials for reuse, and treating their wastes to eliminate toxicity.

Radiation is another pollutant that can pollute the soil and harm plants, and animals. Fallout from tests of nuclear weapons and waste materials from nuclear power plants produce radioactive materials. Scientists have not agreed exactly on what effects small amounts of radiation have. But they are quite positive that large-scale contamination such as occurred from the Chernobyl power plant incident in Russia did extensive damage.

Acid rain that forms when moisture in the air combines with *nitrogen oxide* and *sulfur dioxide* is becoming an increasingly serious problem. It occurs as a result of the combustion of oil or coal. The reaction between the moisture and the chemical compounds produces nitric and sulfuric acids that fall to the earth with rain or snow. It can also occur from the condensation of evaporated surface water containing naturally occurring acids. Acid rain can harm trees, crops, bodies of water, and reduce the fertility of the soil.

Water Pollution

Water pollution comes primarily from industries, farms, and sewage systems. There are many chemical, physical, or biological substances that affect the natural condition of water. Pollution threatens the quality, usefulness, and the availability of water. Water may be considered polluted if it contains an excess or burden of any gaseous, liquid, or solid constituent.

The worldwide population growth has contributed greatly to water pollution through the organic wastes contributed by domestic sewage from both urban and rural areas. Domestic sewage is the most widespread source of degradable organic waste, although industry also contributes a very large amount. The food and pulp industries generate tons of organic wastes that are often dumped into bodies of water. Wastes from farms include animal wastes, fertilizers, and pesticides that drain off farm fields and into nearby streams. New processes in manufacturing, industrial progress, and technological developments have all produced new and complex wastes that contribute to water pollution.

Large volumes of water are used for cooling refineries, petrochemical plants, steel mills, electric power plants, and other industrial operations. Most of this water is hot as it leaves the cooling systems and is returned to the streams, lakes and coastal waters where it raises the water temperature. Unnatural heat is considered a water pollutant because it reduces the solubility of oxygen in the water. It also upsets natural cycles that work to purify water, and can kill animals and plants that are accustomed to living in lower temperatures. This *thermal pollution*, as well as chemical pollution, directly affects all the living organisms in the food chain. Eventually, this affects us, since man is at the top of the food chain.

Natural cycles work to absorb small amounts of waste in bodies of water if it is not heavily laden. Water will undergo "self-purification," a process involving the action of aerobic bacteria. Aerobic bacteria will break down chemicals and other waste and turn them into nutrients or into other substances that will not harm fish or sea plants. If too much waste matter is dumped into the water, the process, of biological degradation becomes *anaerobic*. The bacteria that work to decay waste use excessive oxygen during the decaying process, and less oxygen is available for the animals and plants in the water. The system will begin to break down as the oxygen is depleted and noxious hydrogen sulfide gas, methane, and other gases are produced.

Another pollutant is oil discharged from barges and ships or from tanker accidents. The great oil spill from the Exxon Valdez in Alaska where oil-polluted water resulted in great damage to aquatic life and other wildlife. Waterfowl became so oil soaked that they were unable to fly. Food for fish and other marine life was destroyed so that those who depended on the fishing industry suffered substantial economic losses. At this time it is not possible to assess the environmental damage caused by the war in the Middle East.

In the past, careless handling of crude oil and disposal of drilling fluids in offshore drilling operations created water pollution problems in some coastal areas which did the same harm as tanker spills, except on a smaller scale.

As a result of a rash of oil spills from tankers, pressure is mounting in Congress and in some of the coastal states to ban all leasing and drilling in the U.S. offshore frontiers. Efforts are under way in coastal states and in Congress to block lease sales and buy back old leases. Some would even like to permanently halt drilling, even though in some cases drilling operations have already begun.

New congressional attention to oil tanker safety is receiving both public approval and oil industry support. *It is one of the most pressing environmental issues at the present time in the United States.* In 1988–89, the 43 spills from tankers throughout the world demonstrated very clearly that tanker accidents represent an immediate and potentially devastating threat to our beaches and marine life.

The threat to the United States can only grow because of the increased traffic as a result of growth in oil imports and the absence of deepwater offshore terminals. An increased use of alternate fuels could make a difference in crude oil imports. However, the transition may well be decades away. More deepwater offshore terminals and double-hulled tankers might be the best short term answer.

Learning from Oil Spills

Exxon's massive cleanup effort and the subsequent amount of scientific research, along with environmental studies related to the *Valdez* accident, will provide the industry a wealth of data for dealing with oil spills and their effects in the future. If there were some mistakes made in the beginning, it is evident that BP America had learned from these mistakes in handling the Huntington Beach, California spill that occurred later.

BP was training for an oil spill crisis in the weeks preceding the spill off Huntington Beach. The training involved putting its newly formed crisis management team through drills. Their biggest problem in coping with the spill response and cleanup was assuring that there would be enough personnel with the proper training to perform the manual cleanup while maintaining an optimum force, as mandated by the Occupational Safety and Health Administration (OSHA). BP's cleanup task force totaled more than 2,000 persons at its peak.

In managing such a crisis, top priority is placed on the early hours of a crisis, and centers on an effective individual company contingency plan with a well-trained and organized response team. BP went into action about 30 minutes after the spill happened and the port captain notified the local industry oil spill response cooperatives. The crisis management team asked the communities to determine priorities among the beach sections hit by the spill, and then put its cleanup plan into effect. BP also made an extra effort to obtain input from local citizens as well as advice from state and federal officials. Overall, their decisions proved to be correct, and they felt that the task would have been much more difficult without an organizational approach having been taken.

Planning and response are the two critical objectives that must be met to cope with major spills. Past experience proves that in order to be effective, action to contain and control an oil spill must be taken well within the first 24 hours. No amount of expertise or equipment can be properly organized on the spot after a large spill has taken place.

Regulations

The *U.S. Department of Transportation (DOT)* regulations implement the *Clean Water Act*, which requires contingency plans for marine oil-transfer facilities. Procedures for reporting initial containment of oil discharges must be in these plans that have to be approved by the Coast Guard Captain of the Port. The DOT regulations state that each facility must have ready access to enough oil-containment material and equipment to contain an oil discharge on the water from operations at that facility. EPA regulations require such facilities to list in their plans the equipment to be maintained for use in the event of oil discharge.

Federal, state, and local governments are active in helping to control pollution by providing financial assistance for antipollution programs, and passing laws limiting the amount of pollution that installations such as sewage treatment plants and other facilities can release into the environment. These agencies conduct and support research that leads to a better understanding of environmental problems.

In 1899 the United States passed its first major pollution law which made it a crime to dump any liquid wastes other than sewage into navigable waters. The law was seldom enforced and thus was almost useless in mitigating water pollution. Few other important federal pollution laws were passed in the United States until the 1960s. During and after that period scores of laws dealing with the environment have been enacted.

In 1970 the *EPA* was established as an independent agency of the U.S. government. Its purpose is to protect the nation's environment from pollution. Creation of the agency brought under single management the functions of 15 federal programs dealing with pollution. The EPA is directly responsible to the President of the United States.

The EPA conducts research on the effects of pollution and provides grants and technical assistance to state, local, or other government agencies that strive to prevent pollution. The EPA also establishes and enforces environmental protection standards.

Private Organizations

In the United States, various groups of citizens have formed organizations to control pollution. Primarily, they are concerned with their own local problems. However, some work with national and international organizations as well.

Private groups call public attention to pollution problems and put pressure on elected officials and industry leaders to take action in solving problems. They are responsible for many of the corrective steps taken to control pollution.

Some private organizations achieve their goals through legal action. These organizations sue private companies as well as government agencies to force them to stop polluting the environment. Their cases are based on the belief that all people have a constitutional right to a healthy environment.

Environmental Education

Environmental education deals primarily with the interrelationship of living organisms with one another and with their surroundings. It is an expansion of the study of ecology. This concept certainly is not new, because it is very similar to the one practiced by Native Americans for centuries.

Environmental studies also center around sources of energy, food supply, and the protection and conservation of natural resources. These studies are introduced into the elementary and secondary schools in such a manner as to coincide with science and social studies. University environmental programs have been structured around numerous subject matter areas.

Environmental Documentation

Certain projects, including oil and gas exploration or development, offshore drilling, pipeline or power plant construction or other petroleum-related endeavors involving an agency of the federal government or an entity supported or regulated by the federal government, require environmental documentation before they can proceed. This documentation may be in the form of either an *Environmental Impact Statement (EIS)* or an *Environmental Assessment (EA)*. These are the findings of a thorough investigation into all aspects of the project and the existing environment at the proposed site to determine its probable effect, if any, on the environment. They are most usually called for if the proposed project is to be located on federal land, or on American Indian land held in trust by the U.S. government.

Required under the provisions of the *National Environmental Policy Act (NEPA)*, these documents can be prepared by a government agency such as the U.S. Fish and Wildlife Service, in-house by the organization proposing the project, or by an independent third party who is on the government's list of specialized contractors who are approved to perform such work. Normally, the cost of the EIS or EA is borne by the company or organization requesting environmental clearance. For example, if an oil company wishes to drill on Indian trust land that is under the jurisdiction of the Bureau of Indian Affairs, the BIA will require that the company have an EA or EIS prepared at its expense and approved by the Bureau before the drilling permit can be issued.

The purpose of such research is to make sure that the government agency involved carefully considers whether or not the proposed project will have any detrimental effect on the environment and that possible alternatives to the proposed action are considered. Guidelines for developing these documents are set by NEPA, the *Council on Environmental Quality*, and numerous other federal laws and regulations.

The difference between an EA and an EIS is one of size and scope. For example, a small project such as a single drilling lease on federal land might be initially surveyed by an EA to determine if it could cause any significant impact to the environment. If not, then the federal official in charge could issue a *Finding of No Significant Impact* or "*FONSI*," and the project could proceed. If, however, the EA investigation concluded there might be a significant environmental impact resulting from the proposed action, then a choice would have to be made: either abandon the project or conduct a full-blown EIS.

Some projects by their very magnitude, such as a new refinery or hazardous waste incinerator, may be required to submit the more complex and highly detailed EIS without first completing an EA.

Each document is required to examine and record the current state of the environment at the proposed site and to present evidence as to what, if any, impact the project would have on the environment. The areas of concern include endangered species of plants and animals, air quality, water quality, discharge of solids into navigable waters, visual impacts, the impact of increased noise levels, disposal of wastes generated at the site, impact on the local infrastructure, socio-economic effects, the preservation of historic structures, and the destruction of religious or culturally important sites.

As might be imagined, the preparation of such a document calls for the services of many professionals in many fields. A typical EIS may require a team of mechanical, civil, and petroleum engineers; wildlife biologists, agronomists, soil scientists, historians, anthropologists, archaeologists, mapmakers, illustrators, editors, and other specialists. When all their studies are completed, the reports they write are compiled into a draft copy of the EIS that is submitted to the federal agency in charge of the project.

Notice is then given by the government of a public comment period during which time any interested person or organization may examine the document and make pertinent comments. At the end of this period, the draft EIS is returned to the preparer who includes the comments into a final copy that is sent back to the head of the federal agency overseeing the project. He or she then carefully reviews all the evidence before reaching a decision as to whether or not to issue a FONSI.

In the past, some groups opposing the construction of a particular project have used the environmental laws to delay or prevent actions they viewed as detrimental to the environment.

On the other hand, the research conducted in preparing an EIS in some cases has been used to develop viable alternatives that allowed the project to go ahead with little or no impact on the environment. And in recent years many oil companies have come to realize that performing a thorough environmental investigation before proceeding with a project can bring many long term benefits. For example, the existence of a carefully documented record of the environmental conditions at the outset of a project can defend a company from being accused of causing environmental harm that may have actually happened years before.

Effects on the Petroleum Industry

All of these regulations are having effects on the petroleum industry unheard of a few short years ago, and compliance with them has contributed to the rising cost to oil companies of doing business and providing society with much-needed petroleum products.

What to do with the wastes from refineries and petrochemical plants is a major concern. These must now either be treated before they can be disposed of in a regulated landfill or run through a specially designed incinerator. And the ash and scrubber water from the incinerator must themselves be treated before their disposal.

Drillers are faced with the problem of what to do with used drilling fluids and detritus classified as "deleterious" substances. There are few approved disposal sites, and today few contractors are interested in developing new sites because of possible future environmental liability.

Owners of open pits must now cover them with mesh netting capable of keeping out migratory birds or face severe penalties. The U.S. Fish and Wildlife Service estimates that more than a half-million birds ranging from Bald Eagles to more common species are lost each year in Texas, Oklahoma, and New Mexico alone. The birds become entrapped in oil scum on the surface of tanks and ponds and cannot escape. Federal law provides penalties of up to $10,000 per incident for the death of a single bird, plus a possible six-month jail term.

Even owners and operators of service stations must comply with stringent regulations covering their underground storage tanks and how they dispose of used motor oil, antifreeze, batteries, and solvents. Retail marketers are also faced with the problem of controlling hydrocarbon emissions which occur as they deliver fuel to their customer's gas tanks.

Many oil companies have risen to these challenges and have provided exemplary leadership in their concern for the environment. In the United States in recent years, many wells and pipelines have been constructed both offshore and on land adjacent to such sensitive areas as national parks and bird and wildlife sanctuaries. Not only did most of these pass close government scrutiny, in some cases the companies also earned the approval of national environmental groups.

We must all realize that there is no such thing as a "free lunch" where either energy or the environment is concerned. All areas of society must work together to make the most informed decisions possible.

Chapter References

"The Biocyle Guide to Composting Municipal Wastes," *Biocycle: The Journal of Waste Recycling*, (The J.G. Press, Inc.).

Patrick Crow, "U.S. Industry Refighting Battles on Clean Air Act," *Oil & Gas Journal* (July 23, 1990).

"Drilling, Production, Meet Environmental Challenge," *Oil & Gas Journal* (March 5, 1990).

Environmental Regulations and Technology: Use and Disposal of Municipal Wastewater Sludge. (Washington D.C.: U.S. Environmental Protection Agency Intra-Agency Sludge Task Force, September, 1984).

Nora Goldstein, "Composting Activity: Solid Waste," *Biocycle: The Journal of Waste Recycling.*

"Refiners, Petrochem Plants Focus on New Waste Challenges," *Oil & Gas Journal*, (March 5, 1990).

"Tough Air-Quality Goals Spur Quest for Transportation Fuel Changes," *Oil & Gas Journal*, (June 18, 1990).

Glossary

A

abiogenic Not derived from biological materials.

abrasive drilling The use of a harder mineral or substance than the rock being drilled through as a drilling medium.

absorber A tower or column so constructed that it provides contact between the rising natural gas being processed and the descending absorbent.

absorption oil The absorbent liquid hydrocarbon used to absorb the components from the natural gas being processed.

absorption plant The total plant that uses an absorber for the initial separation of the components of the natural gas.

accumulator A vessel for the temporary storage of a liquid or gas.

acidizing The introduction of acid into a formation to dissolve deposits of alkali and thus open passageways for the fluids to flow through.

acoustic log A measuring device that uses sonic waves to directly measure lithography and porosity. Also used to detect poor cement bonding between wellbore and casing.

AGA American Gas Association.

air balance beam pumping unit A pump jack that uses compressed air to balance the weight of the beam.

air bursts A marine geophysical technique in which bursts of compressed air from a gun towed behind a seismic vessel are used to produce sound waves which were formerly generated by high explosives.

air drilling Rotary drilling system using compressed air instead of liquid as the circulation medium.

air injection A secondary or tertiary means of recovery in which compressed air is introduced into a formation in order to force the oil out.

alcohols Organic compounds containing an OH (hydroxyl) group. Ethanol, C_2H_5OH, is commonly called alcohol, but it is only one of many members of its class.

alkanes This series, derived from petroleum, has its carbon atoms arranged in a straight chain. It includes methane, ethane, butane, propane, pentane, hexane, and heptadecane.

alkylation The reaction of alkenes or olefins with a branched chain alkane to form a branched, paraffinic hydrocarbon with high antiknock qualities.

alkylation process The process of making gasoline-range liquids from refinery gases.

all-levels sample Taken by submerging a stoppered beaker to a point in a tank as near as possible to the draw-off point and then opening it and raising it.

allowable The amount of oil of gas a well is permitted to produce by order of a regulatory body.

alluvial Pertaining to the sediment deposited by water flow.

Alpha 1 A cable-operated surface unit that uses less energy to operate than a conventional beam pumping unit.

angular unconformity Strata lying at an angle across the folded and tilted edges of the beds below it.

annulus Downhole space between the drillstring or casing and the borehole wall, or between the production tubing and casing, or between the surface casing and production tubing and casing.

node The electrode at which oxidation occurs in an electrolytic cell.

anticlines Arches or upfolded rock formations.

API gravity (API°) The standard method of expressing the gravity, or unit weight, of petroleum liquids.

arch A rock formation that folds upward like an inverted trough.

Archean The oldest era of the geologic record; more than 2,500 million years ago.

Archie's equation George Archie pioneered studies in resistivity for log interpretation and determined that water saturation is equal to the square root or the 100% water-wet resistivity, $R°$, divided by the formation resistivity, R_t : $S_w = R_o/R_t$.

aromatics Cyclic hydrocarbons, originally found naturally in aromatic gums or oils.

artificial drives Reservoir drives from other than natural means—in-situ combustion, waterflooding, etc.

artificial lifts Any mechanism other than natural reservoir pressure of sufficient force to make oil flow to the surface.

associated gas Natural gas that occurs with oil, either in solution or as free gas.

asphaltene Any of the dark, solid constituents of crude oils and other bitumens.

atmospheric discharge The release of gases and vapors from pressure relief and depressuring devices to the atmosphere.

average sample Averaging of two or more tank samples.

automatic tank batteries (also see LACT units) Lease tank batteries equipped with automatic measuring, gauging, and recording devices.

B

back pressure The pressure existing at the outlet of the pressure relief device due to pressure in the discharge system.

baffles Plates or obstructions built into a tank or other vessel that change the direction of the flow of fluids or gases.

bailer A bucket-shaped cylinder used in cable tool drilling to remove rock cuttings and mud from the borehole.

barefoot Well completed without casing in a sandstone or limestone formation that gives no indication of caving in.

barite A mineral often used as one of the components of drilling mud to add weight; barium sulphate.

barrel (bbl) 42 U.S. gallons.

basin (sedimentary basin) A depression of the basement rock filled by sediments. An example is the Anadarko Basin of Oklahoma where the deepest part is approximately 45,000 ft.

batch A shipment of a particular product through a pipeline.

batch interface The point where two shipments or product touch in a pipeline, e.g., a batch of gasoline followed closely by a batch of kerosene.

batch separator A device used to keep shipments apart in a pipeline so that different liquids do not intermingle.

batching sequences The order in which product shipments are sent through a pipeline.

B/D (or b.d.) Barrels per day.

bead Deposit of molten filler material laid down during the welding process.

bedding plane A division plane which separates individual strata or beds in rock.

bentonite Highly absorbent rock composed principally of clay materials and silica.

benzene Light petroleum distillates in the gasoline range. The term "benzene" is still used as a synonym for gasoline in some European countries.

Big Inch The 22-in. pipeline built from Texas to the East Coast by the government in World War II.

biogenic Substances derived from biological materials.

bits The cutting tools used on the working end of the drill string.

blind flange A companion flange with a disc bolted to one end to seal off a section of pipe.

block valve A valve in a pipeline used to seal off or block a section of the pipeline.

blowdown stack A vent or stack into which the contents of a processing unit are emptied when an emergency arises.

blowout When excessive well pressure runs wild and blows the string and tools out of the hole.

blowout preventer Device consisting of a series of hydraulically controlled rams that can be triggered instantly to seal off a well.

boiling point Point at which the vapor pressure of a liquid equals the external pressure on the liquid.

bottom drive Pressure from salt water under the oil which forces it upward.

bottom fraction The heaviest components of petroleum—those remaining at the bottom of the barrel after the lighter ends have been removed.

bottom sample Obtained from the material at the lowest point in the tank.

bottom water Water located at the bottom of the reservoir under the petroleum accumulation.

Bourdon Tube A small, crescent-shaped tube closed at one end, connected to a source of gas pressure at the other, used in process recording devices or in pilot-operated control mechanisms; actuates control or recording instruments.

bright spots White areas on seismographic recording strips that may indicate the presence of hydrocarbons.

BS Basic sediment.

BS&W Basic sediment and water.

Btu (British thermal unit) Unit of measurement. The amount of heat needed to raise the temperature of 1 lb of water 1° F.

bubblepoint The pressure and temperature at which gas, held in solution in crude oil, breaks out of solution as free gas and forms a small bubble; saturated pressure.

Burton Process Early process developed by Dr. William C. Burton that increased production of light products by using heat and pressure.

butane A hydrocarbon fraction; it is a gas, but it is easily liquefied at ordinary atmospheric conditions.

bypass valve A valve that controls an alternative route for liquid or gas.

C

cable-operated long stroke Pumping device that uses a cable from a tower instead of a walking beam to lift the sucker rods.

calorimeter An apparatus which determines the heating value of a combustible material.

Cambrian Geologic period from about 600,000,000 B.C. to 500,000,000 B.C.

cannel coal Bituminous coal that burns with a heavy smoke.

cap bead Final welding process on pipeline joints.

cap rocks Dense rocks that obstruct or delay the upward seepage of hydrocarbon fluids. Layers of salt and clay often form cap rocks, as they are dense and are not as likely to develop fractures as more brittle rocks.

capillaries The minute openings between rock particles through which oil and water are drawn.

capillary action The upward and outward movement through the pore spaces in rock.

carbon A natural element found in hydrocarbons.

carbon black A "soot" produced from natural gas used in the manufacture of tires and other products.

carbon dioxide CO_2 a constituent of the atmosphere at a level of 0.003% by volume. Atmospheric carbon dioxide appears to have been the chief source of carbon in the carbonate rocks, and it is the chief source of carbon in plants.

carbon dioxide-injection A tertiary means of recovery that uses compressed carbon dioxide to force oil from a well.

casing Pipe used in a well to seal the borehole to prevent fluid escape and to keep the walls from collapsing.

casinghead gasoline Natural gasoline—actually the condensate from natural gas.

catalysis A process in which the chemical reaction rate is affected by the introduction of another substance.

catalyst A substance that is used to slow or advance the rate of a chemical reaction without being affected itself.

catalytic (cat) cracking The use of catalysis to break petroleum down into its various components.

catalytic reforming Process used to upgrade hydrocarbons, already quite high in octane, to even higher octane.

cathead Spool-shaped hub on a winch shaft connected to the draw works around which a rope may be snubbed.

cathode The electrode at which reduction occurs in an electrolytic cell.

caustic injection Introducing caustic substances into a formation to increase porosity by breaking down particulate matter and also using pressure to increase the flow of oil.

caverns Larger openings between the rocks in a formation.

cellar Area dug out beneath the drilling platform to allow room for installation of the blowout preventer.

cellulose A very high molecular weight polymer.

cement Mixture which is used to set the casing firmly in the borehole. A slurry, it is allowed to set until it hardens.

cementation The natural filling in of the pore spaces in a reservoir by limestone.

cementing Pumping the cement slurry down the well and back up between the casing and the borehole. Once hardened, the cement is then drilled out of the casing.

Cenozoic Geologic era from about 63,000,000 B.C. to the present.

centralizers Devices fitted around the outside of the casing as it is put in place to keep it centered in the hole.

channelization The act of taking a shallow, winding waterway and straightening and deepening it so that it can be used for transportation.

check valve A valve with a free-swinging tongue or clapper that permits fluid to flow in one direction only. A back-pressure valve.

chemical precipitates Material formed in place under the Earth's surface by the action of dissolved salts.

cheese box Early still using a vessel that resembled a cheese container.

christmas tree Array of valves, fittings, and pipes placed atop a free-flowing well.

circulating system That portion of the rotary drilling system which circulates the drilling fluid or mud.

clamshells Hinged, jaw-like digging tools used on a dragline.

clean circulation In rotary drilling, when the drilling fluid returns clean without cuttings coming to the surface. A method of refining that eliminates much of the residuum.

clearance sample A sample taken four inches below the level of the tank outlet.

coking The undesirable buildup of carbon deposits on refinery vessels.

combination drive Two or more natural mechanisms present in a reservoir such as water and gas-cap drives.

combination trap A reservoir formed by folding, faulting, and porosity changes.

combustion The process of burning; the vigorous reaction of a substance with oxygen, giving off heat and light.

common carrier A public, for-hire transport, i.e., bus, train, airline, pipeline, etc., whose regulation is under the jurisdiction of the Department of Transportation.

completion Finishing a well. Preparing a newly drilled well for production.

composite sample Composed of equal portions of two or more spot samples.

composite spot sample A blend of spot samples mixed in equal portions.

compound A substance in which the molecules contain different kinds of atoms.

compressor Mechanical device used in the handling of gases much as a pump is used to increase the pressure of fluids. Also used to increase air pressure.

compression ratio The ratio between suction pressure and discharge pressure; pressure ratio; the ratio of the volume of an engine's cylinder at the beginning of the compression stroke to its volume at the end of the stroke.

compressor station Placed at selected intervals along a gas pipeline, these units maintain the pressure necessary to keep the gas flowing through the lines.

conceptual models By using various types of information, a geologist can illustrate on paper what underground structures probably look like.

Condeep Offshore drilling and production structure designed for use in the North Sea by Shell.

condensate liquid hydrocarbons produced with natural gas which are separated from the gas by cooling and various other means.

conductivity Property of, capacity for, or tendency toward conductance of an electric current, varying according to formation composition.

conductor Outer pipe near the top of the well used to seal off unstable formations or protect ground water near the surface.

connate water Salt water not displaced from the pore spaces which coats the surfaces of the larger

openings and fills the smaller pores.

continental environment Sediments deposited by the wind.

contour map A map on which the elevation in height (or depth) is visually indicated.

convection Transfer of heat by fluid motion such as through the atmosphere.

core Literally a cylinder-shaped plug lifted or cut out of the Earth at a predetermined depth.

core drilling Using a special bit for the purpose of cutting a core.

core sampling Taking out a core for geological examination of the composition of the strata at a particular depth.

coring bit A hollow bit designed to make a circular (cylindrical) cut for a core sample.

correlation markers Indicators used on a map to cross reference one particular feature to another.

corrosion A complex chemical, physical, or electrochemical action destructive to metal.

corrosion inhibitor Additive or agent or treatment used to stop or inhibit corrosion.

cracking Process of breaking crude oil down into its various components.

cradling Lifting of the welded and wrapped pipe into the trench.

Cross Process Cracking process of the 1915–1920 era developed by Gasoline Products Co.

cross-sectional map A vertical slice map illustrating features above and below ground.

crown block Pulley at the top of the derrick that raises and lowers the drillstring. The traveling block and hook are attached to it by lines.

crude Oil, unprocessed, just as it comes from the formation.

crust Outer covering of the Earth.

cut A particular hydrocarbon fraction.

cyclic Hydrocarbons with their carbon atoms arranged in a ring circle.

cycling plant A unit that processes natural gas from the field, strips out the gas liquids, and returns the dry gas to the producing reservoir.

D

daily drilling report Completed every morning by the toolpusher from the records kept by the drillers of the activities of the three previous tours.

darcies (d) Unit of measurement of permeability. Named after its originator, Henry D'Arcy.

day-work basis Contractual arrangement of payment for drilling where the drilling contractor is paid by the day rather than by the foot.

debutanizer Equipment for separating butane from a mixture of hydrocarbons.

deltaic plain Depositional deposit area of sedimentation beneath the alluvial plain but above the normal marine.

decompression chamber A sealed room in which the atmospheric pressure may be varied. Used to isolate divers and gradually adjust their bodies to differences in pressure to prevent or treat the "bends."

dehydrate The removal of water from natural gas by means of a substance and equipment.

dehydration plant A plant which contains vessels, equipment, and apparatus designed to effect dehydration.

demulsifier/demulsifying Chemical/chemical action used to break down crude oil/water emulsions by reducing surface tension of the oil film surrounding water droplets.

Department of Energy U.S. Federal umbrella agency charged with overseeing all aspects of energy supply and demand.

depletion drive A situation where the oil does not come in contact with water-bearing permeable sands, and thus must depend on either solution-gas or gas cap drive as a lifting mechanism.

derrickman Member of the drilling crew who works on the tubing board and handles the pipe joints.

desiccant A drying or dehydrating medium, either solid or liquid.

detritus Fragments of minerals, rocks, and shells moved into place by erosion.

Devonian Geologic period from about 405,000,000 B.C. to 345,000,000 B.C.

dew point The point under pressure at which only a small drop of liquid remains when gas breaks out of solution.

dip log/logging Record of formation dip vs. depth.

direct detection Method of reading the possible underground location of hydrocarbons from the white areas on seismographic record strips.

directional drilling Non-vertical well drilling.

disconformity Situation where the layers above and below the unconformity are parallel.

distillate Liquid hydrocarbons, usually light colored and of high API gravity (above 60°), recovered from wet gas; condensate.

distillation The refining process of separating crude oil components by heating and subsequently condensing the fractions by cooling.

distributing lines Pipelines running from the product line to the various markets.

diverter A bypass system used in drilling at sea that vents natural gas away from the drillship or rig in the event of a blowout.

DOE *See* Department of Energy.

dolomite A sedimentary rock. Possibly formed from limestone through the replacement of some of the calcium by magnesium.

dolomitization Ground-water action causing limestone to change to dolomite. As it changes, it shrinks, causing larger openings in the formation.

dome An upthrust in the Earth's surface caused by the forcing upward of salt or serpentine rock by pressure from below.

dope gang Members of a pipeline crew assigned to applying a protective coating.

DOT Department of Transportation.

double two joints of pipe fitted together.

drain sample Sample obtained at the discharge valve.

draw works The hoisting equipment of a drilling rig.

dredge Vessel designed to scoop or pump out a trench or channel in the bottom of a body of water.

driller The man in charge of the drilling crew on each tour.

drilling mud A fluid consisting of water or oil, clays, chemicals, and weighting materials used to lubricate the bit and flush cuttings out of the hole.

drilling program The planning process for assembling all the personnel, equipment, and supplies for drilling and completing a well.

drilling template Mechanical device placed on the ocean floor with orifices through which the drillstring passes and to which the marine riser is attached.

drillstem testing Obtaining fluid samples from a formation using a tool attached to the drillstem.

drillship Large vessel used for offshore drilling operations.

drip Wellhead device for tapping off natural gasoline.

drive The energy force present in a reservoir that causes the oil to rise toward the surface.

Dubbs process The clean-circulation process of cracking developed by Carbon Petroleum Dubbs that greatly reduced the amount of coke deposits in the refinery vessels.

dry hole A duster; a well that fails to hit oil or gas in commercial volumes.

dry-hole money Money paid to an operator by the lease owner, or the owner of a lease near a well site, if the well fails to strike pay sands. Even though no oil is found, valuable information about the underground structures is gained.

DWT Dead-weight tons.

Dynamic positioning Means of keeping a drillship positioned exactly above the undersea drillsite by transmitting position signals from the ocean floor to the ship'Gs computers and thence to the thrusters.

E

edge water Water around the edges of a reservoir that presses inward.

Ekofisk Phillips Petroleum's North Sea drilling and production complex.

electric log General term for a wireline log that measures and records the electrical properties of rocks and the fluids in the pores in a formation.

electrodrills Rotary drills powered by electricity.

element A pure substance that contains just one kind of atom.

electrolysis The process in which an external voltage is used to force a current through a cell and therefore cause chemical changes.

eminent domain Legal concept which gives government, and those so authorized by government, superceding access to land or property.

enhanced oil recovery Usually refers to tertiary recovery methods that alter oil properties in the reservoir in order to improve recovery.

environmental assessment (EA) An environmental examination of projects to be undertaken, usually on federal, or federally administered Indian land, to determine if the proposed project would cause a significant impact on the environment.

environmental impact statement (EIS) A highly detailed and exhaustive study, required by the National Environmental Policy Act (NEPA) and other federal laws and regulations, that must be completed of certain projects such as refinery, power plant or pipeline construction before permits can be issued.

Environmental Protection Agency U.S. federal agency charged with protecting the environment—air, water, etc.

environments Deposits of sedimentary rock. *See* Continental, transitional, and marine.

erosion The process of wearing away by water, ice, wind, or wave action.

escape capsule Floating watertight escape vessel used at offshore rigs for emergency evacuation.

ethane A gaseous hydrocarbon, CH_3CH_3, occurring in natural gas and also in small amounts in coal gas.

ethanol Also known as ethyl alcohol. *See* alcohol.

explosive fracturing The use of explosive charges to shatter a formation in order to increase fluid flow through the formation. May be fired through the sidewalls of a well.

F

fail safe Equipment constructed to be automatically activated to stabilize or secure the safety of the operation in the event of the failure or malfunction of any part of the system.

fault A fracture in the Earth where the rock on one side moves.

Federal Energy Regulatory Commission (FERC) Government agency that regulates energy in interstate commerce.

feedstock Raw material—crude oil or other hydrocarbon liquids or gases—to be processed into various products in a refinery or petrochemical plant.

filler bead Successive hot passes made at pipeline joints to continue building up the weld.

firing line The part of the pipeline welding crew that makes the finishing welds.

flammable Material which can be easily ignited.

flares Devices that burn off excess natural gas at a well or production site.

flash point The temperature at which a given substance will ignite.

flow Movement of petroleum through the reservoir.

flow test Determination of productivity of a well by measuring total pressure drop and pressure drop per unit of formation section open to a well during flow at a given production rate.

folds Buckling of the Earth's strata caused by movement. See upfolds.

fourble Four joints of pipe fastened together.

fraction Each of the separate components of crude oil, or a product of refining or distilling.

fractionating columns Tall metal column used in processing liquid petroleum into its various components.

fracturing Artificially opening up a formation to increase permeability and the flow of oil to the bottom of the well.

Frasch process Process developed by Herman Frasch using cupric acid to treat sulphur-bearing crude oil.

free gas Gas occurring in a reservoir, but separately from the oil.

fuel oil Any liquid or liquefiable petroleum product burned for the generation of heat in a furnace; or for the generation of power in an engine, exclusive of oils with a flash point below 100°F.

G

gamma-ray log A nuclear log that measures natural radioactivity to determine lithography.

gas Natural gas.

gas cap Gas trapped above the oil in a reservoir.

gas cap drive If a gas cap located above the oil is tapped, the gas continues to expand and force the oil downward to the bottom of the well and then back up the bore.

gas drilling Drilling process using natural gas as the circulating medium; similar to air drilling.

gas lift Inducing gas into the reservoir to force the oil out.

gas pipelines Pipelines designed to transport gases such as natural gas or CO_2.

gatevalve A valve made with a wedge-shaped disc or tongue that is moved from the open to the closed position by the action of a threaded valve stem.

gathering lines Pipelines from lease tank batteries to the crude trunk lines running to the refinery. Natural gas production fields may also be served by gas gathering lines.

gauge ticket Written record kept by a gauger or pumper indicating the amount and quality of production.

gauger Person who measures the amount of oil entering a pipeline.

geophones Microphones plugged into the Earth's surface and used to detect seismic waves (*see* also "jugs").

glycol dehydrator A unit for removing minute particles of water from natural gas using glycol as the medium.

go-devil Cleaning device sent through a pipeline (*see* "pig").

G.P.A. Gas Processors Association.

graben Valleys between high peaks of igneous rocks that become filled with sediment.

gravity meter A device that indicates the density of rock formations, measures the gravitational pull of buried rocks, and provides information about their depth and nature.

gun barrel tank A settling tank placed between the pumping unit and other tanks, normally fitted with a connection at the top to separate the gas. It is usually smaller in diameter and taller than the other tanks in the battery.

H

half-cell A cell, containing a single electrode, in which only an oxidation or only a reduction reaction occurs.

heat content The heat energy stored in a substance during its formation.

heat of formation The heat released or consumed when one mole of a compound is formed from the elements.

heat of solution The heat released or absorbed when a solute dissolves.

heater A refinery furnace. A unit used to heat the stream from gas and condensate wells to prevent the formation of hydrates.

heat exchanger Metallic device used to transfer heat from its source to the material or space to be heated.

heavy crude Thick, sticky crude oil represented by very low API gravity numbers.

holiday Gap left in the protective coating of a pipeline.

holiday detector Electronic device used to detect gaps in pipeline coatings (*see* "jeep").

Holmes-Manley Process Texas Co. cracking process of 1915–1920.

horsehead End of a walking beam to which the polished rod is attached; resembles the shape of a horse's head.

horst High peaks of igneous rock.

hot pass Successive welds made over the stringer bead to fill up the groove between the joints in the pipeline.

Houdry Process Method of catalytic cracking developed by Eugene P. Houdry that revolutionized the refining industry.

hydraulic fracturing Forcing a formation open by pumping in liquid under high pressure.

hydraulic pumping Using crude oil from the reservoir pumped back into the well under pressure to force more crude to the surface.

hydrate A solid material resulting from combining a natural gas with water.

hydrocarbons Petroleum—a mixture of compounds of which the principal chemicals are carbon and hydrogen, crude oil, condensate or natural gas.

hydrocracking Method of cracking using hydrogen and a catalyst.

hydrogen A natural element; one of the essential components of petroleum.

hydrogen sulfide (H_2S) An odorous and noxious compound of hydrogen and sulfur found in sour gas.

hydrolysis A chemical reaction in which a chemical bond is split by the action of water.

hydrophones Waterproof microphones used to detect seismic echoes at sea.

hydrostatic head A difference in height as measured between two points in a body of liquid.

I

IBP or i.b.p. Initial boiling point

ideal gas The volume of an ideal gas is directly proportional to its absolute temperature and inversely proportional to its absolute pressure.

igneous rock Rock formed as molten magma cools.

independent A person or company engaged in one or more phases of the petroleum industry that is not a part of one of the larger companies.

Indian Territory Now incorporated into the eastern part of the state of Oklahoma, it was an area set aside as a home for Indians driven from their homelands in other parts of the United States.

induction-electrical log Can indirectly measure porosity and reveal a well's potential as a producer.

industrial gas Gas purchased for resale to industrial users.

inerts Elements or compounds which do not react in the combustion process. Nitrogen, helium, carbon dioxide, etc.

inland waterways The rivers, canals, and intercoastal waterways maintained by the U.S. Army Corps of Engineers for domestic water transportation.

inorganic theory A theory of the creation of petroleum that states the elements carbon and hydrogen came together under great pressure under the Earth's surface.

integrated company A large company engaged in many phases of the petroleum industry.

interfacial tension Surface tension occurring at the interface of two liquids.

interstitial Water found in the interstices or pore openings of rock.

isomers Two or more substances having the same molecular formula but different structural formulas.

isopach Maps drawn to illustrate the variations in thickness between the correlation markers. This is usually done by shading or coloring.

isotopes Atoms whose nuclei have the same atomic number (same number of protons) but different mass numbers (different numbers of neutrons).

J

jeep Electronic device used to detect gaps in the protective coatings on pipelines.

jet fuel A specially refined grade of kerosine used in jet propulsion engines.

jet sleds Small, highly maneuverable underwater vehicles powered by jets of water; used by divers.

jet tray Contracting device used in a fractionator consisting of a circular plate with half-moon-shaped openings.

joint A single section of pipe.

K

kelly a hollow, 40-ft joint of pipe that has four or more sides with threaded connections on each end to permit it to be attached to the swivel and the drillpipe. It is used to transmit torque from the turntable to the drillstring. It can move vertically as needed during drilling.

kick Pressure surge in the well.

kinetic energy Energy of motion. Defined as one-half the product of the mass multiplied by the square of the velocity of a particle.

L

LACT (lease automatic custody transfer) A tank battery that is fully automated as to recording and shipping into a gathering pipeline.

lateral fault A fault that has lateral movement.

laterolog Logging instrument in which electric current is forced to flow radially through the formation.

lease A legal document giving one party rights to drill for and produce oil on real estate owned by another. The property described in the document.

lens trap Reservoir brought about by abrupt changes in the amount of connected pore spaces. Porous oil-bearing rock is then confined within pockets of nonporous rock.

lifts Various methods of bringing oil to the surface, including pumping units. The pump itself is usually located downhole. The mechanism that operates the pump is often surface mounted and is considered a pumping unit or lift.

light crude Thinner, freely flowing crude of light specific gravity.

lighter ends The more volatile components or fractions of petroleum. Those with a high API gravity.

liquefied petroleum gas (LPG) Butane, propane, and other light ends separated from natural gasoline or crude oil by fractionation or other processes.

limestone Sedimentary rock composed largely of magnesium carbonate and quartz.

line list Pipelining instructions that include the names of the owners of the property, the length of the line to be built on the property, and any special restrictions or instructions.

linewalker A pipeline inspector who walks the length of the pipeline looking for leaks, potentially hazardous situations, or evidence of theft.

lithology Study, description, and classification of rocks.

lithosphere The solid portion of the Earth.

LNG Liquefied natural gas.

logging The lowering of various types of measuring instruments into a well and gathering and recording data on porosity, permeability, types of fluids, fluid content, and lithography.

LPG Liquefied petroleum gas.

lost circulation Loss of substantial quantities of drilling mud into a formation that has been pierced by a drilling bit.

Lufkin Mark II Pumping unit that has the crank mounted on the front instead of on the rear of the beam. It uses an upward thrust instead of a downward one.

M

magma Rock in its molten state.

magnetometer Device which detects minute fluctations in the Earth's magnetic field and show the presence of seimentary rock.

mandrel Device used to bend pipe without deforming it.

manometer A device for measuring pressure. It often consists of a column of mercury or other fluid.

mantle That portion of the Earth about 1,800 mi thick which lies between the core and the crust.

marine environment Sediments deposited in the ocean.

marine riser Assembly placed between the ocean floor and a barge or ship on the surface to help maintain the drilling position of the vessel and protect the drillstring from movement caused by wind and wave action.

McAfee Process Gulf Refining Company process of 1915 that used anhydrous aluminum chloride as a catalyst.

MCF One thousand cubic feet (the standard measuring unit of gas).

Mesozoic Geologic era from about 230,000,000 B.C. to 63,000,000 B.C.

metamorphic rock Created from sedimentary rock subjected to great heat and pressure.

methane The simplest saturated hydrocarbon; a colorless flammable gas.

methanol Methyl alcohol; a colorless flammable liquid derived from methane (natural gas).

microlaterlog A resistivity logging instrument with one center electrode and three circular ring electrodes around the center electrode.

microlog A resistivity logging instrument with electrodes mounted at short spacing in an insulating pad.

microwaves Electromagnetic radiation and a wavelength between that of short-wave radio and infrared radiation.

middle distillates Hydrocarbons in the middle range of refinery distillation; kerosine, diesel, etc.

middle sample Sample drawn from the middle of a tank.

migration Movement of hydrocarbons in the ground; primary migration is from source-bed or rock to permeable rock.

millidarcies (md) Since average permeability is usually less than one darcy, the measurement is expressed in millidarcies (md) or one thousandth of a darcy.

mineral right The ownership of minerals within a property tract.

MMCF One million cubic feet.

mole The mass of a substance in pounds equal to its molecular weight is a pound-mole or mole.

molecule A group of atoms bound together in a particular way.

moonpool Opening in the center of a drillship through which drilling operations are carried out.

mouse hole Shallow hole drilled to just one side of a well in progress and used to store the next joint of pipe to be used.

mud See drilling mud.

mud logger Person who analyzes the cuttings brought up with the drilling mud returning up the hole.

mud program Planning for the supply of and use of drilling fluids in the drilling process.

multiple completion More than one zone completed from the same hole.

N

naphtha One of the petroleum distillates. Used to make cleaning fluid and other products.

natural gas Gaseous form of petroleum occurring underground.

natural gasoline Casinghead gasoline—a natural condensate of natural gas.

Natural Gas Policy Act of 1978 Legislation that set a new price structure and decontrol schedule for natural gas.

neutron log A nuclear log that can reveal porosity and saturation.

nitrogen An inert gas; a natural element (N_2).

natural fault A fault that has vertical movement.

O

OAPEC Organization of Arab Petroleum Exporting Countries.

Occupational Health and Safety Act of 1971 (OHSA) Comprehensive U.S. federal law covering working conditions and health and safety of workers in industry and business.

octane Term used to indicate the antiknock quality of gasoline. A fuel with a high octane rating has better antiknock qualities than one with a low rating.

oil-based drilling fluid/mud A drilling fluid formulated with an oil base.

oil pipelines Pipeline designed to carry liquids such as crude oil.

oil string The casing in a well that runs from the surface down to the zone of production.

oil treater Device between the wellhead and lease storage tank that separates the natural gas and BS&W from the oil.

olefins One of a class of unsaturated hydrocarbons, such as ethylene, that has many chemical potentials.

OPEC Organization of Petroleum Exporting Countries —a group of South American, Middle Eastern, Asian, and African nations with large petroleum reserves that joined together to control the production and pricing of their resources.

orifice meter A measuring instrument that records the flow rate of gas, enabling the volume of gas delivered or produced to be computed.

organic theory The current prevailing theory of the origin of petroleum which holds that it was formed from plant and animal remains under great pressure beneath the Earth's surface millions of years ago.

overburden Earth material overlying a mineral or other useful deposit.

override An additional payment made in excess of the usual royalty.

oxidation 1. In general, the loss of electrons. 2. Reaction of a substance with oxygen or another oxidizing agent.

oxygen A natural element (O_2).

P

packer Expanding plug used to

seal off tubing or casing sections when cementing or acidizing or when isolating a formation section.

Paleozoic Geologic era from about 600,000,000 B.C. to 230,000,000 B.C.

paraffin The paraffin series is a group of saturated aliphatic hydrocarbons. The term *paraffin* is also used to describe a solid, waxy hydrocarbon.

pay sands The zone of production—where the oil is found in commercially feasible amounts.

PDC bits Synthetic polycrystalline diamond compacts (PDC) designed into controlled geometric patterns over the drilling surfaces of rotary drilling bits.

perforating Literally punching holes in the casing so that the oil and gas can flow into the well from the formation.

permafrost Arctic subsoil that remains frozen all year.

permeability The factor of a reservoir that determines how hard or how easy it is for oil to flow through the formation.

petrochemicals Chemicals derived from petroleum.

photosynthesis The process by which plants synthesize glucose, starches, and cellulose from CO_2 and H_2O, using sunlight as the source of energy.

pig Device sent through a pipeline to clean it out. *See* go-devil.

pipe gang Part of the pipeline crew who line up, prepare pipe, and make initial welds.

pipeline gauger Pipeline company employee who measures the amount and quality of oil entering the gathering lines from the lease tanks.

pipe shoe Device for bending, without deforming, small-diameter pipeline pipe.

pipeyard Working area for cleaning and coating pipeline pipe. Also where two or more joints are welded together before being hauled to the pipeline site.

platform The deck or working surface of a rig. An offshore rig that is anchored to the bottom.

platforming Using a catalytic reforming unit to convert low-quality fractions to those of higher octane.

plugged back To plug off a well drilled to a lower level in order to produce from a formation nearer the surface.

plug trap A salt dome or plug of serpentine rock that has been forced upward by the petroleum accumulated under it.

polymer Synthetic compound having many repeated linked units.

pores The void spaces between the rocks in a reservoir.

porosity The capacity of rock to hold liquid in the pores.

potential energy Energy stored in a system (a molecule, for instance) to be released as kinetic energy, electrical energy, heat, light, etc.

pour point The temperature at which a liquid congeals and ceases to flow.

pressure gradient The difference in pressure at two given points.

primary recovery The first period of obtaining oil from a field either by free flow or by pumping.

primary term The period, expressed in time, a lease is written to cover.

prodeltaic plain Geologic environment lying between the normal marine and the lower deltaic plain.

product Term used to cover any of the products of the petroleum industry.

product pipeline Lines that carry finished products from the refinery.

propane (C_3H_8) One of the alkane of paraffin homologenes series of gases.

proppant/propping agent Granular substance used to keep earth fractures open after the fracturing fluid withdraws.

public utility A company that sells service to the public such as water, gas, electricity, etc.

pump Mechanical device for lifting oil to the surface.

pumper Person in charge of production and records for a producing well or well field.

pumping Lifting liquids as from a reservoir to the surface by artificial means.

pumping off Recovering the oil that has flowed to the bottom of the well through the formation, and then stopping for a rest period. If pumping were maintained continuously, the flow could be stretched so thin the oil would become isolated in pockets and much otherwise recoverable oil wasted.

pumping stations Units placed at intervals along a pipeline to maintain pressure and flow.

Q

quantum number A number used to designate a particular energy level of an atom; it is also used in algebraic expressions to calculate the magnitude of that energy.

R

radioactivity The spontaneous emission of radiation.

radiation Transfer of heat by emission of waves from a fixed point or surface.

radioisotope A particle that has natural or induced radioactivity and which can be introduced into a substance or system as a measuring signal.

rat hole Shallow hole drilled next to a well in progress where the kelly is stored during a trip.

rate of penetration The speed with which the drillstring moves downward.

reboiler Device used to put additional heat into the bottoms liquid to make it boil.

recovery Obtaining petroleum from a reservoir and bringing it to the surface.

reforming Rearranging the carbon and hydrogen molecules by use of catalysts and heat.

reforming processes The use of heat and catalysts to effect the rearrangement of certain of the hydrocarbon molecules without altering their composition appreciably; the conversion of low-octane gasoline fractions into high octane stocks suitable for blending into finished gasoline; also the conversion of naphthas to obtain more volatile product of higher octane number.

reid vapor pressure A measure of the vapor pressure in pounds pressure of a sample of gasoline at 100°F.

remote sensing Using infrared photography and television, often from an aircraft or satellite, to detect mineral deposits, salt-water intrusion, or faults.

reservoir A rock formation or trap holding an accumulation of petroleum.

reservoir rock Sedimentary rock formations that contain quantities of petroleum.

reservoir fluids Crude oil, salt water, and natural gas.

residuals (resid) Matter left over in boilers and refinery vessels.

residue gas Gas remaining after natural gas is processed and liquids are removed.

residuum The sticky, black mass left in the bottom of a refining vessel.

resistivity Property of, capacity for, or tendency toward resistance of passage of an electrical current, varying according to formation composition.

reverse fault A fault that also moves vertically in the opposite direction of a normal fault.

rippers Devices mounted on crawler tractors to remove rocks from the pipeline trench.

rock cycle The repetitious process of magma cooling into igneous rock and being eroded into particles, which become sedimentary rock and then may again be melted into magma.

rod The sucker rod of a pump. A unit of measurement consisting of 16 ½ ft.

rod pumping Using solid metal rods to lift oil to the surface.

rotary drilling Using a turning motion to bore into the Earth's surface.

rotational fault Of particular interest to the petroleum geologist, a rotational fault moves with a twisting movement.

roughneck Drilling crew member who assists the driller.

royalty Fee paid to the owner of the minerals based on the production from the lease.

run Transferring or delivering from the lease tank battery to the pipeline or tank truck.

running sample Taken by lowering an unstoppered beaker from the top of the oil to the level at the bottom of the outlet and returning it at a uniform rate of speed so that it is about three quarters full when returned.

run ticket Written record of the amount and quality of the run.

Rural Electrification Administration (REA) Government agency formed during the 1930s to bring electricity to rural areas not served by commercial power companies.

rusting A particular type of corrosion in which iron is converted to hydrated Fe_2O_3 by the combined action of atmospheric oxygen and water.

S

salt domes Salt plug forced upward through strata because of differences in density.

samples Small amounts of oil drawn from a tank to determine the API gravity and amount of BS&W present.

San Andreas Fault Famous fault line in California running parallel to the Pacific coastline.

sandstone Sedimentary rock composed of grains of sand cemented together by other materials.

saturation The actual amount of a fluid available in a given space.

scraper traps Mechanism on a pipeline for inserting or retrieving pigs or go-devils.

scratchers Devices used to clean drilling mud from the walls of the bore so that the cement will adhere better.

screen liner Perforated or wire mesh screen placed at the bottom of a well to keep larger particles from entering the bore.

scrubbing Purifying a gas by putting it through a water or chemical wash; removal of entrained water.

secondary recovery The next attempt at production after all the oil that can be removed by primary means such as pumping, has been extracted from the reservoir.

sediment Particulate matter carried along by water until it settles out to the bottom.

sedimentary deposition The laying down of a layer of sediments in a particular place.

sedimentary rock Rock created from particles of sediment compressed under great pressure.

sedimentation The building up of layers of sediment on the bottom of a body of water.

self-potential/spontaneous potential Amount of electrical voltage exhibited by a natural material.

separator Device placed between the wellhead and lease tank battery to separate crude oil from natural gas and water.

seismograph Extremely sensitive recording device capable of detecting earth tremors as used in oil exploration to record manmade shock waves.

semisubmersible Marine drilling rig that can either be anchored to the bottom or maintained at a given position between the bottom and the surface.

shake out Using a centrifuge to separate any oil that may be present in a sample of BS&W from a well test.

shale Rock composed of clay and fine-grain sediments.

shale shaker Mechanical device used to separate bits of shale and rock from the drilling fluid as it comes out of the well.

shell stills In use by 1870, these horizontal units permitted continuous thermal cracking operations.

sidebooms Crawler tractors with booms mounted on the sides which are used to lower pipe into trenches.

sidetrack well A well drilled out from the side of an existing well. Sometimes used to bypass a blockage or reduce pressure.

sidewall cock Valve placed on the side of a tank for the purpose of obtaining small samples.

sidewall sampler Device used to obtain a sample from the side of a tank.

sidewall tap Same as a sidewall cock.

Single-point mooring system (SPM) Offshore anchoring and unloading or loading point connected to the shore by an undersea pipeline. Used offshore from existing harbors which are too shallow for laden tankers.

slurry Thin, runny mixture of water and other substances such as clay.

solution A mixture, on the molecular level, of two or more substances.

solution gas Gas dissolved in solution with the oil in a reservoir.

solution-gas drive If the gas-oil solution is so great no bubbles can form, once the pressure is relieved bubbles to form. As they expand, their pressure drives the oil up the wellbore.

solvent 1. A liquid capable of dissolving another substance. 2. That compound of a solution (usually a liquid) which is present in the larger amount.

sour gas Natural gas containing substantial amounts of sulfur or sulfur compunds.

source bed/source rock Original site of the deposition of petroleum, not always the site of present accumulation.

specific gravity The ratio between the weight of a unit volume of a substance compared with the weight of an equal volume of another substance, usually water, taken as a standard.

spot sample Sample taken at a particular level of the tank.

spread All of the manpower and equipment necessary to construct a pipeline.

spread superintendent Person in charge of men and equipment for a pipeline construction job.

spudding in To begin a new well.

squeeze To seal off with cement a section of a well where there is a leak that allows water either into or out of the well.

squeeze cementing Process used to fill any large unwanted openings in the sides of the borehole.

stabilizer A bushing used on the drillstring to help maintain drilling as close to vertical as possible.

standard pumping rig Conventional pumping unit using a walking beam to raise and lower the sucker rods.

steam injection The introduction of steam into the field to obtain secondary or tertiary recovery.

steam stripping Injection of superheated steam into bottom liquids to make them boil

Stick welding Arc welding using a single electrode or rod.

stock tank Storage tank for oil on the lease.

straight hole A hole drilled with as little deviation from the vertical as necessary.

strapping Measuring the dimensions of a new tank for the first time in order to determine its exact capacity.

strata A layer, as of rock.

stringer bead The first weld of a pipeline joint.

stripper/stripper well A well that produces a limited amount of oil, e.g., no more than 10 B/D.

submersible drilling-barge A barge-like vessel capable of drilling in deeper water than the smaller and simpler barge platform. The submersible drilling-barge has a drilling deck separate from the barge element proper. When floated into position offshore in water as deep as 100 ft, the barge hull is flooded and as it slowly sinks, the drilling platform is simultaneously raised on jacking-legs at each corner of the barge, keeping the drilling platform well above the water's surface.

Sub-sea completion system A self-contained unit resembling a bathysphere used to carry two men to the ocean floor to install, repair, or adjust wellhead connections. One type of modular unit is lowered from a tender and fastened to a special steel wellhead cellar. The men work in a dry, normal atmosphere. The underwater wellhead system was developed by Lockheed Petroleum Services Ltd. in cooperation with Shell Oil Company.

Substructure A sturdy platform upon which the derrick is erected. Substructures are from 10 to 30 ft high and provide space under the derrick floor for the blowout preventer assembly.

sucker rod That portion of a beam pumping unit that actually lifts the oil. The sucker rods are connected to the pump down inside the well tubing and to the beam on the surface.

surface lift Any mechanism at the surface such as a pumping unit.

surface equipment Lease equipment used to produce hydrocarbons and clean them in the field.

surface rights Conferred to owners of a tract of real property with or without concurrent ownership of the mineral rights for that tract.

surf line The point along a shore where the depth decreases enough to cause the waves to break.

swabbing Cleaning out a well with a special tool connected to a wireline.

sweet gas Natural gas containing little sulfur or sulfur compounds.

swivel A rotating attachment point on the bottom of the traveling block.

syncline A rock formation folded downward.

T

tank table A chart showing the capacity of a given tank at a given level.

tank battery A group of storage tanks located on a lease.

tank strapping Measuring tanks and computing the volume that can be contained in each interval of a tank.

tap sample Sample withdrawn through a sidewall tap or cock.

tariffs Rules and regulations, including rates, of a common carrier or public utility.

temperature log/logging A record of wellbore temperature vs. depth.

tertiary recovery The third attempt at production after all the oil possible has been recovered by primary and secondary means.

Texas Towers World War II offshore radar towers whose design was copied from early offshore oil platforms.

thermal cracking The use of heat to separate petroleum into its various components.

thermal recovery Tertiary recovery methods in which the oil in the reservoir is affected by thermal treatments, including in situ combustion, steam injection, and other methods of reservoir heating.

thief Device used to obtain a sample of crude from a lease tank.

thiefing The act of using a thief to obtain a sample.

thief hatch Opening in the top of a tank through which the thief can be lowered and recovered.

thribble Three joints of pipe fastened together.

thruster A motor and propeller mounted below the waterline of a vessel which can be used to furnish lateral motion.

thrust fault A horizontally moving fault.

thumper truck Large truck equipped to mechanically produce shock waves which were formerly generated by high explosives.

tie-in Connections between pipelines.

tie-in crew Crew that follows pipeline construction crew to connect the new line to existing lines.

toluene Petroleum derivative of many uses including solvents and explosives.

tool pusher Person in overall charge at the drillsite.

torsion balance Device used to measure the gravitational pull of rocks beneath the Earth's surface.

torque Turning or twisting force, as produced by a rotating shaft.

tour (pronounced "tower") Shift of duty at a wellsite.

towboat Powerful vessel used to push a string of barges on an inland waterway.

tow A barge or string of barges.

transitional environment Sediments deposited in a delta at the mouth of a river or between two such deltas.

trap A geologic structure that halts movement of a petroleum accumulation.

traveling block The largest pulley on the drill rig. It has the hook attached to its bottom and it moves up and down on lines running to the crown block.

treater A system for treating gases or liquids usually for the purpose of removing impurities.

trip The process of pulling the string of tools out of the hole (tripping out), or reentering the hole (tripping in).

trunk line A main line; fed by gathering lines.

Tube and Tank Process Standard of New Jersey refining process of about 1920.

tundra Arctic region lying between the permanent ice cap and the more southern forests. The subsoil usually remains frozen and plant life in the area is limited.

turbodrill A rotary drilling method in which a fluid turbine is usually placed in the drillstring just above the bit. The mud pressure turns the turbine. Since the drill string does not rotate, there is no kelly.

turnkey contract A well drilling contract which calls for the drilling contractor to complete the well and prepare it for production.

two-way spot sample Taken between the 10 and 15 ft levels from tanks in excess of 1,000 bbl capacity.

U

ultra large crude carrier (ULCC) The largest vessels afloat.

unconformity A cap of rock laid down across the cut-off surfaces of lower beds.

upfolds Common deformation of strata—the outer edges are compressed inward and the center rises.

upper sample Taken from the midpoint of the upper third of the tank.

upthrust fault A fault that moves vertically upward and may signal the presence of petroleum accumulations.

V

vacuum still   A vessel in which crude oil or other feedstock is distilled at less than atmospheric pressure.

valence   The capacity of an atom to form chemical bonds; also, the number of bonds an atom can form. For example, oxygen (O^2) has a valence of two.

vapor pressure   The pressure exerted by a vapor being held in equilibrium with its liquid state.

very large crude carrier (VLCC) A tanker larger than a conventional vessel, but smaller than an ULCC.

Vibroseis™   Mechanical means of producing shock waves for seismographic exploration without the use of explosives.

viscosity Used to describe the ability, or lack of ability, of a fluid to flow.

volatile The property of being easily evaporated.

volt Unit of measurement of electrical potential.

vug Large opening between the rocks in a reservoir.

W

wagon drills   Battery of pneumatic drills mounted on a cart or wagon. Used to break up rock when opening the pipeline trench.

water and sediment sample Sample taken and "shaken out" in a centrifuge so that it separates into its various components, the amounts of which can be read directly off a scale.

water-base drilling fluid/mud Drilling fluid formulated with a water base.

water drive   As water moves in to occupy the space left as petroleum is removed, its pressure forces the remaining oil toward the surface.

waterflooding The injection of water under pressure into a reservoir to drive out more oil. A process normally used in secondary recovery.

welding   The joining together of metal by using a filler metal at high temperature generated by electricity or a gas flame.

well jacket   A protective structure, topped with navigational warning devices, placed around a completed offshore well.

welding gang   *See* firing line.

wellhead   Casing attachment to the blowout preventer or the production christmas tree and bolted or welded to the conductor pipe or surface casing.

well log   Geological, formation attribute, and hydrocarbon potential data obtained from downhole using special tools and techniques.

well spacing   The geographic spacing of wells according to regulatory requirements and/or reservoir engineering recommendations.

weight indicator Instrument which constantly displays the total weight of the string in the hole as the well is being drilled.

wet gas Natural gas from a well that contains some of the heavier hydrocarbon molecules which, under surface conditions, form a liquid.

whale oil Fine, high-quality oil rendered from the blubber of whales.

wildcatter An operator who drills the first well in unknown or unproven territory.

wireline A rope or cable made of steel wire.

workover Cleaning, repairing servicing, reopening, or perhaps drilling deeper or plugging back a well to secure continued or additional production.

X

xylene An aromatic hydrocarbon, one of a group of hydrocarbons including benzene and toluene, from which many synthetic chemicals are derived.

Z

zeolite Mineral sometimes used as a catalyst in cracking operations.

zones of lost circulations Crevices, caverns, or very porous formations in which the drilling mud is lost and does not return.

Bibliography

Adams, Neal.*Workover Well Control.* Tulsa: PennWell Books, 1981.

"Age Old Technology Finds New Application." On Target, Vol. 3, No 1., Spring 1990.

Anderson, Kenneth E. *All About Oil.* Stillwater: Anderson Petroleum Services, Inc., 1981.

Baldwin, David L. *All You Ever Wanted to Know About Leasing Indian Land But Were Afraid to Find Out!* Denver: Baldwin & Associates, 1984.

Berger, Bill D. *Facts About Oil.* Stillwater: Oklahoma State University, 1975.

_____, ed. *Principles of Drilling.* Stillwater: Technology Extension, Oklahoma State University, 1978.

_____ and Anderson, Kenneth E. *Gas Handling and Field Processing.* Tulsa: PennWell Books, 1980.

_____, translated by Gustavo Pena and the editors of *Petróleo Internacional. Petróleo Moderno Introducción básica a la Industria Petrolera.* Tulsa: PennWell Books, 1980.

_____. *Refinery Operations.* Tulsa: PennWell Books, 1979.

"The Biocycle Guide to Composting Municipal Wastes," *Biocycle: The Journal of Waste Recycling,1989.* The J.G. Press, Inc.

Burdick, Donald L. and Leffler, William L. *Petrochemicals for the Nontechnical Person.* Tulsa: PennWell Books, 1983.

The Canadian Petroleum Association. *A Brief History of the Petroleum Industry in Canada.* Calgary: November 1980.

_____. *The Geophysical Story: Seismic Exploration for Oil and Gas.* Calgary: 1979.

_____. *Focus on Energy.* Calgary: 1979.

_____. *Focus on Energy: Resource Ownership.* Calgary: 1980.

CBI Industries Inc. *CBI Bulletin 3200, Horton Floating Roof Tanks.* Oakbrook: 1981.

_____. *1988 Annual Report.* Oakbrook: 1989.

Crow, Patrick. "U.S. Industry Fighting Clean Air Battles," *Oil & Gas Journal,* July 23, 1990.

"Drilling, Production Meet Environmental Challenge," *Oil & Gas Journal,* March 5, 1990.

Feehery, John, "Spindletop...Birthplace of a New Era," *Amoco Torch,* Vol. 4, No. 6, November/December, 1976.

Gold, Thomas. *Power from the Earth: Deep Earth Gas.* J. M. Dent & Sons Ltd. London: 1987.

Goldstein, Nora. "Composting Activity: Solid Waste," *Biocycle: The Journal of Waste Recycling,*1989

Gray, Forest. *Petroleum Production for the Nontechnical Person.* Tulsa: PennWell Books, 1986.

Houdry, P. and Lang, D.G. "3-D Land Seismic Acquisitions—A Case History," *Oil & Gas Journal,* October 3, 1983.

Intergraph Corporation. *Geological Applications*. Huntsville: 1988.

Johnson, David E. and Pile, Kathryne E. *Well Logging for the Nontechnical Person*. Tulsa: PennWell Books, 1988.

Kennedy, John L. *Fundamentals of Drilling*. Tulsa: PennWell Books, 1983.

_____. *Oil and Gas Pipelining Fundamentals*. Tulsa: PennWell Books, 1984.

McNair, Will L. "Rig Automation," *Oil & Gas Journal*, April 9 and 16, 1990.

McNamara, Anne. *Our Petroleum Challenge: The New Era*. Calgary: Petroleum Resources Communication Foundation, 1986.

Moore, Preston L. *Drilling Practices Manual*, 2nd ed. Tulsa: PennWell Books, 1986.

Moritis, Guntis. "CO_2 and HC Injection Lead EOR Production Increase," *Oil & Gas Journal*, April 23, 1990.

_____. "Horizontal Drilling Scores More Successes," *Oil & Gas Journal*, February 26, 1990.

OPSEIS Publications. *OPSEIS System Overview*. Bartlesville: 1989.

Parcher, Loris A. and Roush, Clint E. *Legal Land Descriptions in Oklahoma*. Extension Fact Sheet No. 9407, Stillwater: Oklahoma State University Cooperative Extension Service, n.d.

Petzet, G. Alan. "Horizontal Drilling Fanning Out as Technology Advances and Flow Rates Jump," *Oil & Gas Journal*, August 23, 1990.

Railsback, Lynn. "High Tech Brings Oil Risk Down," *PhilNews*, July, 1988.

"Refiners, Petrochem Plants Focus on New Waste Challenges," *Oil & Gas Journal*, March 5, 1990.

Silvaraman, S. "Field Tests Prove Radar Tank Gauge Accuracy," *Oil & Gas Journal*, April 23, 1990.

"Tough Air-Quality Goals Spur Quest for Transportation Fuel Changes," *Oil & Gas Journal*, June 18, 1990.

The University of Texas, Petroleum Extension Service. *Crude Oil Tanks: Construction, Strapping, Gauging and Maintenance*. Austin: 1975.

_____. *Oil Pipeline Construction and Maintenance*. Austin: 1975.

_____. *A Primer of Oilwell Drilling*. Austin: 1976.

U.S. Department of the Interior. U.S. Geological Survey, *Annual Report 1976*, "Plate Tectonics and Man," by Warren Hamilton. Washington D.C.: 1980.

U.S. Department of Transportation. Research and Special Programs Administration. *Guidance Manual for Operators of Small Gas Systems*. Washington D.C.: 1985.

_____.Transportation Safety Institute. *Safety Requirements for Gas Pipeline Systems Conducted for Materials Transportation Bureau and State Agencies*, 2nd ed. Washington D.C.: January 1987.

U.S. Environmental Protection Agency. Intra-Agency Sludge Task Force. *Environmental Regulations and Technology: Use and Disposal of Municipal Wastewater Sludge*. Washington D.C.: September, 1984.

Vielvoye, Roger. "Rising Demand Sparks Burst of Tanker Orders," *Oil & Gas Journal*, July 30, 1990.

Weatherford, Jack McIver. *Indian Givers: How the Indians of North America Transformed the World*. New York: Crown Publishers, 1988.

Index

Crust, 20, 21, 22, 23, 58
Cryogene, 288
Cut, 312
Cuttings, 106, 153
Cyclic hydrocarbons, 369

D

Daily Drilling Report, 125
Daimler, Gottleib, 11, 16, 310
d'Arch, Henry, 38
Darcies, 38
Day Work Contract, 119
Dead Line, 103
Dead Sea, 1
Delay rentals, 79
Derrick, 99
Derrick man, 104, 124
Derrine, Irving, 52
Detergent, 376
Deviation, 127
Diaphrams, 401
Diolefins, 369
Dipmeter, 166
Direct detection, 56
Dissimilar Metals, 418
Dissimilar Soils, 421
Distillates, 335
Distillation, 318
Distillation Tower, 318
Dolmite, 25, 28, 36, 152
Dominion Land Survey (D.L.S.), 77
Double, 100
Down Hole Motors, 97
Dowsing Rods, 49
Drake, Col. Edwin, 8, 9, 49, 94, 223, 277, 279, 310
Draw works, 95, 97
Drift Indicators, 164
Driller, 53, 97, 112, 119, 123, 124
Drilling:
  Aerated, 115
  Air, 113

Directional, 105, 107, 135, 308, 385
Foam, 113, 115
Horizontal, 105, 126, 127, 128, 132, 134
Mist, 113, 115
Natural Gas, 113
Drilling barges, 139
Drilling contract, 104
Drilling controls, 380
Drilling lines, 94
Drilling Mud, 15, 76, 104, 106, 111, 112
Drilling pipe, 15, 16, 100, 102, 103, 105, 106, 115, 119, 122, 126
Drilling platform, 141
Drilling program, 91, 111, 117
Drilling ship, 141
Drilling template, 137
Drilling collar, 99, 102, 104, 105, 112
Drill string, 16, 95, 97, 99, 103, 105, 106, 138
Drip gas, 440
Drives:
  Bottom, 45
  Bottom water, 39
  Depletion, 41, 45
  Edge water , 39, 95
  Gas-cap, 41, 45, 95
  Gravity, 46
  Solution-gas, 41, 45, 46
  Water, 41
Dubbs, C. P., 313
Dubbs process, 313
Dusters, 50
DWT, 283
Dynamic positioning, 141

E

Eastman, John, 127
Eastman Whipstock, 127

Modern Petroleum

Gluckauf, 283
Glycerin, 374
Glycol, 374
Glycol Dehydration, 445
Gravity meters, 59
Great Britain, 6, 135, 283
Greece, 2
Greek, 24
Greenhouse Effect, 286, 466
Gulf of Mexico, 135, 145, 146, 287
Gulf Production Co., 52
Guide shoe, 200
Gusher, 7, 16, 17

H
Habendum Clause, 86
Halliburton, Earle, 192
Hamill, Al, 16
Hamill, Curt, 16
Hamilton Gas Co., 6
Hamilton, Ontario, 7
Hardeman Basin, 63
Haserman, William B., 52
Heaters, 444
Heavy Oil, 458
Helium, 27, 363
Heptadecane, 367, 369
Hewitt Field, 192
Hexane, 367
Higgins, Patillo, 12, 13, 14, 16
Hoisting equipment, 97
Holiday, 306
Holiday detector, 306
Horsehead, 223, 224
Horton Double-Deck Floating Roof, 261
Hot Spot, 416
Houdry, Eugene J., 313
Houdry process, 313
Huntington Beach, CA 135, 472
Hydrates, 440
Hydraulic Fracturing, 209

Hydrogen, 169
Hydrometer, 271
Hussein, Saddam, 284

I
Illinois, 11
India, 2
Indians, 10
Indian lands, 71
Indian Meridian, 68
Indian Territory, 11
Initial Boiling Point (IBP), 318
Input well, 238
Integrated Company, 340
International Association of Drilling
  Contractors (IADC), 107, 119
Interstitial, 39
Isobutane, 325
Isomerism, 369
Isomers, 367
Isopropyl alcohol, 374
Isopach maps, 57
Italy, 470

J
Jack-up Rigs, 139
Japan, 301, 350, 451
Jars, 95
Jeep, 306
Jet fuel, 316
Jobber, 340
Johnstone, Nellie #1, 11
Jolliet Field, 146
Jugs, 52
Jupiter, 19

K
Karcher, J. Clarence, 52
Kelly, 100, 101, 104, 112, 115, 126
Kentucky, 11, 68
Kerosene, 6, 9, 16, 283, 312
Kerr-McGee, 135
Kesselring, E. C. No. 1, 9

Methyl-tertiary-butyl-ether (MTBE), 335
Mexico, 3
Microamperes, 424
Micrologs, 157
Middle East, 49, 115, 236, 262, 282, 302, 451
Milliamps (Ma), 418
Millidarcies, 38
Millivolts, 160
Milne, David, 51
Mineral Deed, 75
Mineral Lease, 76, 79
Mineral Owner, 75, 76, 81, 117
Mintrop, Dr. L., 51
Miracle drug, 3
Mississippi River, 279
Missouri, 279
Molecules, 19, 356, 362, 363
Moon pool, 141
Morning report, 124
Most Efficient Recovery (Mer), 219, 220
Motor Valves, 410
Mouse hole, 124, 126
Mud (see Drilling Mud)
Municipal Solid Wastes (MSW), 470
Mutual bond, 423

N

Naphtha, 10, 326, 371
Napoleonic Code, 85
NASA, 58
Natural Gas:
　Dry, 435
　Sour, 435
　Sweet, 435
　Wet, 435
Natural Gas, Physical Properties, 436, 437
Navajo Nation, 97
Navigation Drilling System (NDS), 131

NEPA, 475
Neon, 363
Neptune, 19
Neutrons, 361, 363
Nevada, 115
New Jersey, 282
New Orleans, La., 281, 297
New York, 282
New York State Natural Gas Corp., 96
New Zealand, 96
Ninian Field, 142
North Africa, 282
North Dakota, 129
North Sea, 118, 127, 142, 145, 146
North Slope, 295, 350
Nova Scotia, 6
Nuclear Fission, 359
Nucleus, 366

O

Occupational Health and Safety Act of 1971 (OSHA), 472
Octane, 434
Octane Number, 333
Offset well, 81, 84
Offshore drilling, 135
Offshore Platform, 131
Offshore Wells, 135, 136
Ohio, 9, 11, 13
Ohio Valley, 4, 11
Ohmmeter, 156
Ohm's Law (E=IR), 427
Oil Creek, PA., 8, 9, 49, 310
Oil seeps, 8
Oil Springs, 78
Oilwell Drainhole Drilling Co., 127
Oklahoma, 11, 76, 117, 128, 133, 192, 353
Oklahoma City, 52, 76, 346
Olefin Plants, 376
Olefins, 325, 330, 356, 369, 373

Pumper, 247
Pumps:
  Air balance beam, 226
  Ball, 233
  Cable operated long
    stroke, 226
  Centrifugal, 230
  Continous-Flow Gas
    Lift, 231
  Gas lift, 231
  Plunger, 226
  Sonic, 233
  Surface lifts, 223, 229
Pumping:
  Casing-Type Hydraulic, 231
  Conventional, 223
  Hydraulic, 229
  Rod, 219, 223, 226
  Submersible, 233
Pumping off, 234

Q
Quaker State, 4
R
Radar gauging, 404
Radiation-type thermometer, 398
Radioactivity, 22
Radium, 363
Raffinate, 374
Railway tank car, 279
Rat hole, 101, 124, 126
Recovery:
  Air injection, 238
  Carbon dioxide injection, 240
  Caustic injection, 240
  Enhanced recovery, 46, 47, 58,
    129, 130, 236, 240
  In Situ combustion, 237, 238
  Primary recovery, 129, 234, 236
  Secondary recovery, 47, 58, 173,
    236, 240
  Steam injection, 237, 240

Tertiary recovery, 58, 236, 237
  Waterflooding, 58, 236
Rectifiers, 429
Reformate, 333
Reforming, 316
Reformulated gasoline, 334
Remote sensing, 59, 60, 62, 380
Research Octane Number (RON),
  333, 334
Reservoir, 25, 29, 30, 33, 37, 38,
  39, 40, 46, 56, 58, 59, 128
Right-of-way, 292, 303, 307
Riparian rights, 71
Rocks:
  Igneous, 21, 22, 23
  Magma, 21, 22
  Metamorphic, 23
  Sedimentary, 21, 28, 29, 30
Rock cycle, 22
Rock oil, 15
Rods:
  Polished, 218, 224, 225
  Pony, 225
  Sub, 225
  Sucker, 214, 218, 223, 224, 226
Rotary, 91, 99
Rotary drilling, 15, 17, 111
Rotary pumps, 15
Rotary table, 104
Roughneck, 124
Roustabout, 247
Routine drilling ahead, 126
Rowland, T. F., 135
Royalty, 79
Royalty clause, 86
Ruffner, David, 93
Ruffner, Joseph, 93
Rumania, 94
Run ticket, 270
Russia, 115, 280, 300
RVP, 335

Modern Petroleum

S

Sacrificial Anode, 416
Sahara, 4
Salt domes, 14, 17
Sample cocks, 272
Sandstone, 25, 28, 36, 58, 152, 155
San Juan Basin, 456
Santa Barbara, CA., 135
Saskatchewan, 78
Satellite Navigation System, 137
Saturation, 37
Saunders, J. B., Jr., 280
Scanning electron microscope (SEM),153
Schlumberger, 157
SCR-controlled rigs, 133
Scratchers, 200, 201
Scrubbers, 448
Sedimentation, 22, 25
Sedimentary deposit, 27
Sedimentary rock, 21, 22, 23, 25
Seismograph, 52, 54, 380
Seismology, 52, 136
Self-potential, 151
Semi-submersible, 141
Seneca Oil Co., 8
Separator: 39
   Gravitational, 252
   Horizontal, 244
   LTX, 442
   Spherical, 244
   Vertical, 243, 244
Shale, 25, 28, 36, 112, 152, 167, 168
Shale oil, 458
Sheaves, 103
Shell-Esso, 142
Shell Oil, 56, 64
Shell stills, 312
Shooters, 54
Siberia, 146

Sidewall coring tool, 153
Silicates, 21
Silicon, 21
Silliman, Benjamin, Jr., 310
Single Point Mooring System (SPM), 287, 288
Sinker bars, 95
Skelly, W. G., 192
SIAR, 59
Slips, 124
Smith, William "Uncle Billy," 8, 94, 223
Solar systems, 19, 20, 27
Solid-Desiccant Dehydration, 446
Solomon's Temple, 1
Sonde, 156, 157
Sour crude, 9
South Africa, 457
South America, 115
Spain, 3
Specific gravity, 21, 40, 242, 271
Spindletop, 12, 16, 96, 112, 236
Spontaneous Potential, 156, 158, 160, 162
Spring-pole drilling, 94
Spudding in, 50, 124
Standard Oil Company of Indiana, 10
Steamboats, 10
Strategic Petroleum Reserve (SPR), 266
Strapping, 267
Strata, 30, 31, 34, 52
Stray current, 423
Stripper wells, 234
Stuffing box, 215, 218, 224, 225
Submersible, 139, 140
Subsea blowout preventer stack, 138
Sulfur, 14
Sumerians, 1
Superports, 287
Surface owners, 117

Unsaturated hydrocarbons, 359
Upset tubing, 211
Uranus, 19
U. S. Coast Guard, 286
U. S. Department of Transportation (DOT), 473
U. S. Geological Survey, 76
U. S. Minerals Management Service, 76
Utah, 459
U–Tube manometer, 402

V

Valves:
Ball, 391
Butterfly, 391
Check, 392
Gate, 390
Globe, 393
Lock stopcock, 394
Plug, 390
Relief, 392
Shutoff, 411
Tank shutoff, 394
Vapor pressure, 333
Veer-Root Corp., 346
Venezuela, 2, 3, 49
Venus, 19
Vibroseis, 54, 56
Viscosity, 37, 40, 219, 337
VLCC, 283, 287
Vugs, 36

W

Waiting on cement (WOC), 196
Walking beam, 94, 223

Washington, 286
Washington, George, 4
Washington Monument, 143
Water drive, 39
Water pollution, 470
Welding gang, 304
Welding wire, 305
Wellbore, 95, 102, 112, 151
Wellhead, 214, 241
Wetlands, 62
Whale oil, 6
Wildcat, 180
Wildcatters, 11, 49
Williams, James Miller, 7, 8
Wireline, 151
Workover, 249
World Prodigy, 284
World War I, 279, 283
World War II, 127, 135, 280, 315, 316, 344, 370
Wyoming, 11

X

Xenon, 363
X-Rays, 301, 306, 360

Y

Yale University, 310
Young, James, 5, 310
Yukon, 78

Z

Zones of lost circulation, 112
Zublin, John, 127